Vector fields

Vector fields

Vector analysis developed through its application to engineering and physics

J. A. SHERCLIFF

Professor of Engineering Science, University of Warwick

CAMBRIDGE UNIVERSITY PRESS

CAMBRIDGE

LONDON · NEW YORK · MELBOURNE

Published by the Syndics of the Cambridge University Press
The Pitt Building, Trumpington Street, Cambridge CB2 1RP
Bentley House, 200 Euston Road, London NW1 2DB
32 East 57th Street, New York, NY 10022, USA
296 Beaconsfield Parade, Middle Park, Melbourne 3206, Australia

Library of Congress Catalogue Card Number 76-8153

ISBN hardcovers 0-521-21306-1
ISBN paperback 0-521-29092-9

First published 1977

Printed in Great Britain at the Alden Press, Oxford

Contents

Note. An asterisk denotes a section which is more difficult, and which can be omitted without detriment to the understanding of the following sections.

Preface

This textbook is an expanded version of a course of thirty second-year lectures which the author has given for several years to students of engineering science. It is aimed at those students of the engineering, physical or mathematical sciences who would not be content with a merely mathematical approach to vector analysis. It is complementary to the many books on vector analysis which are primarily mathematical. Throughout, the mathematical concepts are allowed to invent themselves out of the need to describe the physical world quantitatively. The book is intended to provide a central core to the curriculum, a vehicle for a great deal of essential teaching which though mathematical in form is more than mathematical in content; this core course is envisaged as the place where the student learns the fundamental generalised ideas of electromagnetic theory or of fluid dynamics, even though he will also be meeting these subjects elsewhere in courses where they are applied to the design of electric or fluid energy conversion devices and systems. The book also introduces in a coherent way many topics of importance in engineering which are commonly overlooked; two notable examples are seepage flow in porous media and nuclear reactor criticality. The engineering science curriculum is now so crowded that time must be used efficiently; this book shows how basic electromagnetism, fluid mechanics and mathematics can be telescoped.

The book is an *ideas* book rather than a *techniques* book. Mathematical methods are only explored far enough to give the interested student a glimpse of the activities that lie beyond, typically in third-year options. For instance the use of curvilinear coordinates is not seriously pursued and Bessel functions make only minor appearances. For one thing the advent of the computer has reduced the importance of many of the old methods and the norm now is to attack problems well beyond the reach of analytical methods. It is more important nowadays that a third-year student should learn about such things as finite element methods than many of the analytical topics formerly taught. This book does not itself venture far into numerical or computational areas. Simple analysis and familiar standard functions suffice to display the essential nature of the phenomena under discussion. The book does not develop tensor analysis, and for this reason viscous effects are not mentioned, except in passing. It is imagined that applied mathematicians will use the book as a source of case material to supplement the orthodox, formal treatments of vector theory which can be found in any of the many largely mathematical books that are available. It should be stressed that this book is about three-dimensional field phenomena and does not discuss vector spaces, discrete systems with n degrees of freedom or rigid body mechanics.

A feature of the book is its insistence on blending the whole wide range of vector field phenomena so as to bring out illuminating analogies, to save time by avoiding repetition of essentially the same ideas in different contexts on different occasions, and to introduce each new mathematical concept in whatever context makes it most easily grasped because it is then so obviously helpful and necessary. This technique makes it feasible to introduce students progressively to increasingly powerful, generalised ideas up to a level not universally contemplated in engineering courses, even though a proper formulation of electromagnetic or fluid dynamic theory is only possible in such terms. The book best suits a curriculum in which unified teaching of physical or engineering science is undertaken; it is not appropriate to a physics or electrical engineering department which ignores fluid mechanics, for instance. (However, students who are being taught both electromagnetic and mechanical topics separately should find the book provides valuable complementary reading.) The approach obviously has its pitfalls. A reader who picks out only the electromagnetic passages, for instance, will soon stumble unless he is already a master of the mathematical ideas which are taught via fluid mechanical or other applications in the book. To help the reader who might otherwise become disoriented by the proliferation of physical applications, chapter 3 starts by providing an exposition of the structure and rationale of the book, chapters 1 and 2 having been concerned mainly with consolidation of first-year ideas. The chapters range over such a variety of physical topics that succinct yet comprehensive chapter titles are not always possible; instead the titles mainly reflect the progression of mathematical ideas through the book. As this progression is tightly structured, discrimination is needed if the reader or teacher seeks a reduced selection of the material. As an aid to such selection, an asterisk notation is used against those chapter sections or problems which may be found difficult and which can be omitted without detriment to the understanding of the remainder. One way of further lightening the load is to ignore all sections or problems that refer to polarisation or magnetisation, **P**, **D**, **M** or **H**.

Each chapter ends with an ample collection of problems. In accordance with the general spirit of the book, these are mostly of physical interest. There is an abundance of merely mathematical problems in other books. The problems test (and often extend) the ideas in the same order that they appear in the corresponding chapter and they are appropriately referred to in the text. There are many worked examples in the text and guidance is given in many of the problems. Serious attention should be given to the problems, for much of the book's content lies in them. Some of the more advanced ideas are encountered only in the problems, in fact, so that the less advanced student is not troubled by them in reading the main text.

In the choice of symbols for current and the root of minus one it is impossible to satisfy both electrical engineers and others. This book uses i for $\sqrt{(-1)}$ rather than j, with the concomitant choice of J and j (or their vector, heavy-type equivalents) for current and current intensity respectively. Electromagnetic theory is developed on a basis of S.I. units.

Chapters 1 and 3 list the mathematical ideas which the student should have mastered in a previous course. He is also assumed to have encountered the usual elementary solid and fluid mechanics, electrical theory, and thermodynamics.

Appendix I is a note on Taylor's theorem in relation to mathematical modelling and on the way differentials are used in this book, and widely elsewhere.

Appendix II gives a brief explanation of some of the lecture demonstrations and other exercises which the author has found useful in the course on which this book is based.

I am greatly indebted to Dr K.H. Swinden for his comments on the book in draft.

Coventry 1976 J.A. SHERCLIFF

1

Vector algebra revisited

1.1 Prerequisites

Before starting on the book proper in section 1.2, the reader may welcome an outline of the formal vector mathematics which he should already have met, together with some revision exercises. The topics all concern vector algebra or vector functions of one variable. We shall write vectors either in bold type (e.g. **a**) or in the form $\overrightarrow{PQ}$, a directed line segment from P to Q. The corresponding italic type (e.g. a) denotes the magnitude of a vector, as does the 'modulus' notation (e.g. $|\mathbf{a}|$).

Basic vector algebra

If l, m are scalars and **a**, **b**, **c** vectors, then

$$\mathbf{a}+\mathbf{b} = \mathbf{b}+\mathbf{a}, \quad \mathbf{a}+(\mathbf{b}+\mathbf{c}) = (\mathbf{a}+\mathbf{b})+\mathbf{c},$$

vector addition being governed by the triangle or parallelogram laws (see figure 1.1), $l\mathbf{a} = \mathbf{a}l$ is a vector parallel to **a** and of l times its magnitude, $-\mathbf{a}$ is a vector antiparallel to **a** and of the same magnitude,

$$l(m\mathbf{a}) = (lm)\mathbf{a}, \quad (l+m)\mathbf{a} = l\mathbf{a}+m\mathbf{a}, \quad l(\mathbf{a}+\mathbf{b}) = l\mathbf{a}+l\mathbf{b}.$$

A unit vector is a vector of unit magnitude, for expressing direction only.

Fig. 1.1. (*a*) and (*b*): Additions by the triangle law, with vectors head-to-tail, (*c*): addition by the parallelogram law, with both vectors rooted at O (i.e. both tails at O).

The right-hand screw convention

When a vector is used to represent a physical quantity that has a *rotational* quality (e.g. angular velocity) the vector is directed along the axis of rotation in the direction of advance of a right-hand screw under the rotation implied.

Cartesian axes

To avoid ambiguity it is essential to use *right-handed axes* as shown in figure 1.2.

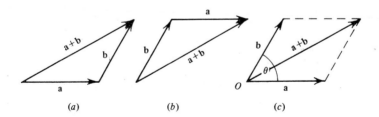

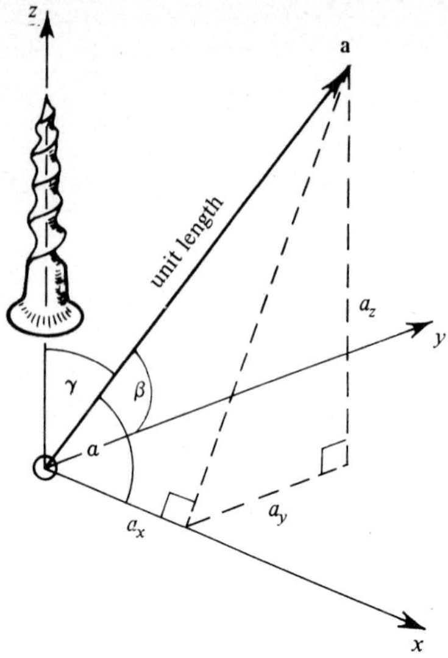

Fig. 1.2. Right-handed axes and the components (direction cosines) of a unit vector.

These are characterised by the fact that a rotation about Oz of Ox towards Oy would cause a right-hand screw to advance in the z-direction. The three axes are mutually perpendicular.

Unit base vectors: **i, j, k**

These are of unit magnitude, oriented respectively in the x, y and z directions. The components (a_x, a_y, a_z) of a vector **a** are such that

$$\mathbf{a} = \mathbf{i}a_x + \mathbf{j}a_y + \mathbf{k}a_z = \sum \mathbf{i}a_x \quad \text{and} \quad a^2 = \sum a_x^2,$$

if we use Σ to denote a sum over three cartesian terms.

Direction cosines

If *unit* vector **a** makes angles α, β, γ with Ox, Oy, Oz, respectively, then $a_x = \cos\alpha$, $a_y = \cos\beta$, $a_z = \cos\gamma$. (See figure 1.2.) The quantities $(\cos\alpha, \cos\beta, \cos\gamma)$ are the 'direction cosines' of the vector and of its direction. Note that $\Sigma \cos^2\alpha = 1$, which expresses the fact that only two of α, β, γ are independent.

Vector multiplication

(i) *The scalar product* of two vectors **a** and **b** is written $\mathbf{a}\cdot\mathbf{b}$ and equals $ab\cos\theta$, where θ is the angle between the two vectors when rooted at the same point as in figure 1.1(c). The product is negative when θ is obtuse. If **b** is a unit vector, $\mathbf{a}\cdot\mathbf{b}$ is the component of **a** in the direction of **b**. If $\mathbf{a}\cdot\mathbf{b} = 0$, then **a** and **b** are perpendicular, if both are non-zero.

Other results: $\mathbf{a}\cdot\mathbf{b} = \mathbf{b}\cdot\mathbf{a}$, $\mathbf{a}\cdot(\mathbf{b}+\mathbf{c}) = \mathbf{a}\cdot\mathbf{b} + \mathbf{a}\cdot\mathbf{c}$, $(l\mathbf{a})\cdot\mathbf{b} = l(\mathbf{a}\cdot\mathbf{b}) = \mathbf{a}\cdot(l\mathbf{b})$, $\mathbf{i}\cdot\mathbf{i} = \mathbf{j}\cdot\mathbf{j} = \mathbf{k}\cdot\mathbf{k} = 1$, but $\mathbf{i}\cdot\mathbf{j} = \mathbf{j}\cdot\mathbf{k} = \mathbf{k}\cdot\mathbf{i} = 0$, $\mathbf{a}\cdot\mathbf{b} = \Sigma a_x b_x$ and $\mathbf{a}\cdot\mathbf{a} = \Sigma a_x^2 = a^2$.

The conventional shorthand $\mathbf{a}^2$ is normally used for $\mathbf{a}\cdot\mathbf{a}$. The point P, (x, y, z), lies on the plane through Q, (x_0, y_0, z_0), perpendicular to the vector $\mathbf{v} = l\mathbf{i} + m\mathbf{j} + n\mathbf{k}$ if the scalar product

$$\mathbf{v}\cdot\overrightarrow{QP} = l(x-x_0)+m(y-y_0)+n(z-z_0) = 0.$$

Hence the plane $lx + my + nz = \text{const.}$ is perpendicular to the vector $\mathbf{v} = l\mathbf{i} + m\mathbf{j} + n\mathbf{k}$.

(ii) *The vector product* of two vectors **a** and **b** is written $\mathbf{a}\times\mathbf{b}$ and is a vector perpendicular to both **a** and **b** having the direction in which a right-hand screw would advance if, with **a** and **b** rooted at the same point, **a** were rotated towards **b**. The magnitude of $\mathbf{a}\times\mathbf{b}$ is $ab \sin\theta$, where θ is the angle (less than 180°) between **a** and **b**, if rooted at the same point. If θ is zero because **a** and **b** are parallel, $\mathbf{a}\times\mathbf{b}$ vanishes and the ambiguity of its direction is immaterial. If $\mathbf{a}\times\mathbf{b} = \mathbf{0}$, then **a** and **b** are parallel, if both are non-zero.

Other results: $\mathbf{a}\times\mathbf{b} = -\mathbf{b}\times\mathbf{a}$, a significant departure from normal algebra.

$$\mathbf{a}\times(\mathbf{b}+\mathbf{c}) = \mathbf{a}\times\mathbf{b}+\mathbf{a}\times\mathbf{c}, \qquad l(\mathbf{a}\times\mathbf{b}) = (l\mathbf{a})\times\mathbf{b} = \mathbf{a}\times(l\mathbf{b}),$$

$$\mathbf{i}\times\mathbf{i} = \mathbf{j}\times\mathbf{j} = \mathbf{k}\times\mathbf{k} = \mathbf{0} \quad \text{but} \quad \mathbf{i}\times\mathbf{j} = \mathbf{k}, \mathbf{j}\times\mathbf{k} = \mathbf{i}, \mathbf{k}\times\mathbf{i} = \mathbf{j},$$

$$\mathbf{a}\times\mathbf{b} = \sum \mathbf{i}(a_y b_z - a_z b_y) = \begin{vmatrix} \mathbf{i} & \mathbf{j} & \mathbf{k} \\ a_x & a_y & a_z \\ b_x & b_y & b_z \end{vmatrix},$$

if the idea of a determinant is generalised for mnemonic purposes.

(iii) *Triple products of vectors*. There are three kinds:

A. $(\mathbf{a}\cdot\mathbf{b})\mathbf{c}$, a vector. This is *not* equal to $\mathbf{a}(\mathbf{b}\cdot\mathbf{c})$ in general.

B. $(\mathbf{a}\times\mathbf{b})\times\mathbf{c}$, a vector. This is *not* equal to $\mathbf{a}\times(\mathbf{b}\times\mathbf{c})$ in general. However, $(\mathbf{a}\times\mathbf{b})\times\mathbf{c} \equiv (\mathbf{a}\cdot\mathbf{c})\mathbf{b} - (\mathbf{b}\cdot\mathbf{c})\mathbf{a}$ is a general identity.

C. $\mathbf{a}\cdot(\mathbf{b}\times\mathbf{c})$, a scalar. This equals $\mathbf{c}\cdot(\mathbf{a}\times\mathbf{b})$ and $-(\mathbf{a}\times\mathbf{c})\cdot\mathbf{b}$, etc. (the sign changes if the cyclic order *abcabc* . . . changes), the very important *transformation property* of the scalar triple product. Note that $\mathbf{a}\cdot\mathbf{b}\times\mathbf{c} = \mathbf{a}\times\mathbf{b}\cdot\mathbf{c}$ (i.e., the dot and cross may be interchanged) and that

$$\mathbf{a}\cdot\mathbf{b}\times\mathbf{c} = \begin{vmatrix} a_x & a_y & a_z \\ b_x & b_y & b_z \\ c_x & c_y & c_z \end{vmatrix}.$$

(The brackets round $\mathbf{b}\times\mathbf{c}$ can be safely omitted.)

Vector functions of one variable such as time t. If $\mathbf{a} = \mathbf{a}(t)$, $\mathrm{d}\mathbf{a}/\mathrm{d}t$ is a vector, not in general parallel to **a**, equal to $\Sigma\, \mathbf{i}(\mathrm{d}a_x/\mathrm{d}t)$,

$$\frac{\mathrm{d}}{\mathrm{d}t}(\mathbf{a}+\mathbf{b}) = \frac{\mathrm{d}\mathbf{a}}{\mathrm{d}t}+\frac{\mathrm{d}\mathbf{b}}{\mathrm{d}t}, \quad \frac{\mathrm{d}}{\mathrm{d}t}(l\mathbf{a}) = \frac{\mathrm{d}l}{\mathrm{d}t}\mathbf{a}+l\frac{\mathrm{d}\mathbf{a}}{\mathrm{d}t},$$

$$\frac{\mathrm{d}}{\mathrm{d}t}(\mathbf{a}\cdot\mathbf{b}) = \frac{\mathrm{d}\mathbf{a}}{\mathrm{d}t}\cdot\mathbf{b}+\mathbf{a}\cdot\frac{\mathrm{d}\mathbf{b}}{\mathrm{d}t}, \quad \frac{\mathrm{d}}{\mathrm{d}t}(\mathbf{a}\times\mathbf{b}) = \frac{\mathrm{d}\mathbf{a}}{\mathrm{d}t}\times\mathbf{b}+\mathbf{a}\times\frac{\mathrm{d}\mathbf{b}}{\mathrm{d}t}.$$

If $\mathbf{r}$ is the *position vector* $\overrightarrow{OP}$ of a point P, then $\mathrm{d}\mathbf{r}/\mathrm{d}t = \mathbf{v}$, its velocity, and $\mathrm{d}^2\mathbf{r}/\mathrm{d}t^2 = \mathrm{d}\mathbf{v}/\mathrm{d}t = \mathbf{a}$, its acceleration, which has components $\mathrm{d}v/\mathrm{d}t$ parallel to v and v^2/R perpendicular to v (along the principal normal) where R is the radius of curvature of P's path.

Problems 1A

1.1 Identify the following as either vectors or scalars: momentum, kinetic energy, pressure, angular velocity, acceleration, mass density, weight, electric charge, frequency, voltage, light beam intensity, equilibrium radiation in a black-body enclosure.

1.2 Figure 1.3 shows a plan of two roads X, Y and two points A, B which must be connected by the shortest sewer, subject to the constraint that the sewer must cross each road at right angles (because of the cost of breaking concrete). Find how to construct the route.

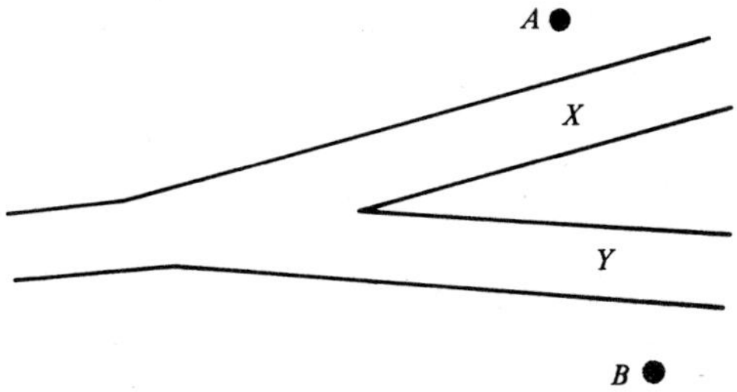

Fig. 1.3

1.3 The normals to two planes have direction cosines $(\cos\alpha_1, \cos\beta_1, \cos\gamma_1)$ and $(\cos\alpha_2, \cos\beta_2, \cos\gamma_2)$. Find an expression for $\cos\theta$ where θ is the angle between the planes. What happens if $\gamma_1 = \gamma_2 = \pi/2$?

1.4 What is the condition for the planes $ax + by + cz = d$ and $kx + ly + mz = n$ to be perpendicular?

1.5 Find the unit vector that is perpendicular to both of the vectors $\mathbf{i} + 2\mathbf{j} + 3\mathbf{k}$ and $4\mathbf{i} + 5\mathbf{j} + 6\mathbf{k}$.

1.6 Find the shortest distance between the line joining the points $(1, 1, 1)$ and $(2, 3, 4)$ and the line joining the points $(3, 1, 1)$ and $(1, 3, 2)$.

1.7 For a parallelogram defined by two adjacent sides $\mathbf{a}$ and $\mathbf{b}$, show that the sum of the squares of the diagonals equals the sum of the squares of the sides, and that the diagonals are orthogonal if the sides are equal in length. Find a simplified expression for the vector product of the diagonals. Is it in the right direction?

1.8 If O_1 and O_2 are fixed points, P is a movable point, $\overrightarrow{O_1P} = \mathbf{r}_1, \overrightarrow{O_2P} = \mathbf{r}_2$, $\mathbf{a}$ is a fixed vector and k is a fixed scalar, find the geometrical constraints on P implied by (i) $\mathbf{r}_1 + \mathbf{r}_2 = \mathbf{a}$, (ii) $\mathbf{r}_1 \times \mathbf{r}_2 = \mathbf{a}$ or (iii) $\mathbf{r}_1 \cdot \mathbf{r}_2 = k$.

1.9 (*a*) Show that the volume of a parallelepiped is $\mathbf{a} \cdot \mathbf{b} \times \mathbf{c}$ if $\mathbf{a}$, $\mathbf{b}$ and $\mathbf{c}$ are three adjacent edges.
(*b*) What is the result if $\mathbf{a}$, $\mathbf{b}$ and $\mathbf{c}$ are coplanar? What is then the angle between $\mathbf{a}$ and $\mathbf{b} \times \mathbf{c}$?
(*c*) Use (*b*) above to find the cartesian equation of the plane defined by the points P_1 P_2 P_3 with position vectors $\mathbf{r}_1, \mathbf{r}_2, \mathbf{r}_3$ (components x_1, x_2, x_3, etc.), by noting that the position vector $\mathbf{r}$ of a point P on the plane is such that $\overrightarrow{P_1P}, \overrightarrow{P_2P_1}, \overrightarrow{P_3P_2}$ are coplanar. Express the result as a determinant. What happens if P_1 P_2 P_3 are collinear?

1.10 From the result $(\mathbf{u} \times \mathbf{v}) \times \mathbf{w} \equiv (\mathbf{u} \cdot \mathbf{w})\mathbf{v} - (\mathbf{v} \cdot \mathbf{w})\mathbf{u}$, show that

$$(\mathbf{a} \times \mathbf{b}) \cdot (\mathbf{c} \times \mathbf{d}) \equiv (\mathbf{a} \cdot \mathbf{c})(\mathbf{b} \cdot \mathbf{d}) - (\mathbf{a} \cdot \mathbf{d})(\mathbf{b} \cdot \mathbf{c}).$$

Interpret this result simply in the case where $\mathbf{a}$ and $\mathbf{c}$ are identical unit vectors and $\mathbf{b}$ and $\mathbf{d}$ are identical unit vectors, inclined at θ to $\mathbf{a}$.

1.11 A bead of mass m slides on a curved wire lying in a plane. The coefficient of friction is μ and gravity is negligible. Show that the speed of the bead is proportional to $e^{-\mu\psi}$, where ψ = angle between tangent and some reference direction. What if there is a point of inflexion?

1.12 (*a*) On a curve or trajectory, the position vector $\mathbf{r}$ can be regarded as a function of s, the distance along the curve from some datum. Show that $d\mathbf{r}/ds$ is a unit vector, $\mathbf{T}$ say, in the tangent direction and that $d\mathbf{T}/ds$ is perpendicular to the tangent, but is zero if the curve is locally straight. (The direction of $d\mathbf{T}/ds$ is called the *principal normal*.)
(*b*) Find the *unit* principal normal vector to the curve $y = x^2, z = x^3$ at the point (1, 1, 1). (Note that $d\mathbf{T}/dx$ is also in the principal normal direction.)
(*c*) If the *radius of curvature* R is defined such that $1/R = |d\mathbf{T}/ds|$, show that, for a particle traversing a curve, the acceleration $d/dt(v\mathbf{T})$ has components $v dv/ds$ and v^2/R along the tangent and principal normal respectively, v being the magnitude of the velocity. (Note that $v = ds/dt$.)
(*d*) A point P moves arbitrarily on a sphere of radius ρ, centred at the origin. By successively differentiating the equation $\mathbf{r} \cdot \mathbf{r} = \rho^2$ show that the velocity is perpendicular to the radius vector $\mathbf{r}$ and that the *radial* component of acceleration is $-v^2/\rho$. Note that ρ will not be the radius of curvature of the locus in general. Examine the case where the locus is a circle of radius smaller than ρ.

1.13 A particle of unit mass, initially moving with unit velocity in the positive z-direction at the origin, suffers a force

$$\mathbf{i}\sin 2t + \mathbf{j}\cos 2t - \mathbf{k}\sin t,$$

expressed in terms of time t. Show that its acceleration and velocity are always perpendicular. Verify that its speed is constant and find the radius of curvature of its path as a function of time. (See problem 1.12(c).)

1.2 Vectors and scalars

Before we move on to our main business of vector *calculus*, it is necessary to spend two chapters extending physical appreciation of vectors and vector algebra.

A major step in the systematisation of knowledge of the three-dimensional world was the realisation that most physical quantities may be classified as *scalars* or *vectors*. Scalars have magnitude only (e.g. mass or temperature) whereas vectors have magnitude and direction (e.g. force or displacement). To specify a vector we must state *three* independent quantities, which might be its three cartesian components or its magnitude and direction cosines (only two of which are independent). One consequence is that a vector equation is really three pieces of information written concisely to look like one. A primitive example is $\mathbf{a} = \mathbf{b}$, which implies the three independent facts that each of the three cartesian components of $\mathbf{a}$ equals the corresponding one for $\mathbf{b}$.

There are physical quantities of practical importance which are neither scalars nor vectors. These are *tensors*, to specify which 6, 9 or even more magnitudes are necessary. A good example is *stress* in a solid or viscous fluid. This book avoids any discussion of tensors.

1.3 Addition of vectors

Not all physical quantities that have magnitude and direction are vectors, however. Vectors must also obey the familiar triangle or parallelogram addition laws for finding the combined physical effect (whatever that may mean in each case) of two vectors acting together. These addition laws function by representing vectors as *directed line segments*, whose lengths represent their magnitudes to some arbitrary scale. Only in the case of vectors that are actual displacements is this representation literal rather than conventional. It does however always lead to the correct addition of vectors. Figure 1.1 shows the addition laws applied to two vectors $\mathbf{a}$ and $\mathbf{b}$, represented by directed line segments, the direction being indicated by an arrowhead.

The statement that velocity, for instance, is a vector therefore includes the fact that the combined physical effect of two velocities acting simultaneously, a familiar notion which is easily interpreted, can be calculated from the triangle or parallelogram laws. One obvious consequence of these laws is that

$$\mathbf{a} + \mathbf{b} = \mathbf{b} + \mathbf{a}.$$

Through the failure of this relation, problem 1.14 reveals a counter-example which shows that there are physical quantities which have magnitude and direction but which are not vectors.

The notion of 'combined effect' must be used with discretion. Like can only be added to like, and the combined effect of a current **J** with a magnetic field **B** at the same point is certainly not **J** + **B**. Later we shall find that a kind of multiplication is appropriate here, instead.

1.4 The essential nature of vectors; 'cosine-quality'

As well as their addition properties, physical quantities that are vectors have another characteristic quality. This emerges when we need to consider the 'contribution', 'effect' or 'influence' that a vector quantity makes or exerts in some *direction of interest*, different from the vector's own direction. If we are crossing the road diagonally, our speed directly across the road is of great relevance to our chance of survival. This leads to the formal idea of *resolution*.

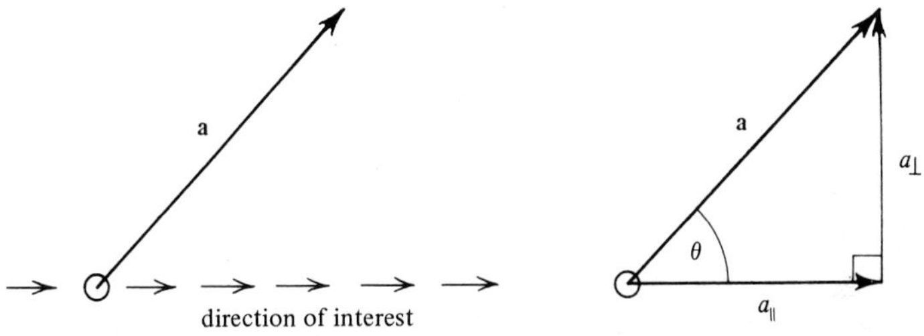

Fig. 1.4. Resolution along and normal to a direction of interest.

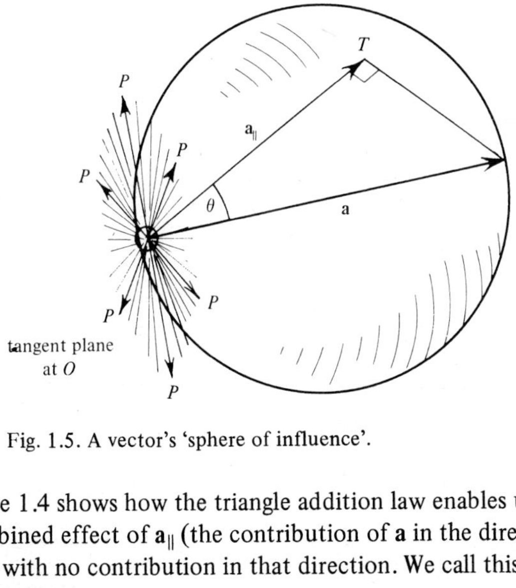

Fig. 1.5. A vector's 'sphere of influence'.

Figure 1.4 shows how the triangle addition law enables us to write a vector **a** as the combined effect of $\mathbf{a}_\parallel$ (the contribution of **a** in the direction of interest) and $\mathbf{a}_\perp$, a vector with no contribution in that direction. We call this *resolving* **a** into *components* $\mathbf{a}_\parallel$ and $\mathbf{a}_\perp$.

The key point to notice is that the magnitude of $\mathbf{a}_\parallel$ is $a \cos\theta$, i.e. *a vector is a* *physical entity whose influence in a particular direction varies like cos θ, where θ is the angle between that direction and the unique direction of maximum influence.* We shall frequently exploit this characteristic *cosine-quality* as a means of recognising vectors.

If we plot in three dimensions how $\mathbf{a}_{\parallel}$ varies with θ, for a given vector $\mathbf{a}$, we get figure 1.5, which shows how the tip T of $\mathbf{a}_{\parallel}$ traverses a sphere based on $\mathbf{a}$ as diameter. The influence pattern is axisymmetric about $\mathbf{a}$. It is characteristic of a vector that it exerts no influence in all directions OP that are perpendicular to it, and makes a maximum contribution in just one direction (its own).

1.5 Localisation of vectors

Can vectors be moved about with impunity? In figure 1.1 we appear to be assuming that the vector $\mathbf{a}$ is still the same entity whether we put it at the bottom or the top of the picture. The normal situation is that the vector is a physical effect which prevails ('is localised at') some particular *point* in space, whereas for visual-aid and addition purposes we habitually represent vectors as arrowed segments of straight lines. Where do we then put the line segments in relation to the location of the effects? The usual convention is to put the *tails* of the arrows there, as at O in figure 1.1(c) – but it is pure convention – and then the parallelogram law seems the more natural one for addition, as for instance when we seek the combined effect of two superposed forces, localised at the same point.

On the other hand, when there is physical meaning in adding two vectors not localised at the same point, as for example with two successive displacements, the triangle law seems more appropriate, indeed obvious. But either law may be used for the addition process if we are prepared to move the vectors about, preserving their magnitude and direction.

There is one case particularly likely to cause confusion, namely, that of the *position vector* $\mathbf{r}$ which is the displacement vector $\overrightarrow{OP}$, used to specify the position of a point P relative to some origin O. The vector is rooted at O (the arrow-tail) but can be regarded as localised at P (the arrow-head) in the sense that it describes P, and is liable to be combined with other vectors or operations localised at P. The same is true of variants of $\mathbf{r}$ such as the inverse-square law field $\mathbf{r}r^{-3}$. So be warned!

There are circumstances where a vector need not be regarded as localised at a point. In rigid dynamics, to determine the instantaneous state of acceleration of a body we need specify only the lines of action along which the forces acting are localised, while any couples acting can be represented as vectors that are not localised at all, as regards their instantaneous effect on the dynamics. But if we are interested in the subsequent motion of the body or its state of stress, or if it is not rigid, then it matters greatly where the forces or couples act and we are back to the normal situation of localisation at a point. Problem 1.15 takes the discussion further.

1.6 Scalar and vector fields

In this book we shall be mostly concerned with phenomena which are distributed in space. The distributions in space of the physical quantities that are relevant we shall call *fields*, and these are usually either vector or scalar fields. Consider the air in a room: it moves about in a manner described by a velocity field (a vector field) under the action of variations in the temperature field (a scalar field) which affects the density (a scalar field) and consequently also the weight distribution (a vector field) in the fluid. In problems of this kind there is no ambiguity about localisation, as all vectors are localised at the points for which they are specifying the conditions that prevail, as are the scalars, also.

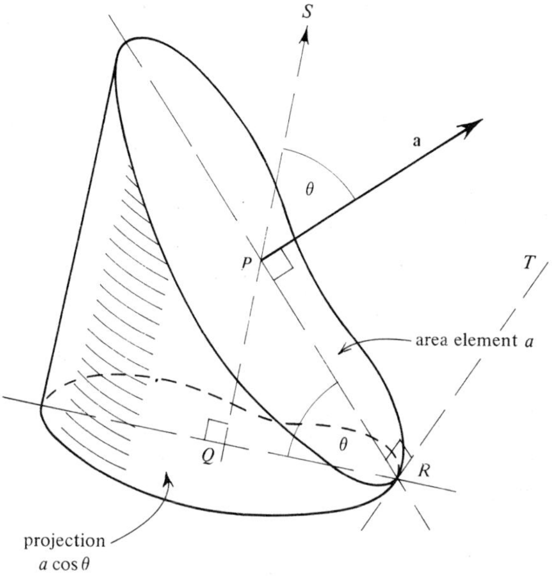

Fig. 1.6

1.7 Vector area

The reader will be accustomed to force and velocity vectors and able to interpret the meaning of 'contribution' or 'influence' in a particular direction in such contexts. But many other kinds of directed quantities will concern us in this book. For example a plane area of finite extent and arbitrary shape has a magnitude and an orientation. Can these aspects (if not the details of the shape) be represented by a vector? Some clue is given by the fact that the vector product $\mathbf{a} \times \mathbf{b}$ of two directed line segments $\mathbf{a}$ and $\mathbf{b}$, localised at a point O, equals in magnitude the area of the parallelogram defined by $\mathbf{a}$ and $\mathbf{b}$, as in figure 1.1(*a*), and its direction is perpendicular to the area in question. This suggests that for representing a plane area *of any shape* by a vector, the natural choice is to make the vector $\mathbf{a}$ normal to the plane and its magnitude a that of the area element, as in figure 1.6. At this stage the direction is arbitrary.

To confirm true vector behaviour, the idea of adding areas to check the addition rule is not obviously meaningful except in the degenerate case of coplanar elements, so instead we examine whether area has the crucial cosine-quality.

If we view the area element in an arbitrary direction *PS*, inclined at θ to the normal as in figure 1.6, we see an effective, or projected, area equal to $a \cos \theta$, for θ is also the angle between the plane *PRT* containing the element and the plane *QRT* perpendicular to *PS*. Thus cosine-quality is indeed exhibited; the component or 'effect' of the vector **a** in an arbitrary direction finds a ready interpretation as the projected area in that direction. (See problem 1.16(*b*).)

This new concept of vector area is particularly convenient in connection with fluid pressure p. The pressure-force on a plane surface element defined by vector **a** can now be succinctly written as $\pm p\mathbf{a}$, where the sign depends on which side the fluid lies.

Example 1.1 Show that fluid pressure is isotropic in the absence of shear stresses.

Let figure 1.6 now represent a small element of fluid, assumed to have a typical linear dimension ϵ. The pressures on the curved face have no component in the direction *PS*. If we denote the pressures on the faces containing *P* and *Q* as p_P and p_Q, then the balance of forces in the direction *PS* requires that

$$\begin{aligned} p_Q(a \cos \theta) - (p_P a) \cos \theta &= \text{the body or inertia forces on the element} \\ &= O(\epsilon^3), \end{aligned}$$

for the terms on the right-hand side must be of the same order as the volume or mass of the element, if accelerations, etc. are to be finite. But a is of order ϵ^2 and so dividing by $a \cos \theta$ gives the result:

$$p_Q - p_P = O(\epsilon)$$

and tends to zero in the limit $\epsilon \to 0$, i.e. the pressure is isotropic.

Just as direction cosines are used to describe the direction of a vector or a straight line, so the direction of a plane or plane area element can be specified by the direction cosines of a vector normal to it. One then speaks of 'the direction cosines of the area'.

1.8 Vector flow intensity

Another important new kind of vector concerns physical phenomena involving the flow, transport or diffusion of some entity such as heat, electric charge, seeping fluid or a foreign species (such as dopant in a semiconductor) through a region of space or a material medium. In the case of heat, we can specify that at a given point it passes at an intensity Q per unit area of a plane perpendicular to the direction of flow, for it is obvious that heat flow does have a direction; heat going from left to right is not going from right to left, or upwards! But is heat flow intensity a true vector **Q** (or, more strictly, a vector *field*, since it is distributed in space)?

It is not meaningful to discuss whether the heat flow intensity can be added by the triangle or parallelogram law so as to qualify for being deemed a vector. Adding

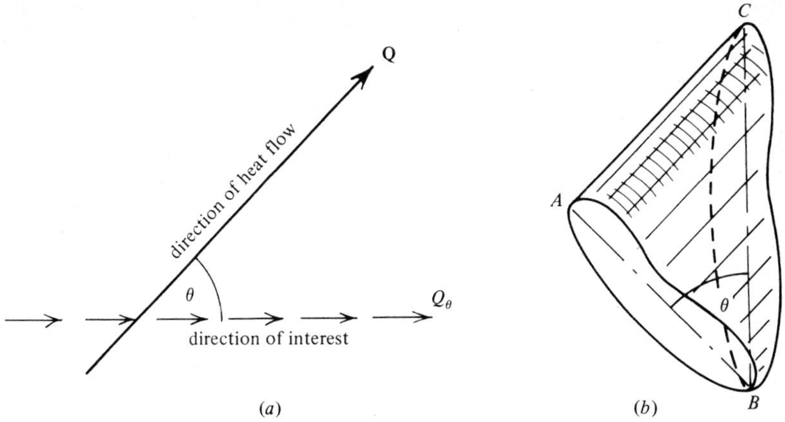

Fig. 1.7

heat flows means nothing physically (although, mathematically, solutions of *linear* heat flow problems can be superposed in a meaningful way). The test we must apply is to see whether heat flow has the characteristic *cosine-quality* of vectors, i.e. whether Q_θ, the heat flow rate across unit area oriented in a particular direction of interest, inclined at an angle θ to the direction of heat flow as shown in figure 1.7(*a*), is equal to $Q \cos \theta$. To verify this, consider the small volume element shown in figure 1.7(*b*). The plane face BC of arbitrary shape faces in the direction of interest, whereas face AB is its projection as seen in the direction of heat flow. The area a of BC is of order ϵ^2, where ϵ is a typical linear dimension of the element. The area of face AB is $a \cos \theta$ because θ is also the angle between the two plane faces.

Then $\quad Q_\theta = (\text{heat crossing } BC) \div a,$

whereas $Q = (\text{heat crossing } AB) \div a \cos \theta.$

As there is no heat flow across the curved face, which consists of lines parallel to the heat flow, the difference between the heat flows across BC and AB must be the heat accumulation rate inside the element, which can only be of the same order as the mass or volume, i.e. of order ϵ^3, if effects such as the rate of rise of temperature are to be finite.

Hence $\quad Q_\theta a - Qa \cos \theta = O(\epsilon^3)$, with $a = O(\epsilon^2)$,

and $\quad Q_\theta - Q \cos \theta = O(\epsilon) \to 0$ in the limit, as $\epsilon \to 0$.

So Q_θ does equal $Q \cos \theta$ at a point, and a quantity such as heat flow intensity does have the characteristic cosine-quality which identifies it as a true vector $\mathbf{Q}$. In future we shall treat all such flow intensities as vectors without more ado.

In this and the previous discussion we have glossed over two points. One is that quantities such as p and $\mathbf{Q}$ have to be derived via the calculus, because in general they vary from place to place and must be defined at a point by taking the limit of force or heat flow rate divided by area as the area tends to zero (but stopping short in the usual way before molecular lumpiness manifests itself). Similar considerations apply to defining scalar densities such as ρ (mass density) or q (net charge density) in non-uniform situations, where the limit of mass, charge, etc. per unit volume is taken as the volume element tends to zero. Densities per unit *length* distributed along a line, or per unit *area* distributed on a surface, are handled similarly.

The other related point is that we have treated p, Q_θ and Q as virtually uniform over the plane faces of elements, and the lines parallel to **Q** as straight (even though **Q** wanders, in general). The terms that we have implicitly omitted are in fact of a higher order of smallness and do not contribute in the limit. Appendix I gives some further discussion of these points. Otherwise in this book we normally ignore, without comment, such higher-order terms.

1.9 Vector multiplication

The reader will have met vector products formally, but may not yet have come to terms with them physically.

Both the scalar or 'dot' product $\mathbf{a} \cdot \mathbf{b}$ and the vector or 'cross' product $\mathbf{a} \times \mathbf{b}$ really invent themselves from the need to describe physical phenomena. What is philosophically intriguing is that these and other concepts, designed to meet one physical duty, then prove to be of service in a seemingly endless series of other, apparently unrelated, physical situations.

1.10 Scalar products

Taking mechanics, for instance, once we realise that the work (a scalar) associated with the displacement **r** of the point of action of a force **F** inclined at θ to **r** is the product $Fr \cos \theta$ of the magnitude of either with the component of the other parallel to the first one, the scalar product concept is born. We write this work quantity as $\mathbf{F} \cdot \mathbf{r}$, spoken as '$F$ dot r'. Problem 1.16(*a*) explores an application.

If the point of action is moving at a speed **v**, the *power*, or rate of doing work, is $\mathbf{F} \cdot \mathbf{v}$. In general, if we have a vector equation expressing a balance of forces, including inertia terms, then scalar-multiplying the whole equation term-by-term by **v** gives a power conservation statement, as we shall see later in discussions of charged particle motion and in fluid dynamics.

Once established, the scalar product idea can be put to use in quite different physical situations. Consider for instance the two less familiar vectors **Q** (heat flow intensity) and **a** (a plane area element, imagined drawn in a conducting medium), **Q** being assumed uniform in the vicinity of the element. We saw previously that the heat flow across the area element is $aQ \cos \theta$ where θ is the angle between **Q** and **a**. Thus $\mathbf{Q} \cdot \mathbf{a}$ is the natural notation for this heat flow. (See also problem 1.17.) Similarly a uniform fluid velocity **v** in the vicinity of a plane area **a** drawn in the fluid gives a volumetric flow rate $\mathbf{v} \cdot \mathbf{a}$ through the area. In all such cases of vector flow, the flow rate, or *flux* (to use the standard technical term) through an area equals the area times the normal component of flow intensity. The concepts of a normal vector to represent area and of a scalar product are just what are needed to express flux succinctly.

Example 1.2 A vector $\mathbf{v}$ is the sum of three vectors $\mathbf{v}_x, \mathbf{v}_y, \mathbf{v}_z$, respectively parallel to the cartesian axes. Show that the component of $\mathbf{v}$ in any direction λ is the sum of the components of $\mathbf{v}_x$, $\mathbf{v}_y$ and $\mathbf{v}_z$ in that direction.

Identify λ by means of a unit vector $\mathbf{I}$. This facilitates taking components:

$$\mathbf{I}\cdot\mathbf{v} = \text{component of } \mathbf{v} \text{ in direction } \lambda,$$

$$\mathbf{I}\cdot\mathbf{v}_x = \text{component of } \mathbf{v}_x \text{ in direction } \lambda, \text{ etc.}$$

But $\mathbf{v} = \Sigma\, \mathbf{v}_x$ and so $\mathbf{I}\cdot\mathbf{v} = \Sigma\, \mathbf{I}\cdot\mathbf{v}_x$, which proves the result (for $\mathbf{a}\cdot(\mathbf{b}+\mathbf{c}\ldots) \equiv \mathbf{a}\cdot\mathbf{b}+\mathbf{a}\cdot\mathbf{c}\ldots$).

1.11 Vector products

Mechanics forces us to invent the vector product concept $\mathbf{a}\times\mathbf{b}$ as well. (This is spoken as '*a* cross *b*'. Some books denote it by $\mathbf{a}\wedge\mathbf{b}$.) From elementary kinematics, for instance, we learn that the velocity $\mathbf{v}$ at a point whose position vector is $\mathbf{r}$ in a rigid body rotating about an axis through the origin at angular velocity $\boldsymbol{\Omega}$ (a vector oriented according to the usual right-hand screw convention) can be concisely written as

$$\mathbf{v} = \boldsymbol{\Omega}\times\mathbf{r}.$$

It is worth taking time to see how the idea of *moment* can be generalised into a vector concept, a vector product in fact. Figure 1.8 is intended to be used as a three-dimensional model. The corner of the page should be folded up along the line XX until it is perpendicular to the rest of the page. Then OQ is an axis normal to the plane OPR containing both the origin O and the force $\mathbf{F}$, acting at P and represented by the directed line segment $\overrightarrow{PR}$. The position of P can be specified by the position vector $\overrightarrow{OP}$, or $\mathbf{r}$. The angle between F and the *forwards* direction of $\mathbf{r}$ is θ and the moment M of $\mathbf{F}$ about the axis OQ is given by $M = rF\sin\theta$, since its moment arm ON equals $r\sin\theta$, $\angle ONP$ being a right-angle.

Next consider the moment M_ϕ about an axis OS inclined at an angle ϕ to OQ. OQ and OS define the plane represented by the folded-up corner of the page. Since the moment of a force is the same wherever it acts along its line of action, the moment of $\mathbf{F}$ may be found by taking it as acting at T, where it intersects the plane OQS. Let the length OT be x and $\angle OTP$ be α. At T we resolve $\mathbf{F}$ into the components $F\sin\alpha$ and $F\cos\alpha$, which respectively do and do not contribute to the moment about OS. The part $F\sin\alpha$ is perpendicular to the plane OQS and acts at a moment arm TU about the axis OS, where $\angle OUT$ is a right angle, and TU equals $x\cos\phi$, since $\angle OTU$ equals ϕ. Since $ON = r\sin\theta = x\sin\alpha$,

$$M_\phi = (F\sin\alpha)(x\cos\phi) = rF\sin\theta\cos\phi = M\cos\phi.$$

The implication is that the moment of F about various axes through O shows cosine-quality, i.e. that the moment is a maximum for one direction of the axis (OQ) and diminishes by a factor $\cos\phi$ for all axes inclined at ϕ to OQ. It is clear

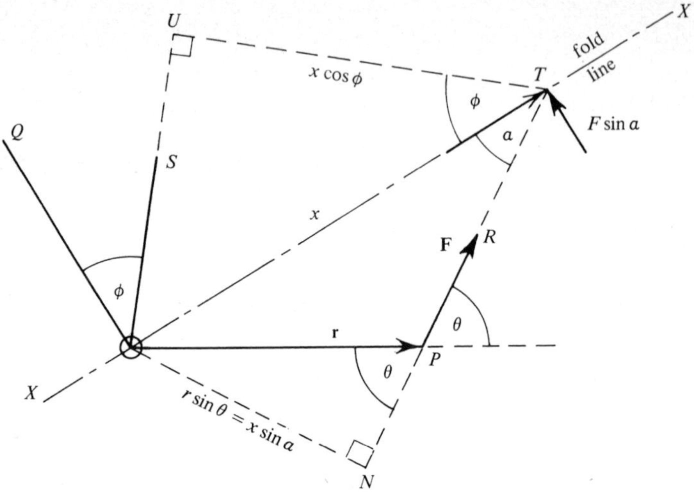

Fig. 1.8. Fold along *XX* until *OQ* is perpendicular to the plane of the page.

that if we define a vector **M** of magnitude *M*, directed along *OQ*, then its component in any direction gives the moment of **F** about an axis through *O* parallel to that direction. We have thus generalised the idea of moment about an axis to *vector moment about a point* (i.e. about all axes through that point).

Moreover, it is clear that this vector moment **M** equals the vector product $\mathbf{r} \times \mathbf{F}$, for its magnitude is $rF \sin \theta$ and its direction is perpendicular to both **r** and **F** and accords with the right-hand screw convention.

The proof in the degenerate case where $\alpha = 0$ is left to the reader. Notice also that the degenerate case where **r** and **F** are parallel causes no trouble: **F** has no moment about any axis through *O* and we are spared the embarrassment of finding a direction for **M** when the plane *OPR* is not uniquely defined. Where there are several forces acting, vector moments may be added (see problem 1.18) and if there is equilibrium, the vector sum is zero. Problem 1.19 explores the moment properties of a *vector couple*, while problem 1.20 shows that vector moment equals $\mathbf{r} \times \mathbf{F}$ by using triple products.

1.12 Multiple products of vectors

Several results concerning triple products of vectors were listed in section 1.1. Of these, quite the most important is the scalar triple product $\mathbf{a} \cdot \mathbf{b} \times \mathbf{c}$, with its remarkable transformation property. (It is a splendid illustration of the old adage that what is most elegant mathematically is often the most significant physically.) One example follows and the next chapter includes further outlets for the transformation property.

Example 1.3 Two equal, opposite and non-collinear forces, constituting a couple of moment **M**, act on a rigid body that is rotating with angular velocity $\mathbf{\Omega}$. Show that their power input is $\mathbf{\Omega} \cdot \mathbf{M}$.

If P and Q are two points in the body separated by the displacement vector $\overrightarrow{PQ}$, then $\Delta\mathbf{v}$, the velocity of Q relative to P, is given by the vector product $\mathbf{\Omega} \times \overrightarrow{PQ}$. Let the two forces $\mathbf{F}$ and $-\mathbf{F}$ be applied at Q and P respectively, constituting the couple of vector moment $\mathbf{M}$ equal to $\overrightarrow{PQ} \times \mathbf{F}$. The rate at which work is being done is

$$\mathbf{v}_Q \cdot \mathbf{F} + \mathbf{v}_P \cdot (-\mathbf{F}) = \Delta\mathbf{v} \cdot \mathbf{F} = (\mathbf{\Omega} \times \overrightarrow{PQ}) \cdot \mathbf{F} = \mathbf{\Omega} \cdot \overrightarrow{PQ} \times \mathbf{F} = \mathbf{\Omega} \cdot \mathbf{M},$$

by the transformation property of scalar triple products. So the scalar product of angular velocity and couple vectors gives *power*, a result pleasingly analogous to the scalar product of velocity and force.

Problems 1B

1.14 Let the arrows ↑ (A) and → (B) indicate a rotation through 90° according to the right-hand screw convention. Take a die (or equivalent object) and check whether $A + B$ (operation A followed by operation B) is the same as $B + A$.

1.15 Can a force applied to a body be regarded as localised in a line rather than a point in relation to its instantaneous rate of doing work on the body?

1.16 (*a*) All points of a plane area element $\mathbf{a}$ of arbitrary shape move a distance $\mathbf{r}$. What is V, the volume swept through? If the element is acted on by a uniform pressure p, show that the work done is pV.
(*b*) An area element of the plane $x + 2y + 3z = 4$ has a projection on the plane $z = 0$ which is a small rectangle, with sides parallel to the x and y axes of magnitude $\mathrm{d}x$ and $\mathrm{d}y$, respectively. Show that the element has an area equal to $(\frac{1}{3}\sqrt{14})\,\mathrm{d}x\mathrm{d}y$.

1.17 Heat is flowing uniformly at an arbitrary inclination, with a vector intensity $\mathbf{Q}$, in a conducting medium. A plane triangular area element ABC is defined on some arbitrary plane, such that BC, CA and AB are respectively the intersections of the plane with planes perpendicular to the x, y and z axes. The three latter planes intersect at D.

If ABC is represented by the vector $\mathbf{a}$, interpret the result

$$\mathbf{Q} \cdot \mathbf{a} = \sum Q_x a_x$$

in terms of the heat flows through the faces of the tetrahedron $ABCD$, there being no heat released or stored in the medium.

1.18 Relative to certain axes, a body is acted on by three forces with the following components and points of application:

$(1, 1, 0)$ at the point $(1, 0, 0)$,
$(0, 2, 2)$ at the point $(0, 1, 0)$,
and $(3, 0, 0)$ at the point $(1, 1, 1)$.

Find the direction cosines of the axis through the origin about which the total moment is a maximum, and determine this maximum value.

1.19 Two equal and opposite forces $\mathbf{F}$ and $-\mathbf{F}$ act at points Q and P respectively, constituting a couple. Show that its vector moment about *any* third point O is independent of the position of O.

1.20 A force $\mathbf{F}$ acts at a point P identified by a position vector $\mathbf{r}$ relative to an origin O. The moment of $\mathbf{F}$ about an axis OQ is to be found. R is the point on OQ nearest to the line of action of the force and S is the point on this line of action nearest to OQ, so that $\mathbf{p} = \overrightarrow{RS}$ is a vector perpendicular to both PS and OR.

If $\mathbf{e}$ is a unit vector parallel to $\overrightarrow{OR}$ show that:

(*a*) the moment of $\mathbf{F}$ about OQ (defined as the magnitude of $\mathbf{p}$ times the component of $\mathbf{F}$ perpendicular to OQ) is equal to

$$\mathbf{p} \cdot (\mathbf{F} \times \mathbf{e}),$$

(*b*) $\mathbf{p} = \mathbf{r} + k\mathbf{e} + l\mathbf{F}$, where k and l are scalars, and hence that

(*c*) the moment of $\mathbf{F}$ about OQ equals the component of $\mathbf{r} \times \mathbf{F}$ parallel to OQ.

2

Vector algebra applied to electromagnetism

2.1 The fundamental forces on charged particles

Vector algebra plays an important role in the concise and accurate description of electromagnetic phenomena, both in charged-particle dynamics and in continuum electrodynamics, where one is interested in the properties of virtually continuous media such as conductors or dielectrics. This goes beyond individual particle behaviour to such concepts as current flow intensity, which, though brought about by millions of individual charged-particle motions, can be 'blurred out' into an effectively continuous, distributed flow, in the same way that in gas dynamics we blur out the individual molecular events into continuous fluid motion.

To revise some important ideas and to prepare for later parts of the book that deal with things electromagnetic, it is worth rehearsing, in vector notation, the basic notions of the subject. It should however be noted that the treatment given here is pre-relativistic and therefore some residual anomalies are left unresolved.

If we leave material magnetisation aside, electromagnetism is the science of charged-particle interactions. Experiment shows that it is necessary to invent the admittedly weird idea of *positive* and *negative charge* to account satisfactorily for the observed attractions and repulsions of 'charged' bodies. (See problem 2.1.) To systematise the description of such interactions, we consider a given array of charges in space, perhaps attached to bulk material, perhaps moving in some way, and then seek a description of the forces which they exert on some small *test charge* in the vicinity. (Strictly, 'small' means vanishingly small in a mathematical limiting process, as otherwise a finite charge may upset the original array, and confusion result.)

The simplest case is electrostatics, where both the array and the test charge are at rest. Then the force on a test charge is found to be proportional to its charge (with attention to sign) and in a definite direction at each point. To describe this we invent the abstraction $\mathbf{E}_s$, a vector defined at each point and equal to the force per unit charge on a test charge at that point, the suffix 's' serving to remind us that this is the *static* phenomenon. The distribution of $\mathbf{E}_s$ is a vector field known as the *electrostatic field*. Just how $\mathbf{E}_s$ is related mathematically to the array which is responsible for it we shall see later.

Events become considerably more subtle when the array includes moving charges, i.e. electric current. We then find that, if the test charge is also moving (with velocity $\mathbf{v}$), it suffers an extra force over and above $\mathbf{E}_s$, the force due to the mere presence of charges. The nature of this extra force at a particular point is quite remarkable. The observed facts are:

(i) When the test charge moves in any particular direction, the force is proportional to its speed v and to its charge.

(ii) For different directions of $\mathbf{v}$, the extra force's magnitude varies. There is one particular direction (*the null direction*) for $\mathbf{v}$ which results in *no* extra force, while if $\mathbf{v}$ is inclined at θ to this null direction, the force varies like $\sin\theta$, reaching a maximum with $\theta = \pi/2$. (Note how this contrasts with 'cosine-quality'.) Thus

$$\text{extra force per unit charge} = kv \sin\theta,$$

in which k and the null direction depend in some way on the moving charge array responsible for these effects.

(iii) The direction of the extra force is found to be perpendicular not only to $\mathbf{v}$ but also the null direction.

These curious and verbose statements can be tersely summed up if we invent another abstraction, $\mathbf{B}$, *the magnetic field*, defining it as a vector which at each point has a magnitude equal to the local value of k and lies in the null direction. Then

$$\text{extra force per unit charge} = \mathbf{v} \times \mathbf{B},$$

in which this order for the vector product implies a precise convention for the direction of $\mathbf{B}$ along the null direction. Had we not already done so in mechanics, it would have been necessary at this point to invent the vector product concept. Indeed, the credit for the invention seems to be more due to mother nature than *Homo sapiens*!

The very odd 'sideways' character of the magnetic effect has been loosely but evocatively summed up by describing magnetism as 'electricity looked at sideways'.

Our tacit assertion that $\mathbf{E}$ and $\mathbf{B}$ are vectors is consistent with experiment and implies that when the effects of several sets of charges are to be combined, the resulting contributions to $\mathbf{E}$ or $\mathbf{B}$ may be added according to the vector addition laws.

There is another important phenomenon to disclose. If the charge array responsible for $\mathbf{B}$ is such that the field $\mathbf{B}$ is changing in time then there is a further extra force, even on a test charge *at rest*. We may denote this extra force per unit of test charge as $\mathbf{E}_i$, where the suffix 'i' refers to the phenomenon of *induction* which we are describing. In an air-cored transformer it is the accelerating electrons in the primary winding which produce a changing magnetic field and stimulate the electrons in the secondary winding into motion. Exactly how $\mathbf{E}_i$ is determined we leave until later.

It is customary, if confusing, to lump $\mathbf{E}_s$ and $\mathbf{E}_i$ together into one vector $\mathbf{E}$, the *electric field*, which is the force due to all causes upon unit test charge, *at rest*. But $\mathbf{E}_i$, being essentially a magnetic phenomenon, has rather more in common with $\mathbf{v} \times \mathbf{B}$ than with $\mathbf{E}_s$, as will emerge later.

To sum up, the total force on a moving particle of charge Q in the vicinity of other charge arrays is

$$Q(\mathbf{E} + \mathbf{v} \times \mathbf{B}).$$

This is called the *Lorentz force*.

Example 2.1 Find an energy equation for a charged particle moving freely in fields **E** and **B**.

Newton's law becomes

$$M\frac{\mathrm{d}\mathbf{v}}{\mathrm{d}t} = Q(\mathbf{E} + \mathbf{v} \times \mathbf{B}),$$

$\mathrm{d}\mathbf{v}/\mathrm{d}t$ being the acceleration and M the mass of the particle. Scalar multiplying by **v**, which equals $\mathrm{d}\mathbf{r}/\mathrm{d}t$ if **r** is the particle's position vector, produces a power equation

$$M\frac{\mathrm{d}\mathbf{v}}{\mathrm{d}t}\cdot\mathbf{v} = \frac{\mathrm{d}}{\mathrm{d}t}(\tfrac{1}{2}M\mathrm{v}^2) = Q(\mathbf{E} + \mathbf{v} \times \mathbf{B})\cdot\mathbf{v} = Q\mathbf{E}\cdot\frac{\mathrm{d}\mathbf{r}}{\mathrm{d}t},$$

since $(\mathbf{v} \times \mathbf{B})\cdot\mathbf{v} = 0$. (The magnetic force is always normal to the motion and so does not participate in the energetics.) The power equation may be integrated into the energy equation:

$$Q\int \mathbf{E}\cdot\frac{\mathrm{d}\mathbf{r}}{\mathrm{d}t}\,\mathrm{d}t = \tfrac{1}{2}M\mathrm{v}^2 + \text{const.},$$

which states how the work done by the electric force creates kinetic energy. Integrals of the kind appearing on the left-hand side will be discussed in chapter 3. The following example has **E** absent and the kinetic energy constant.

Example 2.2 A charged particle moves freely in a plane under a uniform field **B** perpendicular to the plane. There is no electric field. Show that the speed **v** of the particle is constant, as is the radius of curvature of its path, and find the frequency with which it circulates in terms of B, its charge Q and mass M.

The force on the particle $Q\mathbf{v} \times \mathbf{B}$ is normal to its path and in the plane of its path (being perpendicular to **B**). Its magnitude is QvB, for **v** and **B** are perpendicular. The component of acceleration in the direction of **v** (i.e. $\mathrm{d}v/\mathrm{d}t$) must vanish, there being no force component in that direction, and so v is constant. The transverse force QvB is responsible for the centripetal acceleration v^2/R (if R is the radius of curvature) and so

$$QvB = Mv^2/R \quad \text{and} \quad R = Mv/QB = \text{const.}$$

Thus the orbit is a circle, traversed with a frequency f, where

$$f = v/2\pi R = QB/2\pi M,$$

which is independent of v or R. Problem 2.2 explores some other simple aspects of charged-particle dynamics.

2.2 Continuous media, conductors and dielectrics

Though the motion of charged particles is of considerable practical interest – in forming a television picture, for instance – there is greater interest in situations where there are so many particles, charged or otherwise, present that a bulk or macroscopic treatment is called for. In a metal conductor there are roughly 10^{29} mobile electrons per cubic metre and when they move relative to the positively charged host material, very large magnetic forces can result. But overall electric forces are usually very small because the enormous charge of the electrons ($\approx 10^{10}$ coulombs per cubic metre) is normally more or less balanced by the positive charges occupying the same region of space. (If they were not, very large potential differences indeed would result.)

Later on we shall see that these ideas explain why pre-relativistic electrodynamics accounts so adequately for most phenomena involving media. The main point to note at this stage is that usually current is due to relative motion of opposite charges and is therefore independent of the state of motion of the observer (see problem 2.5(*c*)). Its consequence, the magnetic field, is then likewise absolute, i.e. independent of the frame of reference.

The balance between positive and negative charges in a given small region will not in general be exactly maintained. Then there will be a *net* charge density q, measuring the excess of positive charge over negative charge per unit volume. Frequently the imbalance of charge is particularly marked in a thin layer at the surface of bodies and then a density per unit area is appropriate.

In media it is important to distinguish between two kinds of charge: *free* and *bound*. Free charges are able to move indefinitely, as in the case of conduction electrons in a metal or ions in an electrolyte, whereas bound charges have a definite home and only venture small distances which depend in some way on the electric field prevailing, as occurs in a dielectric. There is of course thermal agitation in addition to these effects. The process of creating bound charge migration is called *polarisation*: an overworked word which has several other meanings. The magnitude P of the polarisation is defined as the net amount of bound charge that has migrated across a small surface element $\mathbf{a}$, expressed per unit area, the direction of $\mathbf{a}$ being chosen to intercept the maximum charge migration. P is measured in coulombs per square metre. Polarisation is in fact a vector $\mathbf{P}$, oriented in the direction of $\mathbf{a}$. For other orientations of an area element $\mathbf{a}$, the charge migration intercepted is $\mathbf{P}\cdot\mathbf{a}$, a new use for scalar products. $\mathbf{P}$ is often parallel and proportional to $\mathbf{E}$. It is most easily comprehended in the case of a parallel-faced slab, uniformly polarised normally to its faces. Then no net charge appears inside the slab, because as much charge moves into as moves out of each volume element, but surface charge layers appear of density $\pm P$ per unit area. (See also problem 2.3 for an alternative view of polarisation.)

2.3 Current flow

If, over all the charge particles in a given small element, we sum up the quantities $Q\mathbf{v}$ and divide by the volume of the element we get a vector $\mathbf{j}$ which measures the net intensity of charge flow per unit area and time. (*N.B.* this **j** should not be confused with the unit vector in the y-direction.) The reader should satisfy himself of the correctness of this idea, at the very least in the case of a unit cube in which all charges are uniformly distributed and moving at the same velocity parallel to one edge. (Consider how much charge passes through the end face in time dt.) The vector **j** is analogous to the heat-flow intensity vector which has already been discussed. We shall refer to **j** as the *current intensity*, measured in amperes per square metre.

It can take three main forms, which often co-exist:

(*a*) *Convection current* $q\mathbf{v}$ occurs whenever a net charge density q is attached to a medium that is moving bodily with a velocity **v** (as on the belt of a van der Graaf generator). Here the summing of $Q\mathbf{v}$ terms is particularly simple.
(*b*) *Polarisation current* $\partial\mathbf{P}/\partial t$ occurs whenever the polarisation **P** of a dielectric is changing at a given point. Then the moving bound charges constitute a current intensity equal to $\partial\mathbf{P}/\partial t$, which will be inclined to **P** when **P** is changing its direction.
(*c*) *Conduction current* $\mathbf{j}_c$ occurs whenever free charges drift under the influence of electric or magnetic fields or because of diffusion under thermal, concentration or chemical gradients.

Notice that (*b*) and (*c*) do not require departure from net neutrality in the form of non-zero q, but they do depend on motion of charge relative to neutral particles or charges of opposite sign. They thus involve collisions and are therefore in general thermodynamically irreversible or dissipative, giving rise to the *lossy* dielectric and the conductor with *resistance*, respectively.

Example 2.3 If copper contains 1.14×10^{29} conduction electrons per cubic metre, calculate their mean drift velocity when j has a magnitude of 10^7 A/m^2 (a rather high intensity). (Electron charge $= 1.6 \times 10^{-19}$ coulombs). If the current alternates at 50 hertz calculate the amplitude of travel of the average electron.

Charge q per unit volume $= (-1.14 \times 10^{29}) \times (1.6 \times 10^{-19})$. Mean electron velocity $v = j/q = 0.55$ mm/s, the smallness of which may surprise the reader. If $v = dx/dt = 0.55 \cos \omega t$, with $\omega = 100\pi$, x being the electron displacement in millimetres from its mean position,

$$x = (0.55/\omega) \sin \omega t,$$

and the amplitude of travel equals $\pm\, 0.55/100\pi$ mm $= \pm\, 1.75\ \mu$m, a remarkably short distance. (*N.B.* We have taken the current magnitude as being the peak rather than the r.m.s. value.)

2.4 Body forces

Consider now a region within a continuous medium in which the local macroscopic electric and magnetic fields are **E** and **B**. This is a somewhat sophisticated concept because, at the level of molecular detail, these fields suffer violent variations in the vicinity of each charged particle – indeed these variations are a description of some of the collision forces that we have just mentioned – and so we must allow **E** and **B** to refer to blurred-out average fields over the many particles in each element. The error in this we shall correct by allowing for collisions.

If we add up the Lorentz forces $Q(\mathbf{E} + \mathbf{v} \times \mathbf{B})$ over all the charged particles in a given element, we find that the total body force per unit volume is

$$q\mathbf{E} + \mathbf{j} \times \mathbf{B}$$

because, per unit volume, all the Qs add up to q and the $Q\mathbf{v}$s add up to **j** (and any collisional reactions cancel out between the particles in the element). We have now found another very important manifestation of the vector product idea in the shape of $\mathbf{j} \times \mathbf{B}$, the magnetic force per unit volume due to a current of intensity **j**. Another version of this occurs when we deal with a wire (a conductor long compared with its girth) which can carry a total axial current **J**. Summing $\mathbf{j} \times \mathbf{B}$ over the cross-section leads to the result:

$$\text{magnetic force per unit length of wire} = \mathbf{J} \times \mathbf{B}.$$

How the $\mathbf{j} \times \mathbf{B}$ force is communicated to the bulk material is explored in problem 2.5.

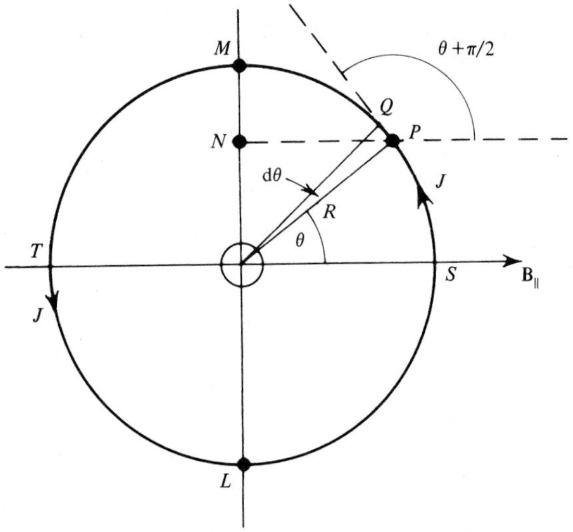

Fig. 2.1

Example 2.4 A plane, circular loop of wire carries a current J in a uniform magnetic field **B**. Relate the magnetic torque on the loop to its area A.

Let the magnetic field have components $\mathbf{B}_\perp$, perpendicular to the plane of the loop, and $\mathbf{B}_\parallel$, in the plane of the loop. Consider the torque, or moment, about the diameter LM, perpendicular to $\mathbf{B}_\parallel$, as shown in figure 2.1. The magnetic force per unit length on the element PQ of the wire has a component $\mathbf{J} \times \mathbf{B}_\parallel$, of magnitude $JB_\parallel \sin(\theta + \pi/2)$ or $JB_\parallel \cos\theta$, perpendicular to and into the paper, and a component $\mathbf{J} \times \mathbf{B}_\perp$, in the plane of the paper and therefore having no moment about LM. Element PQ has a length $Rd\theta$ if R is radius and contributes

$$(JB_\parallel \cos\theta)(Rd\theta)(R\cos\theta)$$

to the moment about LM, for $PN = R\cos\theta$. Hence

$$\text{total moment about } LM = \int_0^{2\pi} JB_\parallel R^2 \cos^2\theta d\theta = JB_\parallel \pi R^2 = JAB_\parallel.$$

The reader should verify that there is no moment about the other diameter ST or about the third axis through O perpendicular to the paper. Note that if we let $\mathbf{A}$ denote a vector of magnitude A directed normally *out of* the paper, then these conclusions can be summarised by saying

$$\text{vector moment about } O = J\mathbf{A} \times \mathbf{B},$$

for $J\mathbf{A} \times \mathbf{B} = J\mathbf{A} \times \mathbf{B}_\parallel + J\mathbf{A} \times \mathbf{B}_\perp$, in which the latter term vanishes because $\mathbf{A}$ and $\mathbf{B}_\perp$ are parallel. $J\mathbf{A} \times \mathbf{B}_\parallel$ lies in the direction $\overrightarrow{LM}$ and is of magnitude $JAB_\parallel$.

Problem 2.4 develops a more general version of this result by different methods.

2.5 Moving media

The rate of change of a magnetic field in general looks different to observers moving relative to each other. This is particularly obvious in the case of a non-uniform magnetic field which is steady in time relative to one observer, but not as seen by a moving observer. Thus the induction field $\mathbf{E}_i$ (and so also the electric field $\mathbf{E}$ which includes it) appears different to different observers, or media, in relative motion. We say that $\mathbf{E}$ is a *relative* (as distinct from an *absolute*) quantity, i.e. dependent on the state of motion of the observer. But in the pre-relativistic electrodynamics that we are studying, the magnetic field $\mathbf{B}$ is an absolute quantity, as was remarked previously. Because a particle's acceleration looks the same if observed from two frames of reference that are moving relatively at a *constant* velocity, we note that the Lorentz force $\mathbf{E} + \mathbf{v} \times \mathbf{B}$ on a travelling unit charge must be *absolute* also, i.e. it takes the same value relative to either frame. But $\mathbf{v}$ and so $\mathbf{v} \times \mathbf{B}$ also are clearly *relative*, which confirms our belief that $\mathbf{E}$ is a relative quantity.

Let the charge have velocity $\mathbf{v}$ relative to a reference frame 1 and let frame 2 move with the particle. Then, equating Lorentz forces, we have

$$\mathbf{E}_1 + \mathbf{v} \times \mathbf{B} = \mathbf{E}_2 + \mathbf{0}, \text{ (suffix denotes frame)}$$

which shows that $\mathbf{E}$ as seen from a frame moving at velocity $\mathbf{v}$ equals $\mathbf{E} + \mathbf{v} \times \mathbf{B}$, where $\mathbf{E}$ is the electric field in the original ('stationary') frame. The change in the $\mathbf{E}_i$ part of $\mathbf{E}$ upon change of frame is exactly compensated by the change in $\mathbf{v} \times \mathbf{B}$

due to the change in **v**. This confirms our earlier assertion that $\mathbf{E}_i$ and $\mathbf{v} \times \mathbf{B}$ are interchangeable aspects of the same basic magnetic phenomenon.

2.6 Ohm's law

This empirical result is the proportionality-relation between conduction current and electric field which applies to most conductors when free charge carriers drift, colliding with the background material. The familiar equation $V = JR$ for the potential difference V across a discrete resistance R bearing a total current J becomes, in terms of current intensity **j** in a stationary conductor,

$$\mathbf{j} = \sigma \mathbf{E} \quad \text{or} \quad \mathbf{E} = \tau \mathbf{j},$$

in which σ = electric conductivity and τ = electric resistivity = $1/\sigma$. This law shows that, in microscopic terms over many particles and many collisions, the average electric force on a drifting charge (proportional to **E**) just balances the average collisional drag force (evidently proportional to **j** and so to the average drift velocity). But note that at sufficiently high frequencies of alternating current, the rates of change of **j** in time may be sufficient for the mean accelerations to be significant, in which case Ohm's law gains an inertia term involving $\partial \mathbf{j}/\partial t$. (Similar effects can occur in dielectric media where a basic linear equilibrium relation making **P** proportional to **E** may gain first a 'drag' term proportional to the current $\partial \mathbf{P}/\partial t$ and then an inertia term proportional to $\partial^2 \mathbf{P}/\partial t^2$.) Problem 2.5 explores another phenomenon, the *Hall effect*, which becomes significant in conductors with relatively few charge carriers (such as semiconductors or tenuous ionised gases) because of the magnetic force on the drifting charges.

Example 2.5 Sea-water has a resistivity of 0.3 ohm-metre and its polarisation **P** may be taken as $79\epsilon_0\mathbf{E}$, where $\epsilon_0 = 8.854 \times 10^{-12}$. Find the ratio of amplitudes of the conduction and polarisation current intensities when E is oscillating at 100 megahertz.

Let $E = A \sin \omega t$. Then conduction current intensity $= E/\tau = (A/\tau) \sin \omega t$ and polarisation current intensity $= 79\epsilon_0(\mathrm{d}E/\mathrm{d}t) = 79\epsilon_0\omega A \cos \omega t$. Ratio of amplitudes of intensities $= 1/(79\tau\epsilon_0\omega) = \frac{1}{7}$ approximately, if $\omega = 2\pi \times 10^8$ and $\tau = 0.3$.

In electrotechnology there is great interest in moving conductors. Ohm's law then becomes

$$\tau \mathbf{j} = \mathbf{E} + \mathbf{v} \times \mathbf{B}$$

because the conductor, moving at velocity **v**, responds to the electric field $(\mathbf{E} + \mathbf{v} \times \mathbf{B})$ observed relative to a frame moving with it, if **E** is still referred to the stationary frame. The quantity $\mathbf{v} \times \mathbf{B}$ is seen to be just as capable of driving electric currents against resistance as is **E** and for that reason is categorised as one of the several kinds of *electromotive force* or *e.m.f.*, for short. (More strictly, $\mathbf{v} \times \mathbf{B}$ is e.m.f. per unit distance, just as **E** is measured in volts per metre.) An important point is that the presence of $\mathbf{v} \times \mathbf{B}$ does not always imply current flow. Often an electrostatic field **E** appears, just sufficient to balance $\mathbf{v} \times \mathbf{B}$. An example is provided by an isolated conducting body moving through a uniform magnetic field. It acquires surface charges which allow $\mathbf{E} + \mathbf{v} \times \mathbf{B}$ to vanish inside it.

As we found with moment and angular velocity earlier, the transformation properties of the triple scalar product often lead to interesting energy conversion statements. The same is true in the present context, for

$$\mathbf{v}\cdot\mathbf{j} \times \mathbf{B} + \mathbf{j}\cdot\mathbf{v} \times \mathbf{B} = 0,$$

in which either term is a *power density* (watts per cubic metre). The first term is mechanical power, done by the $\mathbf{j} \times \mathbf{B}$ force on the moving medium, the second term is electrical power, expressed in the form (current) times (e.m.f. component in the current direction). This equation therefore elegantly and succinctly expresses the basic energy conversion process of electrotechnology. In an electric motor armature winding, the terms are respectively positive and negative, for $\mathbf{v} \times \mathbf{B}$ opposes **j** – hence the expression '*back* e.m.f.'.

2.7 Magnetisation

So far we have treated magnetism as a phenomenon concerning charge-particle motions and electric currents. While not embarking on a full exposition of the involved and extensive subject of material magnetisation, we must for completeness refer to the practically important fact that magnetic fields can have sources other than moving charges. The effect is most marked with ferromagnetic materials, where nearby currents produce greatly enhanced magnetic effects because of the material's own response. With permanent magnetisation, a magnetic field can persist even in the absence of currents.

A somewhat fuller discussion of magnetisation is postponed until chapter 6 when the mathematical tools that we will have then developed will make the task easier. All statements up to that point will assume that magnetisation is absent.

This concludes our review of vector algebra applied to physical phenomena. The emphasis so far has been towards electromagnetism, the field in which the vector product concept finds its finest flowering, but as we move on to consider vector calculus in subsequent chapters, the view will broaden to include a wider range of phenomena of practical importance, ranging from ground water flow to neutron diffusion in nuclear reactors.

Problems 2

2.1 Since the total electrostatic force due to another charge on N identical charges clumped together is N times the force on one of the N charges due to the other (at the same separation), it is reasonable to take charge as an additive entity and say that the force on a charge or collection of charges is proportional to its '*total*' charge. The force is the mutual reaction between *two* charges, however, and so must be proportional to the product of their magnitudes (measured as multiples of some standard charge). For efficient mathematical description, it is desirable to relate the direction of the reaction between two charges to their 'nature', as evidenced by this imagined experiment (which is consistent with physical truth):

Three small charged bodies A, B and C are each arranged in turn so as to be at a standard distance from one other (the third being remote) with the following results: (*a*) A and B repel, (*b*) A and C attract, (*c*) B and C attract, the size of the force being the same in each case.

Is it possible to summarise this experiment with the single formula: force between charges (at a standard distance) = const. × $(\text{charge})_1$ × $(\text{charge})_2$ according to any of the following conventions?

(I) All charges are taken as positive (compare gravity), the sign convention for force being: (i) attraction positive (compare gravity) or (ii) repulsion positive.

(II) The idea that sign can be attributed to charge in the formula is adopted, the sign convention for force being: (i) attraction positive or (ii) repulsion positive. Try taking charge A as positive and negative in turn. Is it sufficient to have just two kinds of charge?

Having decided on a consistent convention (can you tell if it is the one *actually* adopted in standard theory?) examine the force predicted by the formula for the reaction between any one of A, B and C and the other two, clumped together, again at the standard distance from the single charge. (Use the 'total' charge for the clump.) Is this force consistent with the sum of the forces on the individual charges in the clump due to the third charge? (If so, this confirms the validity of the idea of *net* charge.)

2.2 (*a*) A charged particle moves freely at constant velocity $\mathbf{v}$ in a straight line under uniform fields $\mathbf{E}$ and $\mathbf{B}$. Show that $\mathbf{E}$ and $\mathbf{B}$ must be perpendicular to each other and that $\mathbf{v}_\perp$, the component of $\mathbf{v}$ perpendicular to $\mathbf{B}$, must equal $\mathbf{E} \times \mathbf{B}/\mathbf{B}^2$ (which is called 'the drift velocity in crossed fields'), whatever the sign and magnitude of the particle's charge. What is $\mathbf{E}$ as seen by an observer moving with the drift velocity? (Consult section 2.5.)

(*b*) A charged particle is initially at rest in uniform, perpendicular fields $\mathbf{E}$ and $\mathbf{B}$. Find its subsequent motion by first referring the problem to a frame moving with the drift velocity. (Exploit example 2.2.)

2.3 A pair of equal and opposite charges Q and $-Q$, separated by a small distance $\mathbf{r}$, is called a dipole. An example is a polarised molecule. The vector quantity $Q\mathbf{r}$ is called the *dipole strength* $\mathbf{p}$. (It is also called 'dipole moment' but it is *not* a moment of forces.) Polarisation $\mathbf{P}$ can then be regarded as net dipole density per unit volume.

Show that the moment of forces on a dipole in a uniform electric field $\mathbf{E}$ is $\mathbf{p} \times \mathbf{E}$, a couple, but there is no overall force on the dipole.

A dielectric is subjected to a rotating electric field and, because of dissipative losses, **P** is not quite parallel to **E**. Show that there is a torque on each volume element.

2.4 Consider a plane rectangular wire loop of area A, carrying a current under a uniform magnetic field **B**, with components $\mathbf{B}_\perp$ and $\mathbf{B}_\parallel$ respectively perpendicular and parallel to the plane of the loop. One pair of sides of the loop is made parallel to $\mathbf{B}_\parallel$. Show that there is no overall force on the loop and that the torque on the loop can be written as the vector $J\mathbf{A} \times \mathbf{B}$, if we adopt the convention that the vector **A**, representing the plane area spanning the loop, lies in the direction in which **J**'s circulation looks clockwise. Show that the same results hold for a loop that is a right-angled triangle with one shorter side parallel to $\mathbf{B}_\parallel$.

Deduce that the same results hold for a plane loop of any shape, by dividing its area into a large number of small elements of the above types, round each of which J circulates, noting that all current and force contributions cancel except at the periphery, when the elements and their effects are superposed. Referring to problem 2.3, observe the analogy with $\mathbf{p} \times \mathbf{E}$ for an electric dipole; a small current loop $J\mathbf{A}$ is virtually a *magnetic dipole*, at least as regards the force upon it. (This idea plays some part in understanding material magnetisation.)

*2.5 (*N.B.* An asterisk indicates a section which is harder and may be omitted.)

(*a*) Consider a stationary conductor containing n free electrons per unit volume under an electric field **E**. If we treat all the electrons as moving at their average velocity **u** relative to the bulk material, the resulting current density **j** will equal $-ne\mathbf{u}$, if $-e$ is the charge on an electron. The bulk material is composed of neutral particles and/or ions, with a net total charge of ne per unit volume to preserve overall neutrality.

If the average drag force on an electron is $-k\mathbf{u}$ (opposing the relative motion) which balances the electric forces in steady current flow, show that

$$\text{resistivity } \tau = k/ne^2.$$

(*b*) If there is also a magnetic field **B** present, rewrite the balance of forces on the average electron and deduce that Ohm's law becomes

$$\mathbf{E} = \tau\mathbf{j} + \mathbf{j} \times \mathbf{B}/ne.$$

The last term is called the *Hall effect*. It arises because the electrons are moving across the magnetic field and **E** has to include a component *perpendicular* to **j** to balance the sideways magnetic force on them. This component is known as the *Hall field*. It is a very small term in Ohm's law in metals (because n is so big – see example 2.3) but it is not negligible in semiconductors or ionised gas. But even in metals the Hall electric field is physically important because it is responsible for transmitting the magnetic $\mathbf{j} \times \mathbf{B}$ force from the electrons (which feel it direct as $\mathbf{u} \times \mathbf{B}$) to the bulk material in the form of an *electric* force $(\mathbf{j} \times \mathbf{B}/ne)ne$. This force can be big even when the Hall field is weak because *ne* is so enormous. (This is a good example of the important idea that an effect can be negligible from some aspects while not from others, which is often central to the mathematical modelling of physical situations.) Consider the starting process for a current in a wire in a magnetic field. How and why is the Hall field and 'magnetic' force set up?

(*c*) Now consider a conductor moving at velocity $\mathbf{v}$, with the average conduction electron moving at $\mathbf{v} + \mathbf{u}$. The electron current density is $-ne(\mathbf{u} + \mathbf{v})$ but the ion current density is $ne\mathbf{v}$ and so the total current density $\mathbf{j}$ is still $-ne\mathbf{u}$ and depends only on the *relative* motion of the charges.

Rewrite the balance of forces on the average electron and deduce that

$$\mathbf{E} + \mathbf{v} \times \mathbf{B} = \tau\mathbf{j} + \mathbf{j} \times \mathbf{B}/ne.$$

(*d*) Consider the power density equation generated by scalar-multiplying the above equation by $\mathbf{j}$. It relates the electrical energy input $\mathbf{E} \cdot \mathbf{j}$ to the mechanical work and the dissipation. Does the Hall effect involve dissipation?

Consider also the power equations for the electrons and the bulk material separately, deduced by scalar-multiplying the force balance equations by the relevant velocity, and identify the various terms. In particular, note that the mechanical work $\mathbf{v} \cdot \mathbf{j} \times \mathbf{B}$ is actually done on the bulk material by the Hall *electric* field. Thus there is no conflict with the fact that a *magnetic* force cannot do work on a moving charge.

3

Line and surface integrals

3.1 The prerequisite knowledge for later chapters

The remainder of this book is preoccupied with vector calculus, both integral and differential. The reader may therefore welcome a reminder of the main mathematical ideas which he should command before proceeding further. Some revision problems are also provided.

Multiple integrals

The reader should be able to set up and evaluate double integrals such as

$$\int_a^b \mathrm{d}y \int_{x_1(y)}^{x_2(y)} f(x, y)\, \mathrm{d}x, \quad \text{also written} \int_a^b \int_{x_1}^{x_2} f\, \mathrm{d}x\, \mathrm{d}y,$$

over a domain such as that shown in figure 3.1, appreciating that it is a summation first over elements $\mathrm{d}x\mathrm{d}y$ along AB at constant y as x ranges from x_1 to x_2, and secondly a further summation over strips AB as y ranges from a to b, x having vanished from consideration. Note that the inner limits $x_1(y)$ and $x_2(y)$ express the shape of the domain. By taking appropriate inner limits $y_1(x)$ and $y_2(x)$ it is possible to integrate for x and y in the opposite order.

Solid angle

The solid angle enclosed by a pencil of rays through a point O is equal to the area delineated by the pencil on a sphere of radius R centred at O divided by R^2.

The complete solid angle, taking in all directions, is 4π. Solid angles are additive, scalar quantities.

Partial calculus

(i) *Commutative operations*. Provided the independent variable in one operation is constant during the other operation, the order of differentiations and integrations may be revised. (We ignore pathological cases, e.g. where the functions are singular.)

(*a*)

$$\frac{\mathrm{d}}{\mathrm{d}y} \int_{x_1}^{x_2} f(x, y)\mathrm{d}x = \int_{x_1}^{x_2} \frac{\partial f}{\partial y} \mathrm{d}x$$

in the case where x_1 and x_2 are independent of y.

(*b*)

$$\frac{\partial^2 f(x, y)}{\partial x \partial y} \quad \text{which means} \quad \frac{\partial}{\partial x}\left\{\left(\frac{\partial f}{\partial y}\right)_x\right\}_y, \quad \text{equals} \quad \frac{\partial^2 f}{\partial y \partial x}.$$

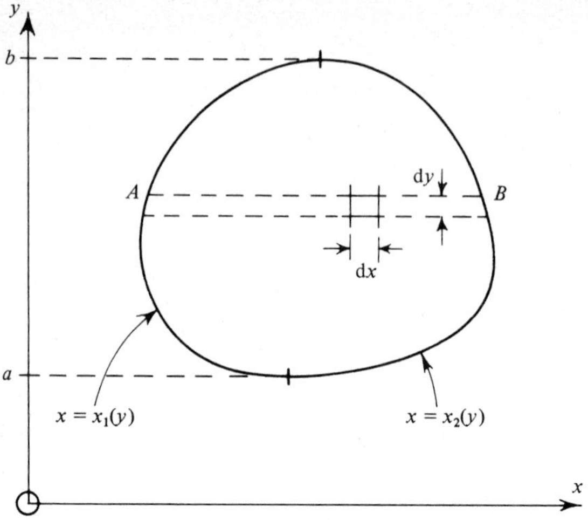

Fig. 3.1

(ii) *Taylor's series* for function of several variables

$$\delta\phi = \phi(x+\delta x, y+\delta y) - \phi(x,y)$$

$$= \delta x\frac{\delta\phi}{\delta x} + \delta y\frac{\partial\phi}{\partial y} + \frac{1}{2!}\left\{(\delta x)^2\frac{\partial^2\phi}{\partial x^2} + 2\delta x\delta y\frac{\partial^2\phi}{\partial x\partial y} + (\delta y)^2\frac{\partial^2\phi}{\partial y^2}\right\} + \text{etc.}$$

(iii) *Total differentials* (See also Appendix I.). If $d\phi$ is written for $\phi(x+dx, y+dy) - \phi(x,y)$ and only first-order small quantities are retained,

$$d\phi = \frac{\partial\phi}{\partial x}dx + \frac{\partial\phi}{\partial y}dy. \quad \text{(Compare Taylor's series above.)}$$

If

$$d\phi = Fdx + Gdy, \quad \text{then} \quad F = \partial\phi/\partial x, \quad G = \partial\phi/\partial y.$$

If

$$x = x(r,\theta), \quad y = y(r,\theta),$$

$$\frac{\partial\phi}{\partial r} = \frac{\partial\phi}{\partial x}\frac{\partial x}{\partial r} + \frac{\partial\phi}{\partial y}\frac{\partial y}{\partial r}, \quad \text{etc.,}$$

the convention being that θ is constant in $\partial\phi/\partial r$, y is constant in $\partial\phi/\partial x$, etc. If

$$\phi = \phi(z) \quad \text{where} \quad z = z(x,y),$$

$$\frac{\partial\phi}{\partial x} = \frac{d\phi}{dz}\frac{\partial z}{\partial x}.$$

(iv) *The change to polar coordinates* (r,θ). The identity

$$\frac{\partial^2\phi}{\partial x^2} + \frac{\partial^2\phi}{\partial y^2} \equiv \frac{\partial^2\phi}{\partial r^2} + \frac{1}{r}\frac{\partial\phi}{\partial r} + \frac{1}{r^2}\frac{\partial^2\phi}{\partial\theta^2}$$

is used in some of the more advanced problems.

Problems 3A

3.1 A sheet of metal is bounded by the positive x and y axes and a quadrant of a circle of unit radius centred on the origin. Its thickness is kxy, where k is a constant. Show that its volume equals $k/8$ by using (a) polar coordinates, (b) cartesians.

3.2 An atomic bomb at height h radiates heat equally in all directions. What fraction of the heat reaches the flat ground beneath within a circle of radius h centred beneath the bomb? (Use the solid angle concept.)

3.3 Verify that

$$\frac{\mathrm{d}}{\mathrm{d}t}\int_0^\infty \phi(x,t)\mathrm{d}x = \int_0^\infty \frac{\partial \phi}{\partial t}\mathrm{d}x$$

in the case where $\phi = \mathrm{e}^{-xt}$, t positive.

3.4 (a) If ϕ and ψ are functions of x and y such that

$$\frac{\partial \phi}{\partial x} = \frac{\partial \psi}{\partial y} \quad \text{and} \quad \frac{\partial \phi}{\partial y} = -\frac{\partial \psi}{\partial x}$$

show that

$$\frac{\partial^2 \phi}{\partial x^2} + \frac{\partial^2 \phi}{\partial y^2} = 0.$$

(b) If ϕ, u and v are functions of x and y, and $\mathrm{d}\phi = u\mathrm{d}x + v\mathrm{d}y$ show that

$$\frac{\partial u}{\partial y} = \frac{\partial v}{\partial x}.$$

3.5 If ϕ is a scalar field, show that its change $\mathrm{d}\phi$ associated with a small change $\mathrm{d}\mathbf{r}$ in the position vector $\mathbf{r}$ is equal to $\mathbf{G}\cdot\mathrm{d}\mathbf{r}$ where $\mathbf{G}$ is the vector

$$\mathbf{i}\frac{\partial \phi}{\partial x} + \mathbf{j}\frac{\partial \phi}{\partial y} + \mathbf{k}\frac{\partial \phi}{\partial z}.$$

(*N.B.* $\mathrm{d}\mathbf{r}$ has components $\mathrm{d}x$, $\mathrm{d}y$, $\mathrm{d}z$.)

3.6 If $\partial\phi/\partial t = \partial^2\phi/\partial x^2$ and $\phi = f(z)$, where $z = x/\sqrt{t}$, show that

$$2\frac{\mathrm{d}^2 f}{\mathrm{d}z^2} + z\frac{\mathrm{d}f}{\mathrm{d}z} = 0.$$

3.7 If $v = f(u)$ where u is a function of x and y, show that

$$\frac{\partial v}{\partial x}\frac{\partial u}{\partial y} = \frac{\partial v}{\partial y}\frac{\partial u}{\partial x}.$$

3.8 (a) The variable depth z of the water in a channel is given by some expression $z(x, t)$ where x = distance downstream from some fixed point and t is time. Show that dz/dt, the rate of change of depth as observed from a boat travelling downstream at an *absolute* velocity v, equals

$$\frac{\partial z}{\partial t} + v\frac{\partial z}{\partial x}.$$

(b) If in (a) $z = f(w)$, where $w = x - vt$, what is dz/dt? Discuss your result.

3.2 Applied vector field theory – an introduction

Before becoming immersed in detail, the reader will benefit from receiving an overall view of the structure of the subject as expounded in this and subsequent chapters. Figure 3.2 categorises it in two ways: mathematical and practical.

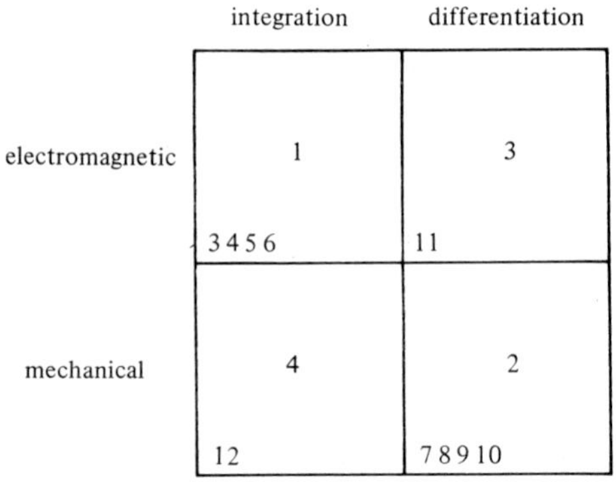

Fig. 3.2. 'The card of the course'.

From the *mathematical* standpoint, we shall at each stage be preoccupied either with *integration*, the summing of contributions over space and time, or with *differentiation*, the analysis of phenomena in terms of variation in space and time. Cutting across this division from the *practical* point of view, we shall turn sometimes to *electromagnetic* phenomena and at other times to *mechanical* ones such as fluid mechanics or heat flow, many of which are of great interest to civil engineers also.

At each stage the area of application first encountered is that where it is easiest to see the need for the mathematical concepts and to acquire a physical feel for them. Though this approach means that the treatment of each area of application is somewhat episodic, the intention is that understanding of each piece of mathematics applied to an appropriate physical topic should by analogy help the understanding of another topic involving the same mathematical concepts.

As the chain of ideas reaches its end and closes on itself, interlinking in several pleasing ways, the reader should find that what at times appeared to be an unmanageably large collection of discrete concepts and facts becomes finite, coherent and mutually supporting. But this is no book for the dyed-in-the-wool electrical or mechanical enthusiast who will have no truck with the other subject.

Returning to figure 3.2, we see that it is divided mathematically and practically so as to yield four zones, numbered in the order in which we visit them. The corner numbers refer to the relevant chapters. We start in Zone 1 (with some forays into Zone 4) because students find vector integration easier than differentiation. Integration suits electromagnetism, moreover, because the subject deals with action at a distance – a charge *here* affects another charge *there* – and so we are naturally interested in adding up all the contributions at a point due to causes elsewhere.

On the other hand, mechanical phenomena such as fluid motion depend mainly on purely local effects and there is not action at a distance, apart from body forces such as gravity. A fluid element accelerates because of pressure variations in the vicinity. (Even when fluid dynamicists talk about induced velocities 'due to' a vortex, this is not a statement of cause and effect but a metaphor based on the close mathematical analogy with action-at-a-distance subjects like electromagnetism.) Notice the emphasis on rates of change in space or time which firmly locates fluid dynamics in Zone 2 under differentiation; the fluid's velocity is unsteady (in time) because the pressure is non-uniform (in space). We might indeed subtitle the chapters in Zone 2 as 'a study of non-uniformity'. We visit Zone 2 as soon as possible because there the need to systematise the physical ideas calls forth a whole new arsenal of mathematical weapons of great power and versatility.

Zone 3 is where speculatively we apply these new weapons to a reformulation of electromagnetism, with astonishingly fruitful results, and in particular realise belatedly that electromagnetism is not really about direct action at a distance at all, but about localised phenomena, feeding off non-uniformity, which propagate around as waves.

Finally, in Zone 4, we discover that the integral approach can be very productive in mechanical subjects, and in particular find that the very abstract and apparently unnatural concept of the *circulation integral* enables us to formulate the most profound statement (Kelvin's theorem) about the nature of fluid motion.

It is suggested that the reader returns to figure 3.2 and the above paragraphs at later stages if ever he begins to feel disoriented.

3.3 Line integrals

In example 2.1 we met the charged particle energy equation

$$Q\int \mathbf{E}\cdot\frac{\mathrm{d}\mathbf{r}}{\mathrm{d}t}\,\mathrm{d}t = \tfrac{1}{2}M\mathbf{v}^2 + \text{const.}$$

This leads straight to Zone 1 because it provides a good illustration of the practical importance of generalising the idea of integration to three-dimensional vector situations. If $\mathrm{d}\mathbf{r}$ is the small displacement undergone by the particle in time $\mathrm{d}t$, then

$$\frac{\mathrm{d}\mathbf{r}}{\mathrm{d}t}\,\mathrm{d}t = \mathrm{d}\mathbf{r} \quad \text{and the integral becomes} \quad \int \mathbf{E}\cdot\mathrm{d}\mathbf{r}.$$

This new kind of integral deserves scrutiny and interpretation, since $\mathbf{E}$ is in general a function of three space coordinates and time. It is one of the various possible kinds of *line integral*. This name is used because the integral is defined along a particular line, trajectory or route in space. Note that *two* statements are necessary to define a curve in space. (One statement, e.g. $z = f(x, y)$, merely defines a surface.) Alternatively x, y and z may each be given in terms of a parameter u.

Ordinary integrals like $\int f(x)\mathrm{d}x$ consist of sums of contributions composed of the product of a small increment $\mathrm{d}x$ of the independent variable x with some weighting function $f(x)$ (the integrand) in the limit as the increments become vanishingly small and infinite in number. In vector analysis we generalise such an excursion along the x-axis into one along an arbitrary curve in space, joining two points A and B as in figure 3.3. This inevitably shows a plane curve but in general the curve

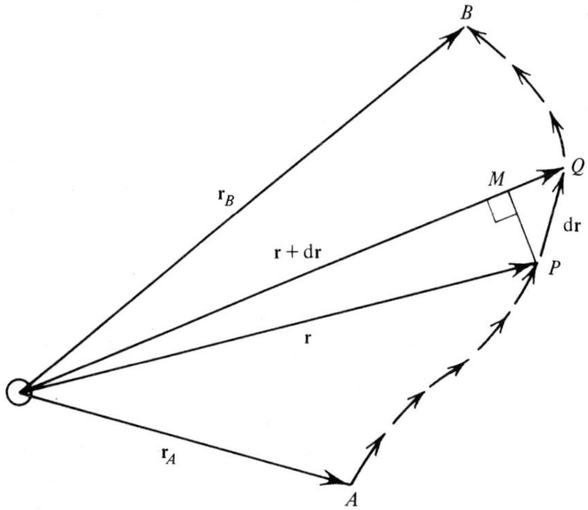

Fig. 3.3

will not lie in a plane. The increment factor becomes either $\mathrm{d}\mathbf{r}$ (i.e. $\overrightarrow{PQ}$), the small change in the position vector $\mathbf{r}$ of a point P moving along the curve, if we require a *vector* independent variable, or a small increment $\mathrm{d}s$, $\mathrm{d}t$ or $\mathrm{d}u$ if we want a *scalar* independent variable, where s = distance along the curve, t = time and u = some general parametric variable. The reader should observe that

$$PQ = \mathrm{d}s = |\mathrm{d}\mathbf{r}| \neq \mathrm{d}|\mathbf{r}| = \mathrm{d}r = MQ.$$

It is noteworthy that the operations 'd' and 'mod' are not commutative.

Just as there are several kinds of products of vectors and scalars, it is possible to have several kinds of line integral, if f and $\mathbf{v}$ are taken as arbitrary scalar and vector variables, defined along the curve:

(*a*) $\int f\mathrm{d}s$, a *scalar*, e.g. the mass of a wire of density f per unit length or the integral $\int \mathrm{d}s/t$ (t = wall thickness) which occurs in the theory of torsion of thin-walled tubes.

(*b*) $\int \mathbf{v}\mathrm{d}s$, a *vector*, e.g. the total electric force on a wire with net charge q per unit length under a field **E** if $\mathbf{v} = (q\mathbf{E})$ (see example 3.1 below), *or* the total electric field at the origin due to a line charge array. (See problem 3.10(*b*).)

(*c*) $\int f\mathrm{d}\mathbf{r}$, a *vector*, which is really a case of (*b*), for if **f** is a tangential vector parallel to $\mathrm{d}\mathbf{r}$, of magnitude f, then $f\mathrm{d}\mathbf{r} = \mathbf{f}\mathrm{d}s$, because $\mathrm{d}\mathbf{r}/\mathrm{d}s$ and $\mathbf{f}/f$ are the same unit vector. Problem 3.9(*a*) is an example.

(*d*) $\int \mathbf{v} \cdot \mathrm{d}\mathbf{r}$, a *scalar*, as in the example which began this discussion and problem 3.9(*b*).

(*e*) $\int \mathbf{v} \times \mathrm{d}\mathbf{r}$, a *vector*, examples of which are discussed later under the heading 'applications to magnetism'.

Types (*b*), (*c*) and (*e*) involve the concept of the limit of a vector sum of a sequence of vectors added nose-to-tail in three dimensions by repeated applications of the triangle law, as the vectors become vanishingly small and infinite in number. A simple degenerate example of this idea is available: the net displacement achieved by the arc of the curve along which these integrals are being taken can be written $\int_A^B \mathrm{d}\mathbf{r}$ (a case of type (*c*) with $f = 1$) equal of course to $\mathbf{r}_B - \mathbf{r}_A$. If the limits are interchanged, the route being unchanged, the values of all five types of integral change sign in the usual way.

It is legitimate to expand types (*b*) and (*c*) into the vector sums of three integrals,

$$\mathbf{i}\int v_x \mathrm{d}s + \mathbf{j}\int v_y \mathrm{d}s + \mathbf{k}\int v_z \mathrm{d}s \quad \text{and} \quad \mathbf{i}\int f\mathrm{d}x + \mathbf{j}\int f\mathrm{d}y + \mathbf{k}\int f\mathrm{d}z$$

respectively, because $\mathrm{d}\mathbf{r}$ has components $(\mathrm{d}x, \mathrm{d}y, \mathrm{d}z)$. Expanding (*d*) gives the sum of three scalars $\int v_x \mathrm{d}x + \int v_y \mathrm{d}y + \int v_z \mathrm{d}z$. The reader should satisfy himself that (*e*) can also be expanded in terms of components.

To define the route and allow evaluation of the integrals, f, v, $\mathrm{d}s$, $\mathrm{d}\mathbf{r}$ must be available in terms of some single parameter, which might be s, t or one of x, y and z. If for instance **v** is given as a function of x, y and z and the curve is specified by the necessary *two* relations in the form $y = f(x)$ and $z = g(x)$, defining the z-wise and y-wise projections of the curve, then on the curve we can in principle eliminate y and z to get **v** and its components purely in terms of x, and also express $\mathrm{d}y$ and $\mathrm{d}z$ as $(\mathrm{d}f/\mathrm{d}x)\mathrm{d}x$ and $(\mathrm{d}g/\mathrm{d}x)\mathrm{d}x$, so that all parts of the integral can be evaluated using x as the parameter. Sometimes polar or other coordinates may be more convenient. (Example 2.4 gave a simple case.)

If the route of integration between two fixed end points is altered by changing $f(x)$ and $g(x)$, then $\mathrm{d}y$, $\mathrm{d}z$ and the integrands f, v_x, v_y, v_z (which depend on x, y and z) will also change. The values of these line integrals between two given points will therefore depend on the route chosen, in general.

The integrand f or **v** need not be a field, defined at all points of a domain in x, y, z space, although it often will be in practice. The integrals are still meaningful if f or **v** is definable only along the curve of integration at the instant at which each point is reached. An example of this is the work done by the forces on a bead sliding on a rough wire in problem 1.11. If **v** is an unsteady field, the line integral may refer to one instant, or may follow a particular time history along the curve.

There is no doubt that type (*d*) is the most important form from a physical standpoint, because of the wide significance of energy equations in which some force **v**, moving its point of application along the curve, does successive amounts of work $\mathbf{v} \cdot d\mathbf{r}$ which accumulate to the integral $\int \mathbf{v} \cdot d\mathbf{r}$. A case of type (*b*) follows.

Example 3.1 The straight line from the origin to the point (1, 1, 1) carries unit charge, uniformly distributed, in the presence of the electric field $\mathbf{E} = x\mathbf{i} - y\mathbf{j} + 0\mathbf{k}$. Calculate the resultant force on the charges and find the point where its line of action cuts the plane $y = 0$.

The length of the line is $\sqrt{3}$ and the charge q per unit length is $1/\sqrt{3}$. On the line, $x = y = z$, $ds = \sqrt{3}\, dx$ and $\mathbf{E} = x\mathbf{i} - x\mathbf{j}$. The resultant force,

$$\mathbf{F} = \int_0^{\sqrt{3}} q\, \mathbf{E}\, ds = \int_0^1 \mathbf{E}\, dx = \left[\frac{x^2}{2}\mathbf{i} - \frac{x^2}{2}\mathbf{j}\right]_0^1 = \tfrac{1}{2}\mathbf{i} - \tfrac{1}{2}\mathbf{j}.$$

The vector moment about 0,

$$\mathbf{M} = \int_0^{\sqrt{3}} \mathbf{r} \times q\, \mathbf{E}\, ds = \int_0^1 \mathbf{r} \times \mathbf{E}\, dx = \int_0^1 (x^2\mathbf{i} + x^2\mathbf{j} - 2x^2\mathbf{k})dx,$$

when $\mathbf{r} = x\mathbf{i} + x\mathbf{j} + x\mathbf{k}$ has been inserted in the vector product, and so

$$\mathbf{M} = \tfrac{1}{3}\mathbf{i} + \tfrac{1}{3}\mathbf{j} - \tfrac{2}{3}\mathbf{k}.$$

If the point of intersection of **F** with the $y = 0$ plane has the position vector

$$\mathbf{R} = X\mathbf{i} + Z\mathbf{k}$$

then

$$\mathbf{M} = \mathbf{R} \times \mathbf{F} = \tfrac{1}{2}Z\mathbf{i} + \tfrac{1}{2}Z\mathbf{j} - \tfrac{1}{2}X\mathbf{k}.$$

Comparing the two values for **M** gives $X = \frac{4}{3}$, $Z = \frac{2}{3}$ (twice) for the required point of intersection.

(*Note.* The fact that this point exists at all depends on the circumstances that this system of forces does reduce to a single resultant force, and not a *wrench*, because **M** has no component parallel to **F** (for $\mathbf{M} \cdot \mathbf{F} = 0$). This allows the equations for X and Z to be consistent. For a contrary case, which *does* reduce to a wrench, take $\mathbf{E} = x\mathbf{i} - y\mathbf{j} + \mathbf{k}$.)

3.4 Closed line integrals, circulation and conservative fields

Great interest attaches to the case where the curve along which we integrate is a *closed loop*, i.e. A and B coincide. It is then immaterial where A, the starting point, lies. We denote such integrals with a special integral sign $\oint$ and, in the case of type (*d*), with a special name, the *circulation* $\oint \mathbf{v} \cdot d\mathbf{r}$.

It is important to realise that, in general, closed line integrals do not vanish in the way that an ordinary scalar integral $\int f(x)dx$ vanishes when the limits are made the same, provided $f(x)$ is single-valued. Even the ordinary scalar integral may not vanish when f is multi-valued and the integral is taken round a closed loop in the f, x plane as in figure 3.4. (The reader will no doubt have met such cases in thermodynamics with $f(x)$ replaced by $p(v)$ or $T(s)$ (where p, v, T, s are pressure, volume,

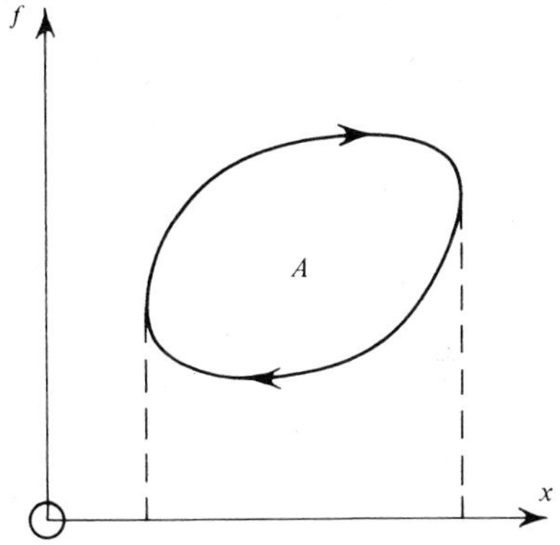

Fig. 3.4

temperature, and entropy respectively) and will know that the integral equals the area A of the loop, with due attention to sign depending on the direction in which the loop is traversed.)

Although in general $\oint \mathbf{v} \cdot d\mathbf{r} \neq 0$, nevertheless there are situations where $\oint \mathbf{v} \cdot d\mathbf{r} = 0$ round *all* loops, and then the field **v** is described as *conservative*. In the case where **v** is a force field like **E** or gravity, it is easy to see why the question whether $\oint \mathbf{v} \cdot d\mathbf{r}$ does or does not vanish is of great interest to the engineer. When **v** is conservative, it is possible to extract limited work from the field by making a finite excursion (not forming a loop) but it is not possible to do this indefinitely on a profit-making basis by repeatedly traversing the same closed loop. It is the fact that under some circumstances the electric field **E** is not conservative which puts the electrical power engineer into business, while in contrast the would-be gravity power engineer goes bankrupt! (The hydroelectric cycle is thermodynamic.)

3.5 Potential

A closed loop may be regarded as an excursion L from A to B followed by a return from B to A by another route M. If **v** is conservative

$$_L\int_A^B \mathbf{v} \cdot d\mathbf{r} + {}_M\int_B^A \mathbf{v} \cdot d\mathbf{r} = 0 \quad \text{and} \quad {}_L\int_A^B \mathbf{v} \cdot d\mathbf{r} = {}_M\int_A^B \mathbf{v} \cdot d\mathbf{r},$$

i.e. the integral takes the same value for all routes $A \to B$. The converse is easily demonstrated.

If we identify one particular end point (the *datum* D) as fixed but allow the other end point P to wander in x, y, z space, then, for a conservative field, $\int_D^P \mathbf{v} \cdot d\mathbf{r}$ depends only on x, y, z (the coordinates of P) and we designate its value as $\phi(x, y, z)$, entitled the (scalar) *potential* of **v** (at P). Thus the vector field **v** has begotten a scalar field ϕ. Going between two points P and Q via D gives

$$\int_P^Q \mathbf{v} \cdot d\mathbf{r} \quad \text{by any route} = \int_P^D \mathbf{v} \cdot d\mathbf{r} + \int_D^Q \mathbf{v} \cdot d\mathbf{r} = \phi_Q - \phi_P.$$

In particular, if PQ is a small excursion over which the variation of $\mathbf{v}$ is negligible this becomes

$$\mathbf{v} \cdot \mathbf{dr} = \mathrm{d}\phi,$$

where $\mathrm{d}\phi$ denotes the increment of ϕ observed in going along the small vector element $\mathbf{dr}$, which has components $(\mathrm{d}x, \mathrm{d}y, \mathrm{d}z)$. If $\mathbf{v}$ has components (F, G, H), each a function of x, y and z, then evidently

$$\mathrm{d}\phi = F\mathrm{d}x + G\mathrm{d}y + H\mathrm{d}z.$$

We may note in passing that, conversely, the condition for an expression like the right-hand side to be an *exact* or *perfect differential* of some function $\phi(x, y, z)$ is that the vector field with components (F, G, H) should be conservative. Other ways of identifying a conservative field or a perfect differential will be encountered in chapter 10.

3.6 Applications to electrostatics and other central force fields

In referring to the electrostatic field $\mathbf{E}_s$ in chapter 2, we postponed all consideration of its relationship to the charge arrays responsible for it. The basic statement of this relation is *Coulomb's inverse-square law*, to the effect that the electrostatic field of a single point charge Q at the origin is observed to be

$$\frac{Q}{4\pi\epsilon_0}\frac{\mathbf{r}}{r^3}.$$

Notice that the magnitude of $\mathbf{r}/r^3$ is indeed $1/r^2$, of inverse-square law form. The curious factor $4\pi\epsilon_0$ and the even odder value of $\epsilon_0(8.854 \times 10^{-12})$ should be regarded as the detritus of the chequered history of units in electromagnetism. On its own it has no particular significance. To call it the 'permittivity of empty space' is perverse; one might as well talk about the colour of a vacuum!

All that concerns us for the moment is that $\mathbf{E}_s$ is of the form $\mathbf{v} = f(r)(\mathbf{r}/r)$, a so-called *central force field*, directed radially to or from its source at the origin and with a single-valued magnitude $f(r)$, dependent only on distance from the source. Gravity is another example. We shall show that any such field is conservative.

Referring to figure 3.5, which shows a part PQ of a loop of integration, we observe that $OP = r$, $OQ = r + \mathrm{d}r = OM$ in the limit, and so $PM = \mathrm{d}r = |\mathbf{dr}| \cos\theta$, the component of $\mathbf{dr}$ in the direction of $\mathbf{v}$. Thus $\oint \mathbf{v} \cdot \mathbf{dr} = \oint f(r)\mathrm{d}r = 0$, i.e. $\mathbf{v}$ is conservative, since f is single-valued (in contrast to the case shown in figure 3.4).

In particular the electrostatic field $\mathbf{E}_s$ of a single charge Q is conservative. This needs generalising to the more usual case where $\mathbf{E}_s$ is associated with an array of several charges Q_1, Q_2, etc., located at several points O_1, O_2 etc. relative to which the position vectors of a further point P are $\mathbf{r}_1, \mathbf{r}_2$, etc. respectively. As P moves along the loop of integration $\mathbf{dr} = \mathbf{dr}_1 = \mathbf{dr}_2$, etc., while $\mathbf{E}_s = \Sigma\, \mathbf{E}_1$, the combined vector sum of $\mathbf{E}_1, \mathbf{E}_2$, etc., the individual, conservative coulomb fields of each charge. Hence, summing over all the charges, we see that

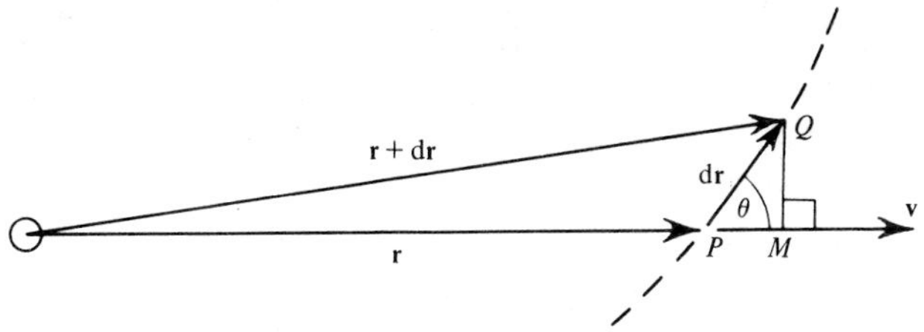

Fig. 3.5

$$\oint \mathbf{E}_s \cdot d\mathbf{r} = \oint \sum \mathbf{E}_1 \cdot d\mathbf{r} = \sum \oint \mathbf{E}_1 \cdot d\mathbf{r}_1 = 0$$

and $\mathbf{E}_s$ is indeed conservative and has a potential, equal to the sum of the potentials due to the individual charges. The universal convention in electrostatics is to set $\int_D^P \mathbf{E}_s \cdot d\mathbf{r}$ equal to *minus* the electric potential V at P so that $\mathbf{E}_s \cdot d\mathbf{r} = -dV$. The energy conservation equation for a charged particle discussed in section 3.3 then becomes

$$QV + \tfrac{1}{2}m\mathbf{v}^2 = \text{const.},$$

in which the term QV is seen to behave like other potential energy terms which occur in mechanics; it falls as the kinetic energy rises.

Example 3.2 A particle is acted upon by two repulsive central force fields, each of constant magnitude $f = k$. One field is centred at the point O_1 $(-c, 0, 0)$ and the other at O_2 $(c, 0, 0)$. Show that all points having a common value for the potential ϕ lie on ellipsoids of revolution.

The combined field is conservative and has a potential, from arguments similar to those used above in connection with electric potential. The potential ϕ, for the force field centred at O_1 (where $\mathbf{r}_1$ denotes the position vector relative to O_1), is

$$\phi_1 = \int \mathbf{v}_1 \cdot d\mathbf{r}_1 = \int k dr_1 = kr_1,$$

if we note again that the component of $d\mathbf{r}_1$ in the direction of $\mathbf{v}_1$ is dr_1. We have used the *plus* convention for potential and taken the datum at O_1. Similarly, the potential ϕ_2 for the field centred at O_2 is $\phi_2 = kr_2$, r_2 being the distance from O_2. The total potential is therefore $k(r_1 + r_2)$.

For all points P where $\phi = \text{const.}$, $r_1 + r_2 = \text{const.}$ If ρ is the distance of P from the x-axis, $r_1 = \sqrt{\{(x+c)^2 + \rho^2\}}$, $r_2 = \sqrt{\{(x-c)^2 + \rho^2\}}$ and $\sqrt{\{(x+c)^2 + \rho^2\}} + \sqrt{\{(x-c)^2 + \rho^2\}} = \text{const.}, 2a$ say.

Squaring, isolating the root and squaring again gives

$$x^2/a^2 + \rho^2/b^2 = 1, \quad (b^2 = a^2 - c^2),$$

the equation of an ellipsoid of revolution with O_1, O_2 as foci. Such a surface is an *equipotential surface*, as all points on it have the same potential.

Problem 3.10 gives some important examples of the calculation of electric fields and potentials from Coulomb's law.

3.7 Applications to magnetism

Magnetism provides various roles for the integral form $\int \mathbf{v} \times d\mathbf{r}$ of type (e). Consider a curved wire carrying a current of uniform magnitude J from A to B in the presence of a magnetic field $\mathbf{B}$. From section 2.4 the magnetic force on an element $d\mathbf{r}$ of the wire, of length ds, is $(\mathbf{J} \times \mathbf{B})ds$, where $\mathbf{J}$ is the vector current, i.e. $\mathbf{J}ds \times \mathbf{B}$. This equals $Jd\mathbf{r} \times \mathbf{B}$ since $\mathbf{J}ds = Jd\mathbf{r}$. ($\mathbf{J}/J$ and $d\mathbf{r}/ds$ are the same unit vector.) The total force on the wire is

$$J \int_A^B d\mathbf{r} \times \mathbf{B}, \quad \text{of type } (e),$$

since the constant scalar factor J can be taken outside the integral.

Example 3.3 A section of a circuit carrying a current J is a wire which consists of one turn of a helix of pitch p wound on a cylinder of radius a. If it is subjected to a uniform field $\mathbf{B}$, parallel to its axis, find the resultant force on it.

Let the ends P, Q of the helix be the points $(a, 0, 0)$ and $(a, 0, p)$ in cartesian terms, and let the z-axis be the axis of the helix. (See figure 3.6, which shows two projected views of the helix.) The helix is specified by the *two* equations

$$x = a\cos(2\pi z/p) \quad \text{and} \quad y = a\sin(2\pi z/p).$$

Total force on the helix

$$= J\int_P^Q d\mathbf{r} \times \mathbf{B} = J\int_P^Q \begin{vmatrix} \mathbf{i} & \mathbf{j} & \mathbf{k} \\ dx & dy & dz \\ 0 & 0 & B \end{vmatrix}$$

$$= J\int_P^Q \{\mathbf{i}Bdy - \mathbf{j}Bdx\} = 0,$$

because y and x are unchanged in going from P to Q. There may however be a resultant *couple* on the helix.

Total vector moment about 0

$$= J\int_P^Q \mathbf{r} \times (d\mathbf{r} \times \mathbf{B}) = J\int_P^Q \begin{vmatrix} \mathbf{i} & \mathbf{j} & \mathbf{k} \\ x & y & z \\ Bdy & -Bdx & 0 \end{vmatrix}$$

(in which we have inserted the components of $d\mathbf{r} \times \mathbf{B}$)

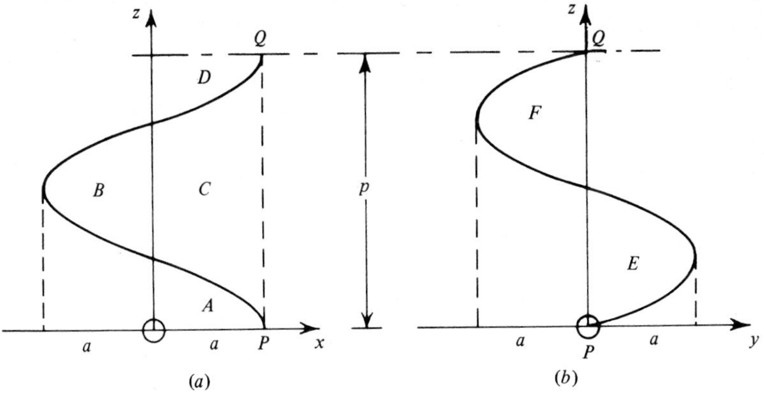

Fig. 3.6. $\int_P^Q z\mathrm{d}x$ = area B + area C = (area A + area D) + area $C = ap$.
$\int_P^Q z\mathrm{d}y$ = − area E + area F = 0.

$$= JB \int_P^Q \{\mathbf{i}z\mathrm{d}x + \mathbf{j}z\mathrm{d}y - \mathbf{k}(x\mathrm{d}x + y\mathrm{d}y)\}.$$

From figure 3.6 it is readily seen that

$$\int_P^Q z\mathrm{d}x = ap \quad \text{and} \quad \int_P^Q z\mathrm{d}y = 0.$$

Also

$$\int_P^Q (x\mathrm{d}x + y\mathrm{d}y) = \tfrac{1}{2}[x^2 + y^2]_P^Q = 0,$$

again because x and y do not change between P and Q. The wire is therefore subject to a couple $JBap$ about an axis parallel to the x-axis.

Another important application of the integral form $\int \mathbf{v} \times \mathrm{d}\mathbf{r}$ to magnetism provides a statement as to how a magnetic field is related to the currents responsible for it. The experimentally observed relation is known as the *Biot–Savart law*, which expresses the field at a point O in terms of currents J flowing in line elements $\mathrm{d}\mathbf{r}$ at points P located by position vectors $\mathbf{r} = \overrightarrow{OP}$. The contribution $\mathrm{d}\mathbf{B}$ to the field at O due to the element $\mathrm{d}\mathbf{r}$ is

$$\frac{\mu_0 J}{4\pi}\left(\frac{\mathbf{r}}{r^3}\right) \times \mathrm{d}\mathbf{r},$$

in which the inverse-square law factor $\mathbf{r}/r^3$ finds a new application, as does the vector product concept. $\mathrm{d}\mathbf{B}$ is seen to be perpendicular to the plane containing $\mathbf{r}$ and $\mathrm{d}\mathbf{r}$ and to vary like $\sin\theta$, where θ is the angle between $\mathbf{r}$ and $\mathrm{d}\mathbf{r}$. The factor $(\mu_0/4\pi)$ is yet more historical detritus, like ϵ_0 of no physical significance on its own, although the product $\mu_0\epsilon_0$ is very significant indeed and is far from arbitrary. Problem 3.11(d) begins to reveal its significance. μ_0 is sometimes meaninglessly called 'the permeability of empty space'. Its value is $4\pi \times 10^{-7}$.

For a wire AB we may integrate to get the total magnetic field at O:

$$\frac{\mu_0 J}{4\pi}\int_A^B \left(\frac{\mathbf{r}}{r^3}\right) \times \mathrm{d}\mathbf{r}$$

(if J is uniform), another integral of the form $\int \mathbf{v} \times \mathrm{d}\mathbf{r}$.

Example 3.4 A wire carrying a current J takes the form of a parabola. Calculate the magnetic field at its focus, if the distance from focus to apex is d.

For any plane wire, the Biot–Savart law indicates that all contributions to the field at a point P in the same plane are perpendicular to that plane, which contains $\mathrm{d}\mathbf{r}$ and $\mathbf{r}$. Moreover, if we use plane polar coordinates as in figure 3.7, the magnitude

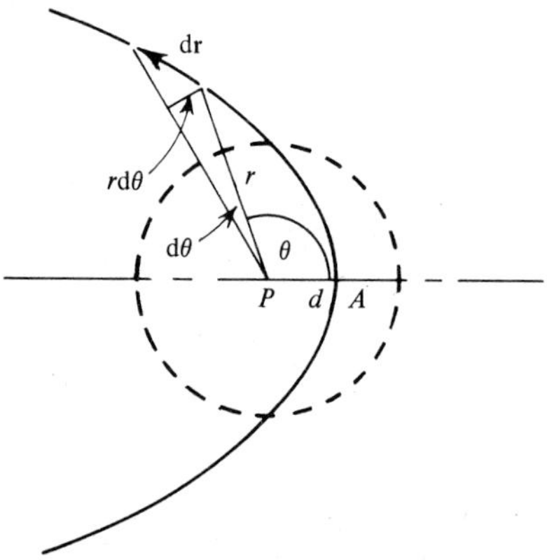

Fig. 3.7

of $(\mathbf{r}/r^3) \times \mathrm{d}\mathbf{r}$ is $(1/r^2)r\mathrm{d}\theta$ or $\mathrm{d}\theta/r$, for $r\mathrm{d}\theta$ is the component of $\mathrm{d}\mathbf{r}$ perpendicular to $\mathbf{r}$. Hence

$$\text{field at } P = \frac{\mu_0 J}{4\pi}\int \frac{\mathrm{d}\theta}{r},$$

evaluated along the wire. (This is a general result.) If P is the focus of the parabola and the apex A is the point $r = d$, $\theta = 0$, then $r(1 + \cos\theta) = 2d$ (a standard result in coordinate geometry)

$$\text{and the field at } P = \frac{\mu_0 J}{8\pi d}\int_{-\pi}^{\pi}(1 + \cos\theta)\mathrm{d}\theta = \frac{\mu_0 J}{4d},$$

the same value, incidentally, as at the centre of the circular wire loop of radius $2d$, shown dotted in figure 3.7.

Problems 3.11(b) and 3.12 provide further examples on the Biot–Savart law. For all but the most simple configurations, the calculation soon becomes a matter for computers.

The $\mathbf{j} \times \mathbf{B}$ or $\mathbf{J} \times \mathbf{B}$ force law and the Biot–Savart law are by no means independent of each other. Problem 3.13 shows how they must be related by the need for the mutual forces between two current-carrying loops to be equal and opposite to each other, in accordance with Newton's third law of action and reaction.

The relation of a magnetic field to its sources, in the form of Biot–Savart current elements $J\mathrm{d}\mathbf{r}$, is seen to be quite different from the Coulomb law that relates the electrostatic field to its sources. It has again that 'sideways' (vector product) quality that magnetism is always revealing. The question whether $\mathbf{B}$ is ever conservative like $\mathbf{E}_s$ we leave until the next chapter.

3.8 Surface integrals

The reader should be familiar with the idea of a double integral $\iint f(x,y)\mathrm{d}x\mathrm{d}y$ over some domain in the x, y plane. Problem 3.1 provided a simple example. The integral is the limit of a sum of contributions over a sequence of small second-order area elements $\mathrm{d}x\, \mathrm{d}y$, with a 'weighting factor' $f(x,y)$ as integrand. Experience with line integrals strongly suggests that generalisation to three-dimensional, vector situations should be possible and fruitful.

If we were trawlermen we might be interested in the rate of volume flow of sea-water through the net, a curved surface S spanning a loop L as in figure 3.8 (knowing how many cod there were per unit volume on the average), or in the total drag that the sea-water exerts on the net, both presumably calculable by integrating over the net's surface. The first is a scalar quantity, the second a vector.

To formalise these ideas we shall apply the concept of vector area developed in section 1.7 to small area elements denoted vectorially by $\mathrm{d}\mathbf{a}$, of scalar magnitude $\mathrm{d}a$. Notice that $\mathrm{d}a$ is *not* the increment of some single variable a that varies over the surface, in contrast to the scalar variable s in line integrals. If f and $\mathbf{v}$ are scalar and vector quantities defined all over the surface, again five kinds of integral are conceivable:

(*a*) $\iint f\mathrm{d}a$, a *scalar*, e.g. the mass of a thin shell of density f per unit area.
(*b*) $\iint \mathbf{v}\mathrm{d}a$, a *vector*, e.g. the drag on the trawler net, if $\mathbf{v}$ is drag per unit area.
(*c*) $\iint f\mathrm{d}\mathbf{a}$, a *vector*, e.g. the force due to a normal pressure f on a submerged surface, discussed further below.
(*d*) $\iint \mathbf{v} \cdot \mathrm{d}\mathbf{a}$, a *scalar*, the very important *flux integral*, e.g. the volumetric flow rate through the trawler net if $\mathbf{v}$ is the velocity of the water relative to it.
(*e*) $\iint \mathbf{v} \times \mathrm{d}\mathbf{a}$, a *vector*, e.g. the vector moment of pressure forces on a submerged surface, discussed further below.

In (*b*), (*c*) and (*e*) the idea of the limit of an infinite sum of vanishingly small vectors is invoked yet again. The order in which the contributions accumulate is immaterial because vector addition is commutative. All but (*a*) may be expanded into three parts to assist evaluation by using the cartesian or other components of $\mathbf{v}$ and/or $\mathrm{d}\mathbf{a}$ in the obvious way. Successful evaluation of such integrals depends crucially on a good choice of two coordinates on the surface in order to identify location and express $\mathrm{d}a$ or $\mathrm{d}\mathbf{a}$ in terms of increments of these coordinates. Problem

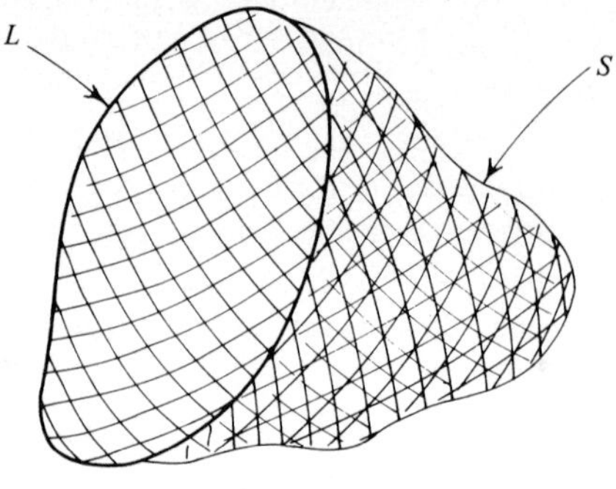

Fig. 3.8

3.15(a) gives a general exercise on this, but often inspection of a well-drawn diagram is a better way of arriving at expressions for the area element's size, position and orientation. It may often be convenient to express everything in terms of x and y, say, and then evaluate the integral as an ordinary double integral over a domain in the x, y plane (see problem 3.14). The equation for the surface $z = f(x, y)$ would enable z to be eliminated. Problem 3.15(b) goes into details.

Example 3.5 Find the cartesian components of a vector area element $\mathbf{da}$ on the Earth's surface expressed in terms of changes in latitude and longitude.

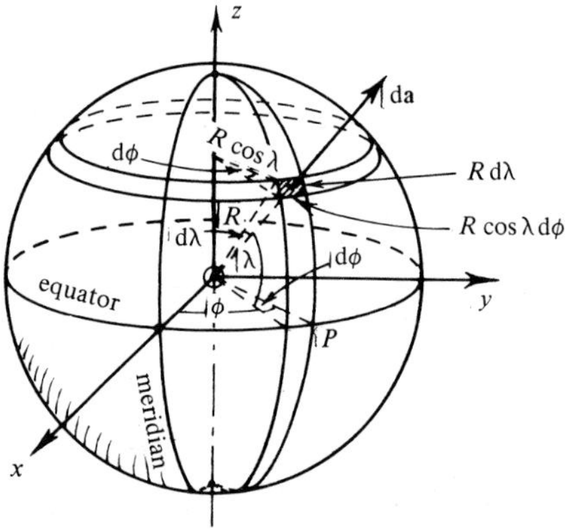

Fig. 3.9. Latitude and longitude coordinates on a sphere.

The familiar latitude λ and longitude ϕ, shown in figure 3.9, are often the most convenient pair of coordinates for defining position on a spherical surface. The magnitude $\mathrm{d}a$ of the area element generated by increments $\mathrm{d}\lambda$ and $\mathrm{d}\phi$, a plane rectangle in the limit, is $(R \cos \lambda \, \mathrm{d}\phi) \times (R\mathrm{d}\lambda)$ or $R^2 \cos \lambda \mathrm{d}\lambda \mathrm{d}\phi$, if we ignore higher order small quantities and R is the sphere's radius.

Let us take cartesian axes as shown in figure 3.9. The *vector* area element $\mathbf{da}$ is radial and has a component da sin λ in the z-direction and a component da cos λ in the x, y plane in the direction $\vec{OP}$. This in turn may be resolved into an x-component and a y-component, using the angle ϕ. Inserting the value of da gives

$$(R^2\, d\lambda d\phi \cos^2\lambda \cos\phi,\ R^2\, d\lambda d\phi \cos^2\lambda \sin\phi,\ R^2\, d\lambda d\phi \cos\lambda \sin\lambda)$$

for the cartesian components of d**a**. They are as usual equal to the projections of da in the x-, y- and z-directions.

Some tactical considerations to observe in the choice of coordinates include the following:

(i) Orthogonal coordinates yielding rectangular area elements are preferable.
(ii) The integration is eased if the choice of coordinates simplifies the integrand (e.g. into the product of a function of one coordinate with a function of the other, or even into a function of just one, because of certain symmetries).
(iii) The integration is eased if the choice of coordinates simplifies the inner limits of the double integral (e.g. because one coordinate is constant along certain boundaries).
(iv) The integration is eased if the choice of coordinates simplifies the vector geometry (e.g. because all the area elements are oriented in the same direction during one stage of the integration) or
(v) The problem is made more simple if the choice of coordinates allows the area to be covered by *monotonic* variation of the coordinates (e.g. for covering a sphere, latitude and longitude are preferable to cartesians).

3.9 Closed surfaces

Some of the most important applications of these ideas are to closed surfaces, without edges, which can be usefully thought of in either of two ways:

View (*a*): The limit of the surface S in figure 3.8 when the loop L is caused to shrink and vanish like a noose on the neck of a bag.
View (*b*): A surface composed of two different surfaces S_1 and S_2, each of which spans the same loop L as in figure 3.10.

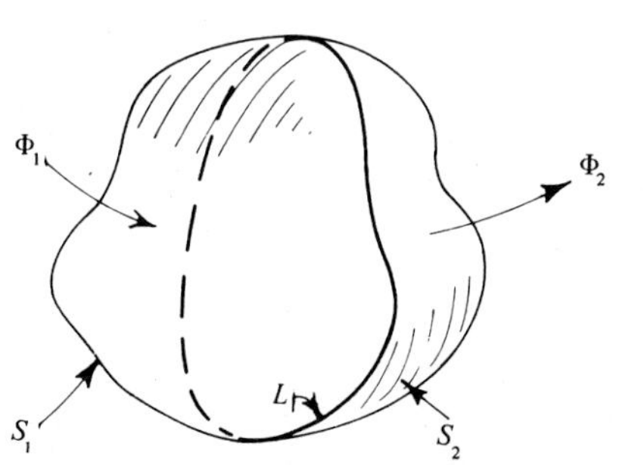

Fig. 3.10. Two alternative surfaces spanning one loop.

An integral over a closed surface is distinguished by the special integral sign $\oiint$. Once a surface is closed it has an inside and an outside and the direction of $\mathbf{da}$ can be unambiguously defined. The standard convention is to make it *outwards*. (Think of the hairs on a gooseberry.) If the region has internal voids, not connected to the outside, they must be deemed to be bits of 'trapped outside', with $\mathbf{da}$ pointing into the void.

3.10 Applications to fluid mechanics

The study of fluid mechanics demands the use of surface integrals. This takes us into Zone 4 of figure 3.2. Consider fluid adjoining a surface, which may be either a material discontinuity (such as the exterior of a hatch on a submarine or the atmosphere/ocean boundary) or an imagined surface drawn in a homogeneous fluid. If the fluid stress consists only of pressure p at each point, shear stress being absent or negligible, then the force exerted on an element of the surface $\mathbf{da}$ (i.e. on the matter beyond it) by the adjoining fluid is $-p\mathbf{da}$, if $\mathbf{da}$ points into this fluid. The surface element exerts an opposite force $p\mathbf{da}$ on the fluid.

Thus the total force exerted by the surface on the adjoining fluid is $\iint p\mathbf{da}$ integrated over the surface, an integral of type (c). To get the component in a particular direction we must integrate the resolved component of $p\mathbf{da}$. The moment of the forces about an axis λ (e.g. the torque about the hinge-line required to open a submarine hatch) is the component along λ of the vector moment integral $\iint \mathbf{r} \times p\mathbf{da}$, i.e. $\iint (p\mathbf{r}) \times \mathbf{da}$, which is of type ($e$). The origin must be on λ. Problems 3.16 and 3.17 provide illustrations of these ideas. Another follows.

Example 3.6 A right circular cone with semi-angle α and slant height L floats half-immersed in water with its axis horizontal. Show that the resultant of the water pressures acting on the curved surface has a line of action which intersects the axis of the cone at a distance $3L/4 \cos\alpha$ from the apex. Take the pressure as zero at the water surface and proportional to depth below it.

There are many ways of doing this problem. One way is to replace the immersed half of the cone by water, on which the same pressures would act. The combined forces on the curved face and on the plane base must then balance the weight of the water. If Z is the point where the resultant force on the curved face cuts the axis of the cone, the moments about Z of the force on the plane base and of the weight of water (acting at its centre of gravity) must be equal and opposite. This fact locates Z. Here we give a more direct method.

A point P on the curved surface can be located by the coordinates θ and l, shown in figure 3.11, θ being a kind of longitude, an angle of rotation about the axis AB of the cone, and l being slant-distance from the apex A. The first stage of the double integration will be conducted along AQ, at constant θ with l varying. The area element at P is a rectangle $\mathrm{d}l$ by $l\epsilon$, where $\epsilon = \angle QAR$. (ϵ equals $\mathrm{d}\theta \sin\alpha$, for $QR = L\epsilon = (L \sin\alpha)\mathrm{d}\theta$, but we shall not use this fact.) The pressure forces on all the elements that compose the virtually plane area AQR are parallel and their

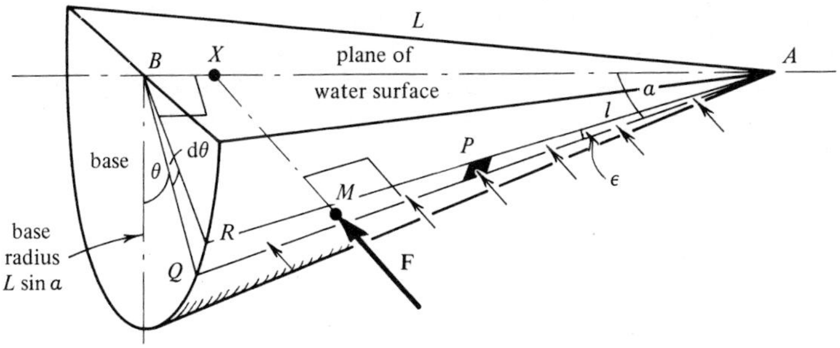

Fig. 3.11

resultant **F** is therefore in the same direction, and obviously intersects the axis AB of the cone, at X say. The point M where **F** intersects the area AQR can be found by taking moments about an axis through A parallel to QR, taking pressure $p = kl$, where $k =$ const. (along AQ), for pressure is proportional to depth.

Hence

$$AM \int_0^L (kl)(l\epsilon)\,\mathrm{d}l = \int_0^L l(kl)(l\epsilon)\,\mathrm{d}l \quad \text{and} \quad AM = \tfrac{3}{4}L.$$

But $\angle XAM = \alpha$ and $\angle XMA = \pi/2$, so that $AX = \frac{3}{4}L \sec \alpha$. This length is independent of θ, i.e. the resultant force on each elementary triangle AQR goes through X and therefore the resultant of all the pressures on the curved surface passes through X, which is at a distance $3L/(4 \cos \alpha)$ from the apex.

A second stage of double integration with respect to θ turns out to be unnecessary, if this approach is adopted.

Closed surfaces play an important role in fluid mechanics. When a closed surface is defined in space so as to delineate a region through which fluid may be passing, it is called a *control surface* (which should not be confused with an aeronautical 'control surface' such as an aileron or rudder) and the space enclosed is called a *control volume*. For a control surface, $-\oiint p\mathbf{da}$ is the total force on the fluid inside, exerted by the surrounding fluid. The minus sign comes from the fact that **da** points outwards. This force, together with any other forces acting, can be related to the momentum changes, as we shall see in chapter 12. In this book control surfaces will be stationary – not moving with the fluid – unless the contrary is explicitly stated.

*3.11 The addition of area elements

(*N.B.* An asterisk indicates a section which is more difficult and may be omitted.)

It was suggested in section 1.7 that addition of non-coplanar area elements was not obviously meaningful geometrically or physically, even though $\iint \mathbf{da}$ over a surface can certainly be interpreted mathematically. However, this integral can now be seen to have at least one meaning, namely, the force due to unit fluid pressure acting on the surface in question. But we can go further than this.

Consider the integral $\iint \mathbf{da}$ taken over a surface S whose edge is the loop L, as in figure 3.12(*a*). Its z-component is $\iint \mathrm{d}a_z$, where $\mathrm{d}a_z$ is the z-wise projection of the element $\mathbf{da}$. Provided $\mathrm{d}a_z$ is of constant sign everywhere as in figure 3.12(*a*), $\iint \mathrm{d}a_z$

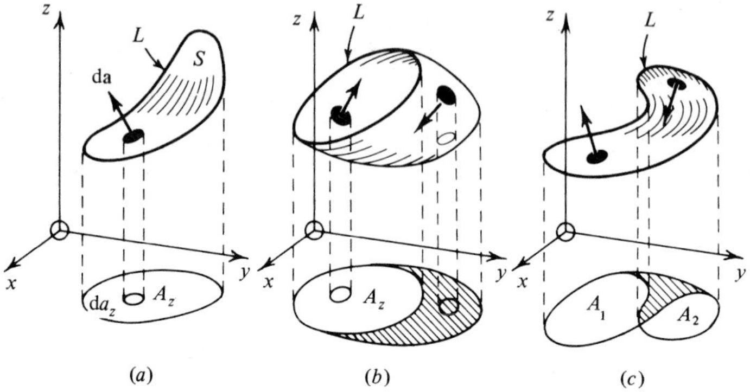

Fig. 3.12

is evidently the area A_z enclosed within L's projection in the z-direction. But this is still true even when $\mathrm{d}a_z$ changes sign, as in figure 3.12(*b*). The shaded part of the total of positive contributions $\mathrm{d}a_z$ is cancelled by the negative contributions. Another mild complication is illustrated in figure 3.12(*c*), in which A_z obviously has to be interpreted as the difference $A_1 - A_2$, and the shaded area cancels out.

The key point to notice is that $\iint \mathrm{d}a_z$ is determined purely by the loop L and is therefore the same for all surfaces S spanning that loop. The same is true of $\iint \mathrm{d}a_x$ and $\iint \mathrm{d}a_y$ and therefore also the vector sum $\iint \mathbf{da}$. We conclude that $\iint \mathbf{da}$ is a property of the loop, being independent of which surface spanning L is used to evaluate it. It can be described as the *vector area of the loop*. Note that the loop will not in general lie in a plane and the direction of its vector area will not be obvious by inspection. Because of the 'cosine-quality' of vectors, however, the direction of $\iint \mathbf{da}$ will be that in which the loop appears to an observer to embrace maximum area, whereas an observer looking in *any* direction normal to $\iint \mathbf{da}$ will see zero net area, with due attention to sign. This discussion glosses over some of the finer topological points such as the alternative of spanning each loop with a one-sided Möbius surface. (See problem 4.11.)

If a closed surface S is regarded as one which spans a loop L that has contracted to a point and vanished, it is immediately clear that $\oiint \mathbf{da} = \mathbf{0}$. The reader should satisfy himself that regarding S as the sum of two surfaces S_1 and S_2, spanning the same loop L as in figure 3.10, leads to the same conclusion once the sign convention for $\mathbf{da}$ on a closed surface is correctly observed. The converse is not valid; $\iint \mathbf{da} = \mathbf{0}$ does not imply that a surface is closed. Non-closed surfaces for which $\iint \mathbf{da}$ vanishes are easily generated.

In a fluid-mechanical context, these results imply that a uniform pressure exerts the same force on any surface spanning a given loop, and no net force on a closed surface, e.g. the complete surface of a body. Problem 3.18 reveals the utility of these ideas and problem 3.22(*b*) goes further. The vector area of a loop also serves a purpose in electromagnetism, illustrated in problem 3.19(*b*).

3.12 The flux integral $\iint \mathbf{v} \cdot \mathbf{da}$

Fluid mechanics provides the most obvious examples of flux integrals. If **v** is fluid velocity, $\mathbf{v} \cdot \mathbf{da}$ is the volumetric flow rate through an element **da** of some stationary surface drawn in the fluid (cf. section 1.10). In other words, **v** can be regarded as the volumetric flux intensity, just as **Q** is heat flux intensity and **j** is current flux intensity. Similarly, $(\rho\mathbf{v})$ is the mass flux intensity vector measured in mass per unit area per unit time, if ρ is the fluid's mass density, which may be non-uniform.

Summing contributions over the surface, we have

$$\iint \mathbf{v} \cdot \mathbf{da} = \text{volume flow rate through the surface,}$$

$$\iint \rho\mathbf{v} \cdot \mathbf{da} = \text{mass flow rate through the surface,}$$

and similar results for **Q** and **j**. It should be apparent by now just how convenient it is that both flow intensity and the area element can be treated as vectors. If a vector **v** is uniform its flux is $\mathbf{v} \cdot \iint \mathbf{da}$.

Since **E** and **B** seem to be essentially abstractions, invented to systematise observations on charged particles, and appear in no sense to represent the actual flow of anything, it might seem perverse to take any interest in the concepts of electric flux $\iint \mathbf{E} \cdot \mathbf{da}$ or magnetic flux $\iint \mathbf{B} \cdot \mathbf{da}$. But they are in fact very important, as will emerge.

Problem 3.20 gives some exercises on evaluating flux. An example follows.

Example 3.7 A unit cube has one corner at the origin O and $QRST$ is one face of the cube, Q being nearest to O. Find the flux of the vector $r^2\mathbf{r}$ through the triangle RST.

Figure 3.13 shows suitable cartesian axes, and an area element $\mathrm{d}y\mathrm{d}z$ at P within RST. As $\overrightarrow{OP}$ is **r**, the component of $r^2\mathbf{r}$ normal to the area element (i.e. parallel to Ox) is r^2OQ, which equals r^2 as OQ is unity. Also $r^2 = 1 + y^2 + z^2$. Since $y = 1 - z$ along RT,

$$\text{flux through } RST = \int_0^1 \mathrm{d}z \int_{1-z}^1 (1 + y^2 + z^2)\mathrm{d}y$$

$$= \int_0^1 \mathrm{d}z \, \{(1 + z^2)[1 - (1 - z)] + \tfrac{1}{3}[1 - (1 - z)^3]\} = 1.$$

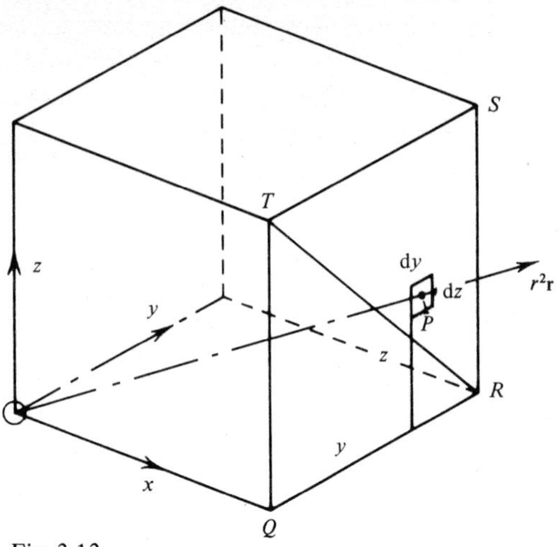

Fig. 3.13

3.13 Closed surfaces and solenoidal fields

If the surface is closed, the integral $\oiint \mathbf{v} \cdot \mathrm{d}\mathbf{a}$ is called the *net flux*, or, more strictly, the *net efflux*, the total outwards flux minus the total inwards flux, for $\mathbf{v} \cdot \mathrm{d}\mathbf{a}$ is negative at points where the flux is inwards as $\mathrm{d}\mathbf{a}$ is directed outwards everywhere.

The very important special class of vector fields for which the integral $\oiint \mathbf{v} \cdot \mathrm{d}\mathbf{a}$ is zero for all closed surfaces is known by the epithet *solenoidal*. For such a field, the fluxes into and out of any closed region always balance. Such ideas will be very important later on in connection with statements of *conservation* for various physical quantities such as energy flow (using the word 'conserve' in another sense).

Problem 3.21 is an interesting example where the concepts of solenoidal and conservative vector fields are brought together, while problem 3.22(*c*) is a proof that magnetic field is solenoidal.

Example 3.8 For the inverse-square law field $\mathbf{r}/r^3$, show that the flux through a surface S spanning a loop L equals the solid angle subtended by L at the origin (the source of the field) and that its flux through a closed surface is 4π or zero, depending whether the source is inside or outside the surface.

The flux through an area element $\mathrm{d}\mathbf{a}$ of S (see figure 3.14) is $\mathrm{d}\mathbf{a} \cdot \mathbf{r}/r^3 = \mathrm{d}a_r/r^2$, if $\mathrm{d}a_r$ is the radial component of $\mathrm{d}\mathbf{a}$, the projection of the element on to a sphere Z of radius r, for the angle θ between $\mathbf{r}$ and $\mathrm{d}\mathbf{a}$ is also the angle between Z and S. But $\mathrm{d}a_r/r^2 = \mathrm{d}\Omega$, the elementary solid angle subtended by $\mathrm{d}\mathbf{a}$ at O, and solid angles are additive. Therefore for a finite surface S, the flux of $\mathbf{r}/r^3$ is

$$\iint_S \frac{\mathbf{r}}{r^3} \cdot \mathrm{d}\mathbf{a} = \Omega,$$

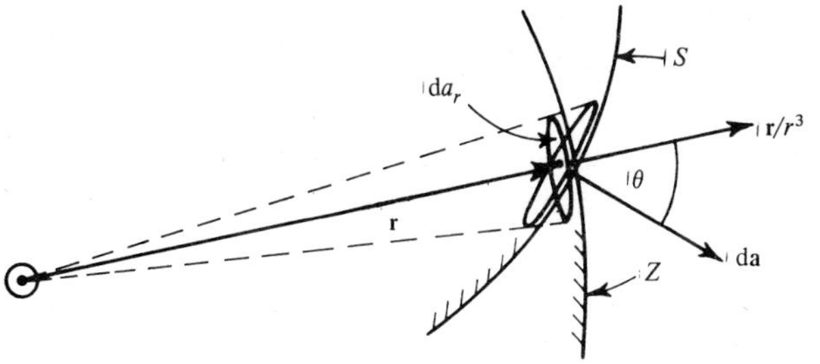

Fig. 3.14

the total solid angle subtended at O by L (with due attention to sign if the surface appears to fold back on itself as seen from O), irrespective of distance from the source O. This result reveals why the inverse-square law is so common in Nature. A torch-beam of given solid angle shines the same light flux on to any surface it encounters, and this implies the inverse-square law for light intensity, in the absence of absorption.

If the surface is closed the familiar argument about letting L shrink to a point might lead one to conclude that $\oiint (\mathbf{r}/r^3)\cdot d\mathbf{a} = 0$ always, but such a conclusion would be too hasty. There are two distinct cases:

(*a*) *O outside the closed surface*. Then $\oiint (\mathbf{r}/r^3)\cdot d\mathbf{a}$ does vanish because each positive contribution to Ω is cancelled by a negative one, even in complicated cases as in figure 3.15.

(*b*) *O inside the closed surface*. Then $\oiint (\mathbf{r}/r^3)\cdot d\mathbf{a}$, the total flux emanating from the source, is not zero but equals 4π, the complete solid angle. The source is said to have a *strength* 4π.

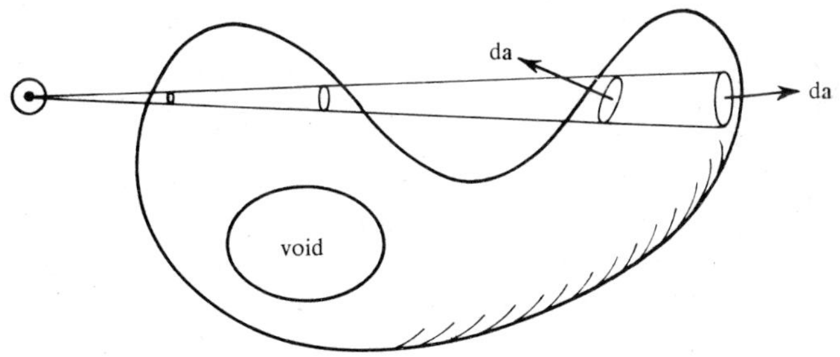

Fig. 3.15

The above example shows that, provided we keep away from the source, which is obviously a singularity of the field, an inverse-square law field is solenoidal. Though any central force field is *conservative*, it is *solenoidal* only if its magnitude has inverse-square dependence on r.

In the case of the inverse-square law electrostatic field $\mathbf{E}_s$ due to a point charge Q, the flux through a loop L subtending solid angle Ω at the charge is $Q\Omega/4\pi\epsilon_0$, and the total flux leaving the charge is Q/ϵ_0. These ideas are applied in problem 3.23.

When $\mathbf{E}_s$ is caused by many charges, it is still solenoidal in regions that do not include these charges, because the flux of the sum of several vector fields is equal to the sum of the fluxes of the individual fields.

3.14 Vector field lines

A vector field has a direction at each point in space. It is possible to construct a series of lines through space which have the property that at each point the tangent gives the direction of the field at that point. The complete pattern of lines refers to one instant in time, even though the field may be changing. We call these lines *field lines.* It is a convenient habit to picture vector fields with the aid of these field lines. Figure 3.16 shows a well known example.

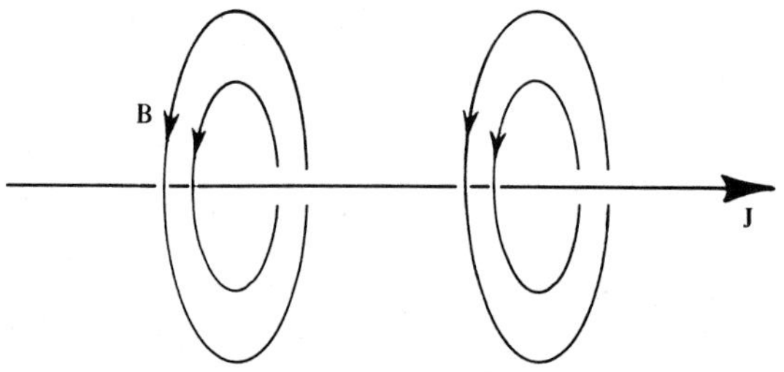

Fig. 3.16. Magnetic field lines round a long straight wire.

Notice that there is no reason why field-line pictures should tell us anything whatever about the *strength* of the field; they describe only its *direction*. This cannot be too strongly stressed. If the field lines run closer together somewhere this does not necessarily indicate whether the field strength is greater there. This is obvious if one considers several different vector fields sharing the same field lines. A good example is the fluid velocity $\mathbf{v}$ and the mass flux intensity field $\rho\mathbf{v}$, where ρ could vary quite freely. Clearly the changing spacing of the field lines could not represent the variations in intensity of both $\mathbf{v}$ and $\rho\mathbf{v}$ if ρ varied. The question how to give precise meaning to the idea of 'closeness of field lines' is discussed later.

Even more hazardous is to consider field lines as *moving* (which implies giving them a persistent identity from one moment to the next) unless the validity of such thought processes has been clearly proven. The field lines are our invention and we must not give them physical properties that they may not possess. It is unfortunate that the treatment of magnetic field lines as physical entities used in most elementary teaching of electromagnetism leads students to believe that all field lines have the same alleged properties.

The problem of finding the form of field lines is essentially mathematical. They cannot in general be found empirically. Even in fluid mechanics, a line of dye continuously released into the stream from a fixed point generates a field line of the instantaneous velocity field (known as a *streamline*) only in the special case of *steady flow*, in which the flow pattern is not changing in time. A released electron does not follow an electric field line; its inertia causes it to 'skid' on the bends. The time-honoured process of putting compass needles nose to tail round a bar-magnet is only a crude version of a finite-difference mathematical integration process.

The mathematical statement that defines a field line consists of two differential equations. Two statements are needed to define a line in space, e.g. as the intersection of two surfaces. To an observer looking in the z-direction, the vector $\mathbf{v}$ appears to have an inclination θ_1 to the x-axis given by $\tan\theta_1 = v_y/v_x$, whereas for the tangent to the field line, the corresponding apparent inclination θ_2 is given by $\tan\theta_2 = \mathrm{d}y/\mathrm{d}x$. The field line is defined by making $\tan\theta_1 = \tan\theta_2$. We treat the other projected views similarly. The result is most tersely written

$$\frac{\mathrm{d}x}{v_x} = \frac{\mathrm{d}y}{v_y} = \frac{\mathrm{d}z}{v_z},$$

which is effectively two differential equations, sufficient in principle to determine a family of field lines if v_x, v_y, and v_z are available as functions of x, y and z at a particular instant. An alternative form of the equations is $\mathrm{d}\mathbf{r} \times \mathbf{v} = \mathbf{0}$, which makes $\mathrm{d}\mathbf{r}$ and $\mathbf{v}$ parallel. (It is a common error to think that, just as the slope of the field line is expressed by differentiation, so the slope of the vector field must also involve some differential coefficient such as $\mathrm{d}v_y/\mathrm{d}v_x$, whereas it actually only requires *ratios* of components, such as v_y/v_x.) The analytical solution of the equations is in general difficult, but problem 3.24 gives some simple examples. Another example follows.

Example 3.9 A vector field $\mathbf{v}$ is given as $\mathbf{i} + \mathbf{j}\sin x + \mathbf{k}\cos x$. For the field line through the origin find y and z as functions of x. Describe the shape of this field line.

Here

$$\frac{\mathrm{d}y}{\mathrm{d}x} = \frac{v_y}{v_x} = \frac{\sin x}{1}, \quad y = 1 - \cos x,$$

and

$$\frac{\mathrm{d}z}{\mathrm{d}x} = \frac{v_z}{v_x} = \frac{\cos x}{1}, \quad z = \sin x,$$

in which the constants of integration have been chosen so as to make $y = z = 0$ when $x = 0$.

The curve is evidently a helix. Its axis is the line $y = 1, z = 0$.

3.15 Vector field tubes

Another useful concept is a *field tube*, which consists of a tubular surface defined by taking all the field lines through a given loop in space. A windsock at an airport provides a good image; a pyjama leg with stripes for field lines is even better. In fluid mechanics, a field tube of a velocity field is called a *stream tube*.

An important function of the field tube concept is that it gives us a precise way of saying what we mean when we describe field lines in three-dimensions as 'getting closer together' and of relating this to the field intensity under certain conditions. The field lines 'get closer together' when the cross-sectional area of a particular field tube becomes smaller. Elementary treatments of electromagnetism often describe this by saying that the 'density' of field lines per unit area (of field tube cross-section) is increasing, the field tube being thought of as a bundle of field lines.

3.16 Solenoidal field tubes

For a solenoidal field the fluxes into and out of any closed surface balance. We apply this fact to a closed surface consisting of a length of field tube that has small transverse cross-sections $\mathrm{d}A_1$ and $\mathrm{d}A_2$ at its ends, as shown in figure 3.17. Let

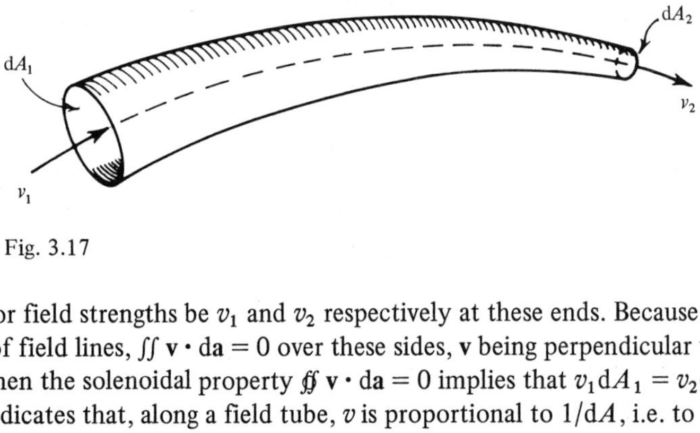

Fig. 3.17

the vector field strengths be v_1 and v_2 respectively at these ends. Because its sides consist of field lines, $\iint \mathbf{v} \cdot \mathrm{d}\mathbf{a} = 0$ over these sides, $\mathbf{v}$ being perpendicular to $\mathrm{d}\mathbf{a}$ there. Then the solenoidal property $\oiint \mathbf{v} \cdot \mathrm{d}\mathbf{a} = 0$ implies that $v_1 \mathrm{d}A_1 = v_2 \mathrm{d}A_2$, which indicates that, along a field tube, v is proportional to $1/\mathrm{d}A$, i.e. to the closeness or 'density' of the field lines. So there is indeed a class of vector fields for which the closeness of the field lines does indicate field strength, namely solenoidal fields. An obvious example is the inverse-square law field, for which v and $1/\mathrm{d}A$ both vary like r^{-2} in a conical field tube with its apex at the origin.

Conversely, any field for which $v \propto 1/\mathrm{d}A$ in all slender field tubes is solenoidal, for we may divide up any closed surface S into elementary field tubes AB as in figure 3.18. The contribution $\mathbf{v} \cdot \mathrm{d}\mathbf{a}$ at B to the total flux through S may be replaced by $v\mathrm{d}A$, in which $\mathrm{d}A$ is the transverse cross-section of the tube at B, i.e. the projection of $\mathrm{d}\mathbf{a}$ in the $\mathbf{v}$-direction. If $v \propto 1/\mathrm{d}A$, the efflux at B is cancelled by the influx at A, and so on for all tubes. Hence $\oiint \mathbf{v} \cdot \mathrm{d}\mathbf{a}$ over S vanishes and we have shown the field to be solenoidal. (The epithet 'solenoidal' comes from the Greek word for 'tube'.)

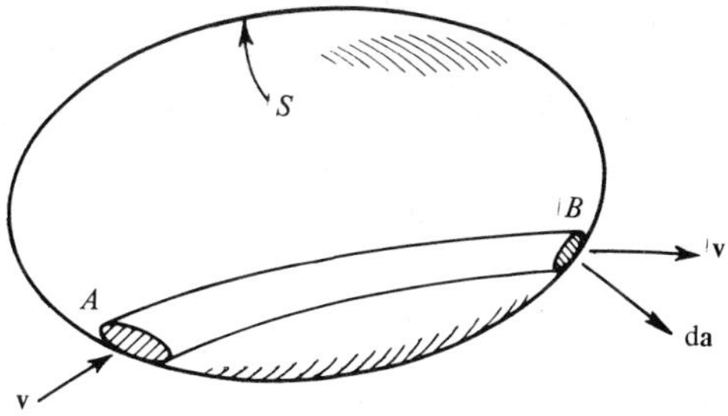

Fig. 3.18

Example 3.10 Show that any magnetic field is solenoidal.

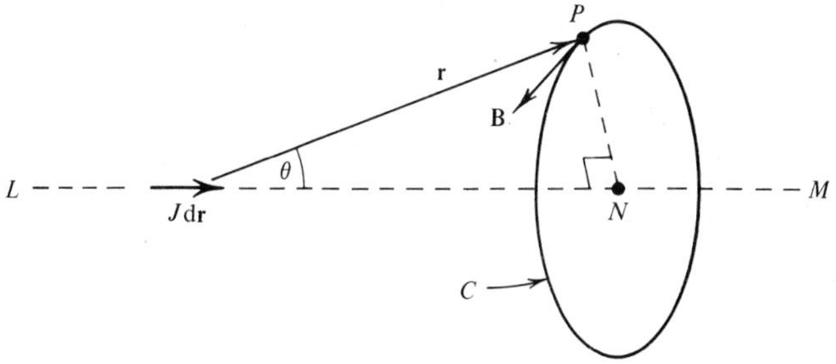

Fig. 3.19

First consider the field of a Biot–Savart current element $J\mathrm{d}\mathbf{r}$, lying in the line LM as in figure 3.19. The magnetic field $\mathbf{B}$ at P is perpendicular to the plane PLM and has magnitude $\mu_0 J\mathrm{d}r \sin\theta/4\pi r^2$, by the Biot–Savart law. This magnitude is constant all round the circle C centred at N in a plane perpendicular to LM. Moreover, the direction of B at all points such as P on C is tangential to C and so C is a field line. A field tube will clearly be a torus of constant cross-section, along which B is constant. This meets the condition $B \propto 1/\mathrm{d}A$ along a field tube (albeit in a degenerate way) and the argument used in the previous paragraph shows that the magnetic field $\mathbf{B}$ is solenoidal.

When there are many Biot–Savart elements, jointly responsible for the total magnetic field, it is still solenoidal because the sum of the fluxes of several vector fields is the flux of the sum of those fields. (Problem 3.22(*c*) explores an alternative proof.)

The familiar old arguments in terms of field line counting used in elementary treatments are valid for magnetic fields only because $\mathbf{B}$ is solenoidal. It is the solenoidal property of $\mathbf{B}$ which also enables magnetic flux to play such an important part in electromagnetic theory.

Remember, however, that in general vector fields are not solenoidal and so field-line spacing may carry no information about field strength. Problem 3.25 exhibits an important example drawn from fluid mechanics where the field gets stronger as the field lines *diverge*, the exact opposite of solenoidal behaviour!

3.17 Flux-linkage

This much-exploited concept in electromagnetism deserves critical scrutiny at this point. 'Flux linked' is loosely thought of as 'flux passing through' a given loop, but flux is defined only relative to a surface spanning that loop and an infinity of surfaces can span each loop. Alternatively we can regard 'flux linked' as 'flux cut' as we tighten the loop like the noose of a lassoo down to vanishing point, but this does not really constitute a different view because the shrinking loop traverses and delineates a surface for which there is an infinity of choices. Flux-linkage for a loop is therefore a meaningless concept unless it is unique for that loop and independent of the choice of surface chosen to span the loop so as to define the flux. Figure 3.10 shows two alternative surfaces S_1 and S_2 spanning the same loop L. A vector field has fluxes Φ_1 and Φ_2 through these surfaces in the same direction. If flux-linkage is to be unique for loop L, then $\Phi_1 = \Phi_2$ and, for the closed surface $S_1 + S_2$,

$$\oiint \mathbf{v} \cdot d\mathbf{a} \;=\; \text{flux out } (\Phi_2) - \text{flux in } (\Phi_1) \;=\; 0.$$

The situation is quite simple therefore: for flux linkage to be meaningful, i.e. uniquely defined for every loop, the vector field must be solenoidal. The converse is easily proved.

In the next chapters these ideas prove to have the most profound consequences for electromagnetic theory.

Problems 3B

3.9 (*a*) A thin steady fluid jet is projected horizontally at velocity V with mass flow rate $\dot{m}$ and falls under gravity g into a lake a distance h beneath. Show that the total momentum of the jet has horizontal and vertical components equal to $\dot{m}V\sqrt{(2h/g)}$ and $\dot{m}h$ respectively.
(*b*) Evaluate $\int_0^P \mathbf{F} \cdot d\mathbf{r}$ where P is the point (1, 1, 1) along the curve $y = x^2$, $z = x^3$ with $\mathbf{F} = xy^2\mathbf{i} + x^2y\mathbf{j} + z\mathbf{k}$. Repeat the calculation by other routes (e.g. $x = y = z$, or three stages, each parallel to an axis). If you think that $\mathbf{F}$ is conservative, so that $\mathbf{F} \cdot d\mathbf{r}$ is a perfect differential $d\phi$ of a potential function $\phi(x, y, z)$, find ϕ by inspection and thence check the values of your integrals.

3.10 (*a*) Show that, at a point P at a distance r from a point charge Q, the electric potential V equals $Q/4\pi\epsilon_0 r$, if V is taken as zero at great distances.
(*b*) Show by integration that the electric field due to charges uniformly distributed at density q per unit length along a virtually endless straight

line equals $q/2\pi\epsilon_0 r$ in magnitude at a point at a distance r from the line and that it is directed normally away from the line. Deduce that the electric potential equals

$$-(q \log r)/2\pi\epsilon_0 + \text{const.}$$

(*c*) If one such line of charges lies in the *z*-direction at $x = a, y = 0$ and another parallel line with equal and opposite charges lies at $x = -a, y = 0$, express the total potential at the point (x, y, z) in terms of x and y, taking it as zero at the origin.

Show that the equipotential surface $V = V_0$ is a cylinder of radius $2ab/(b^2 - 1)$ with axis at $x = a(b^2 + 1)/(b^2 - 1), y = 0$, where $b = \exp(2\pi\epsilon_0 V_0/q)$.

What happens if we let a tend to zero while keeping $p = 2aq$ constant? This configuration is called a *doublet of strength p*.

Sketch the equipotentials in both cases.

(*d*) Point charges Q and $-Q$ lie at points R_1 and R_2 where $\overrightarrow{R_2R_1} = \boldsymbol{\delta}$. Show that the potential at a point P at distances r_1 and r_2 respectively from R_1 and R_2 equals $Q(r_2 - r_1)/4\pi\epsilon_0 r_1 r_2$.

If the vector $\mathbf{p} = Q\boldsymbol{\delta}$ is kept constant as $|\boldsymbol{\delta}| \to 0$, the resulting entity is known as a *dipole of strength* **p**. Show that the potential at P then becomes $\mathbf{p} \cdot \mathbf{r}/4\pi\epsilon_0 r^3$.

3.11 (*a*) Like electric charges are distributed at uniform density q per unit length along two parallel lines at a distance r apart. Show (using problem 3.10(*b*)) that there is a *repulsive* electric force between the lines equal to $q^2/2\pi\epsilon_0 r$ per unit length.

(*b*) If one line of charges moves axially with a velocity v, so constituting a current $J = qv$, show by using the Biot–Savart law that the magnetic field strength is $\mu_0 qv/2\pi r$ at a distance r from the line. What is its direction?

(*c*) If *both* lines of charge move axially with a velocity v in the same direction show that there is an *attractive* magnetic force between them equal to $\mu_0 q^2 v^2/2\pi r$ per unit length.

(*d*) What value of v would make the two forces in (*a*) and (*c*) balance? What happens if a frame of reference moving with the charges is used instead? Is there a magnetic force or is there not?

(Note that $(\mu_0 \epsilon_0)^{-1/2}$ = velocity of light c. This problem shows that at normal velocities ($\ll c$) electric forces vastly exceed magnetic forces unless the interacting charge arrays each consist of almost equal numbers of *opposite* charges in relative motion (the most common situation) whereupon the magnetic forces can become much larger than the electric forces. This problem also indicates that a completely consistent treatment of electromagnetism must be relativistic in order to resolve the anomaly that the force between the lines appears to depend on the observer's velocity.)

3.12 (*a*) From the Biot–Savart law show that the magnetic field of an *N*-turn circular loop of radius a bearing a current J is axial and equal to

$$\tfrac{1}{2}\mu_0 NJa^2(a^2+z^2)^{-3/2}$$

at a point on the axis at a distance z from the plane of the loop.

(*b*) A plane wire loop carries a current J. It consists of two straight sections, each of length d, and two semicircles of diameter d, as shown in figure 3.20, $ABCD$ being a square. Show that the magnetic field at O, the centre of the square, equals

$$\frac{\mu_0 J}{\pi d}\,[\log(\surd 2+1)+\surd 2].$$

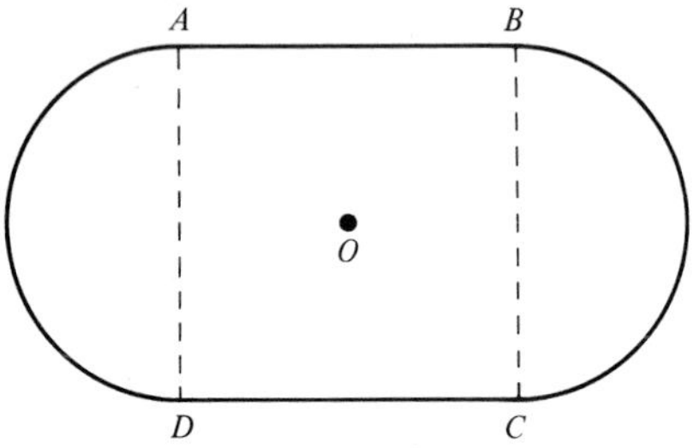

Fig. 3.20

(*c*) Show that the z-wise component of the magnetic field at the origin due to the current J itself along the helix referred to in example 3.3 is equal to $\tfrac{1}{2}\mu_0 J(a^2+p^2)^{-1/2}$. (The rest of the circuit is such that it produces no field at O.)

*3.13 (*N.B.* An asterisk indicates a section which is harder and may be omitted.) Consider two current loops L_1 and L_2 bearing uniform currents J_1 and J_2 and let two line elements at P_1 and P_2 on the loops be respectively $d\mathbf{r}_1$ and $d\mathbf{r}_2$. Denote the vector $\overrightarrow{P_1P_2}$ as $\mathbf{R}$, of magnitude R.

Find an expression for the force on $d\mathbf{r}_1$ due to the magnetic field of L_2 and show that the total force on L_1 due to L_2 is

$$\frac{\mu_0 J_1 J_2}{4\pi}\oint\oint\frac{d\mathbf{r}_1\times(\mathbf{R}\times d\mathbf{r}_2)}{R^3},$$

integrated over both loops in turn. The order of integration is immaterial. Hence show that the force equals

$$\frac{\mu_0 J_1 J_2}{4\pi}\oint\oint\frac{(d\mathbf{r}_1\cdot d\mathbf{r}_2)\mathbf{R}}{R^3},$$

noting that, with P_2 and $d\mathbf{r}_2$ fixed during integration round L_1,

$$d\mathbf{r}_1=-d\mathbf{R}\quad\text{and}\quad\oint_{L_1}\frac{d\mathbf{R}\cdot\mathbf{R}}{R^3}=0,$$

as an inverse square law field is conservative.

Deduce that the total force on L_2 due to L_1 is equal and opposite, in line with Newton's third law. This confirms the consistency and mutual dependence of the Biot–Savart and $\mathbf{j} \times \mathbf{B}$ force laws in this context. (It is Newton's third law which also requires a current-bearing loop to exert no net force on itself, for no possible site for an equal and opposite reaction then exists.)

3.14 The portion of the plane $x + y + z = 1$ that lies in the octant in which x, y and z are all positive is acted on by a force per unit area equal to $\mathbf{r}$. By means of double integrals with respect to x and y find the x- and z-components of the total force on the plane. Note that $\mathrm{d}a$ is greater than, not equal to, the corresponding $\mathrm{d}x\mathrm{d}y$ product (compare problem 1.16(b)).

*3.15 (a) The position vector $\mathbf{r}$ of a point P on a surface may be expressed as a function of two parameters u, v (e.g. latitude and longitude on the Earth's surface) which are curvilinear coordinates defining position on the surface. The vectors $\mathrm{d}\mathbf{r}$ and $\partial\mathbf{r}/\partial u$ taken at constant v are tangential to a constant-v line on the surface. What is a condition for the constant-u and constant-v lines to be orthogonal? Do latitude and longitude satisfy this condition?

Show that, for the evaluation of surface integrals as double integrals with respect to u and v, the area element $\mathrm{d}\mathbf{a}$ may be written as

$$\frac{\partial \mathbf{r}}{\partial u} \times \frac{\partial \mathbf{r}}{\partial v} \quad \mathrm{d}u\mathrm{d}v,$$

and use this result to check example 3.5.

(b) Apply (a) to the case where u and v are x and y and $\mathbf{r} = \mathbf{i}x + \mathbf{j}y + \mathbf{k}z(x, y)$. Show that $\mathrm{d}\mathbf{a}$ has x, y, z components equal to $-(\partial z/\partial x)\mathrm{d}x\mathrm{d}y$, $-(\partial z/\partial y)\mathrm{d}x\mathrm{d}y$, $\mathrm{d}x\mathrm{d}y$, respectively.

3.16 A hemispherical igloo of radius R suffers a wind pressure p given by $p = A - B\sin^2\theta$ at a point P such that θ is the angle between the radius vector to P and the (horizontal) wind direction. Find the total vertical wind force on the igloo. A and B are constants, and the ground is horizontal.

*3.17 A hemispherical undersea shelter lies on the plane seabed, which is inclined at 30° to the horizontal. The depth of the highest point of the shelter equals the radius of the hemisphere. The air pressure inside is atmospheric and the water pressure is proportional to depth. Show that the total pressure force on the shelter is inclined at $\tan^{-1}(\sqrt{3} - \frac{2}{3})$ to the horizontal.

3.18 The roof of a moon station consists of two planes at right angles. From it protrudes an observation dome which consists of part of a sphere of radius r whose centre is on the line of intersection of the planes. The air pressure inside the dome is p. By considering an alternative surface spanning the loop defined by the intersection of roof and dome, find the total pressure force on the dome. (This is equivalent to considering the equilibrium of the dome plus some of the air inside it.)

3.19 (*a*) Show that $\iint$ **da** evaluated over any surface spanning a loop L equals the line integral $\frac{1}{2}\oint \mathbf{r} \times d\mathbf{r}$ round the loop where **r** is expressed relative to any origin.
*(*b*) A wire loop of any shape carries a uniform current J in a uniform magnetic field **B**. Show by generalising the argument used in problem 2.4 that there is no net force on the loop and that the couple on the loop is $J\mathbf{A} \times \mathbf{B}$, in which **A** denotes $\iint$ **da** evaluated over any surface spanning the loop.

3.20 (*a*) Calculate the circulation of the vector $\mathbf{k} \times \mathbf{r}$ round a circle of unit radius in the x, y plane, tangential to the y-axis at the origin.
(*b*) Calculate the flux of the vector **r** through the sphere of unit radius tangential to the x, y plane at the origin.
(*c*) Find the flux of the vector $\mathbf{r}/r^3$ through one face of unit cube centred on the origin.

3.21 Consider the velocity field **v** which corresponds to rotation at a uniform angular velocity $\mathbf{\Omega}$ about the x-axis, and a vector field **w** which is axisymmetric about the x-axis and has no component in the **v**-direction at each point. Any two points A and B are joined by any curve L that does not cut the x-axis, and a cylindrical surface S is generated by rotating L about the x-axis. By considering the elementary strips of S generated by rotating elements $d\mathbf{r}$ of L about the x-axis, show that $\int_A^B \mathbf{v} \times \mathbf{w} \cdot d\mathbf{r}$ along L equals $\Phi\Omega/2\pi$, where Φ is the total flux of **w** through the surface S.

If **w** is solenoidal, show that $\mathbf{v} \times \mathbf{w}$ is conservative. Discuss and interpret this result in the case where **w** is a magnetic field and L is one of several possible routes for an armature conductor joining A to B.

3.22 (*a*) A closed surface S is divided into strips by its intersections with a series of spheres, all centred at P. Consider a particular strip lying between closely adjoining spheres of radius r and $r + dr$, and divide it into approximate parallelogram area elements by a series of closely spaced cuts. Let two adjoining cuts be QL and MN, where Q and M are on the sphere of radius r. Denoting the vectors $\overrightarrow{QL}$, $\overrightarrow{QM}$ and $\overrightarrow{PQ}$ by $d\boldsymbol{\rho}$, $d\mathbf{l}$ and **r** respectively, show that $d\boldsymbol{\rho} \cdot \mathbf{r} = r\,dr$. What is $d\mathbf{l} \cdot \mathbf{r}$?

If **da** represents the area $QLNM$, write **da** in terms of $d\boldsymbol{\rho}$ and $d\mathbf{l}$, and find $d\mathbf{a} \times \mathbf{r}$ in terms of $d\mathbf{l}$, r and dr. Deduce that $\oint d\mathbf{a} \times \mathbf{r}$ over the whole strip is zero (r and dr being constant) and hence that $\oiint d\mathbf{a} \times \mathbf{r} = \mathbf{0}$ and $\oiint d\mathbf{a} \times \mathbf{r}/r^3 = \mathbf{0}$ over the whole surface S.
(*b*) Show that the vector moment of the forces on a surface S about any origin O due to a uniform pressure p depends only on the shape of the rim of the surface.
(*c*) If a Biot–Savart current element is placed at P, deduce that its

magnetic field is solenoidal. (Note that $\mathbf{r}$ in (*a*) corresponds to $-\mathbf{r}$ in the Biot–Savart formula.) Since the flux of the sum of several vector fields is the sum of their fluxes, it follows that the magnetic field of a complete array of current elements forming a circuit or current flow field is also solenoidal.

3.23 (*a*) Point charges Q_i $(i = 1, 2, 3, \ldots, n)$ lie at points O_i upon a straight line XY. The resulting electric field pattern is axisymmetric about XY with the field lines lying in planes that pass through the line XY. For the annular strip produced by rotating a length $\mathrm{d}s$ of an electric field line at a point P about XY let the inner and outer edges respectively subtend solid angles Ω_i and $\Omega_i + \mathrm{d}\Omega_i$ at O_i. Noting that the flux of the sum of vector fields is the sum of their fluxes, and that the flux of $\mathbf{E}$ through the strip is zero, deduce that $\Sigma\, Q_i \mathrm{d}\Omega_i = 0$, summing over the charges, and hence that $\Sigma\, Q_i\Omega_i = \text{const.}$ along the field line. If the cone traced by O_iP as it rotates about XY has a semi-angle θ_i, show that $\Sigma\, Q_i \cos\theta_i = \text{const.}$ along a field line.

(*b*) For the case of two like charges Q and $2Q$, find the inclination at infinity of a field line which leaves the charge Q at right-angles to XY.

3.24 Sketch the two-dimensional vector-fields specified below, by first drawing little arrows in the direction of the field at various points in the x, y plane and then drawing curves which are tangential to a series of such arrows. One way of doing this is to draw the arrows in families, located along curves upon each of which the inclination of the arrows is constant. (Such a curve is called an *isocline*. It should not be confused with a field line.) Then check your sketches by finding the exact shape of the field lines analytically.

(i) $-y\mathbf{i} + x\mathbf{j}$, (ii) $x\mathbf{i} - y\mathbf{j}$,
(iii) $(1/y)\mathbf{i} + x\mathbf{j}$, (iv) $(1/x)\mathbf{i} + y\mathbf{j}$.

3.25 A gas flows steadily at high speed along a conical, diverging passage. All the streamlines are straight lines which if extrapolated pass through the apex of the cone and all properties (density ρ, pressure p, speed v) are uniform over spherical surfaces centred at the apex. The equation relating changes in pressure and velocity is $\mathrm{d}p + \rho v \mathrm{d}v = 0$ and the equation relating changes in pressure and density is $\mathrm{d}p = a^2 \mathrm{d}\rho$, where a is the speed of sound. Conservation of mass requires the mass flux vector $\rho\mathbf{v}$ to be solenoidal.

By considering $d(\rho v)/\mathrm{d}v$ show that if $v > a$ (supersonic flow) the velocity field $\mathbf{v}$ has the property that broadening of the stream tubes is associated with increase, not decrease, of velocity.

Contrast this with the case where $\rho = \text{constant}$ and $\mathbf{v}$ is solenoidal.

4

Further applications to electromagnetism

4.1 Ampère's law

Chapter 3 revealed that Coulomb's law implies that the electrostatic field $\mathbf{E}_s$ is conservative, with $\oint \mathbf{E}_s \cdot \mathbf{dr} = 0$ for all loops. This prompts the query whether the Biot–Savart law similarly implies that $\mathbf{B}$ is conservative, with $\oint \mathbf{B} \cdot \mathbf{dr} = 0$. The fact is that it does *not* make $\oint \mathbf{B} \cdot \mathbf{dr}$ zero in general but instead yields

$$\oint_L \mathbf{B} \cdot \mathbf{dr} = \mu_0 \iint_S \mathbf{j} \cdot \mathbf{da},$$

a result which is known as *Ampère's law*. Note that in one equation it brings together the two main mathematical concepts of chapter 3, the *circulation* type of line integral and the *flux* type of surface integral; the left-hand side is an integral round a closed loop L and the right-hand side is an integral over any surface S which spans L, the total current flux through S. In electrical engineering, electric current is often caused to flow in a wire winding, composed of many turns in series, and then the current flux $\iint \mathbf{j} \cdot \mathbf{da}$ through a surface intersecting these turns equals the current in the wire times the number of turns. The flux is then described as '*ampere-turns*'.

For the signs to be correct, a precise convention is necessary to relate the directions of the line and surface integrals. It is as usual a right-hand screw convention: looking at S in the positive direction for $\mathbf{da}$, $\mathbf{dr}$ positive must give a *clockwise* circuit of L. Note that L need *not* be itself a field line, although it is in the example below.

Example 4.1 Check the right-hand screw convention in relation to figure 3.16 and repeat problem 3.11(*b*) by using Ampère's law. Also calculate the magnetic field inside the wire if its radius is R.

Figure 3.16 is correct because, to an observer looking in the $\mathbf{J}$-direction, $\mathbf{B}$'s circulation appears clockwise. Provided one knows in advance that the field lines *are* circular, with B uniform along each line, Ampère's law, applied along a field line of radius r, makes the calculation of the field strength very simple:

$$\oint \mathbf{B} \cdot \mathbf{dr} = 2\pi r B = \mu_0 J, \quad \text{and} \quad B = \frac{\mu_0 J}{2\pi r} \quad \text{as before,}$$

for J is the current through the loop. If $r < R$, however, and the current intensity is j (equal to $J/\pi R^2$),

$$\oint \mathbf{B} \cdot \mathrm{d}\mathbf{r} = 2\pi r B = \mu_0 \pi r^2 j \text{ instead}$$

and $B = \frac{1}{2}\mu_0 j r$, proportional to r, not its inverse.

Problem 4.2 exploits this result in an interesting way.

---

*4.2 Ampère's law and the Biot–Savart law

It is an instructive exercise in vector analysis to demonstrate that Ampère's law follows from the Biot–Savart law. They are equivalent, alternative ways of relating the magnetic field **B** to its sources (the currents **j**).

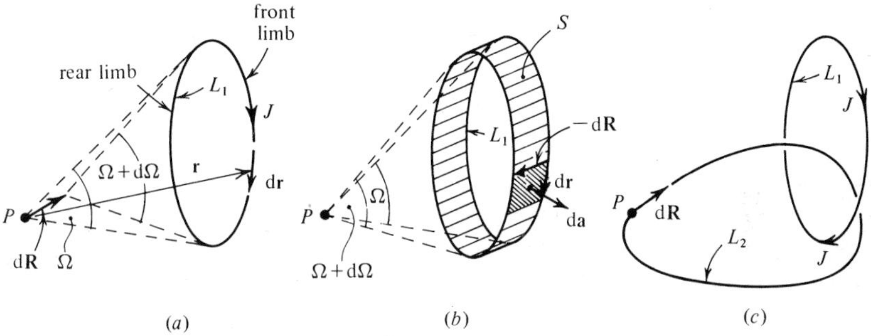

Fig. 4.1

We shall restrict the proof first to the case where a uniform current J flows in a closed loop L_1 as shown in figure 4.1(a). According to the Biot–Savart law, the field **B** at point P is given by an integral round L_1:

$$\mathbf{B} = \frac{\mu_0 J}{4\pi} \oint \frac{\mathbf{r}}{r^3} \times \mathrm{d}\mathbf{r},$$

in which $\mathrm{d}\mathbf{r}$ is an element of L_1 and **r** runs *from P to* $\mathrm{d}\mathbf{r}$ as shown. Ampère's law involves the motion of P along some other closed loop L_2, of which $\mathrm{d}\mathbf{R}$, say, is an element, as shown in figure 4.1(c). It is necessary to use $\mathrm{d}\mathbf{R}$ instead of $\mathrm{d}\mathbf{r}$ for the present to avoid confusion. As $\mathrm{d}\mathbf{R}$ is constant during the integration round L_1,

$$\mathbf{B} \cdot \mathrm{d}\mathbf{R} = \frac{\mu_0 J}{4\pi} \oint \frac{\mathbf{r}}{r^3} \times \mathrm{d}\mathbf{r} \cdot \mathrm{d}\mathbf{R} = \frac{\mu_0 J}{4\pi} \oint \frac{\mathbf{r}}{r^3} \cdot (-\mathrm{d}\mathbf{R}) \times \mathrm{d}\mathbf{r},$$

if we exploit the properties of the triple scalar product. This equation can be interpreted fruitfully in terms of the change $\mathrm{d}\Omega$ of the solid angle Ω subtended by L_1 at P as P moves along $\mathrm{d}\mathbf{R}$. (See figure 4.1(a).) It is simpler to view this change in solid angle relative to P, i.e. to consider the effect on Ω of moving L_1 bodily through $-\mathrm{d}\mathbf{R}$ so that it traces out a strip S while P is kept fixed, as in figure 4.1(b). The product $(-\mathrm{d}\mathbf{R}) \times \mathrm{d}\mathbf{r}$ is the parallelogram area element $\mathrm{d}\mathbf{a}$ shown, and so

$$\oint \frac{\mathbf{r}}{r^3} \cdot (-\,\mathbf{dR}) \times \mathbf{dr}$$

equals the flux of the inverse-square-law field through S, which is $\mathrm{d}\Omega$, the solid angle subtended at P by S. Hence

$$\mathbf{B} \cdot \mathbf{dR} = \frac{\mu_0 J}{4\pi} \mathrm{d}\Omega$$

which integrates to

$$\int \mathbf{B} \cdot \mathbf{dR} = \frac{\mu_0 J}{4\pi} \times (\text{increase in } \Omega).$$

When P traverses a complete loop L_2 there are two cases:

(*a*) If L_2 links L_1 as in figure 4.1(*c*), the increase in Ω is 4π. (Think of an umbrella blowing inside out.) The conclusion here is that

$$\oint \mathbf{B} \cdot \mathbf{dr} = \mu_0 J$$

which is Ampère's law, as J is the total current flowing through any surface spanning L_2 in this case. Figure 4.1(*c*) confirms that the right-hand screw convention is correctly observed; to an observer looking through L_2 in the J-direction, $\mathbf{dR}$ positive gives a clockwise circuit of L_2.

(*b*) If L_2 does not link L_1 there is no net increase in Ω when L_2 is completely traversed and $\oint \mathbf{B} \cdot \mathbf{dR} = 0$, which confirms the at first surprising implication of Ampere's law that current that is not linked does not contribute to $\oint \mathbf{B} \cdot \mathbf{dr}$, however near it is. Although each bit of such a current loop does contribute to $\mathbf{B}$, the contributions to $\oint \mathbf{B} \cdot \mathbf{dr}$ all cancel out and there is no paradox.

Finally, we may note that it is possible to generalise this proof to the case of a general distribution of currents $\mathbf{j}$ in space if they can be regarded as the superposition of many individual current loops such as the one we have just treated.

4.3 Multi-valued potentials; magnetic potential

Ampère's law throws up an important general idea which also has implications for other subjects, including fluid mechanics. It raises the question as to whether a potential can be defined for a field such as $\mathbf{B}$ for which $\oint \mathbf{B} \cdot \mathbf{dr} = 0$ *sometimes* (rather than *always*, as in sections 3.4 and 3.5). Consider a situation where a toroidal region R (shown in part and sectioned in figure 4.2) contains circulating currents which total J. Then $\oint \mathbf{B} \cdot \mathbf{dr} = 0$ round any loop which does not link R, and B begins to look conservative; but $\oint \mathbf{B} \cdot \mathbf{dr} = \mu_0 J \neq 0$ round any loop such as $L_1 + L_2$ which does link R. Thus

$$\int_{L_2\,P}^{Q} \mathbf{B} \cdot \mathbf{dr} + \int_{L_1\,Q}^{P} \mathbf{B} \cdot \mathbf{dr} = \mu_0 J \quad \text{and} \quad \int_{L_2\,P}^{Q} \mathbf{B} \cdot \mathbf{dr} = \int_{L_1\,P}^{Q} \mathbf{B} \cdot \mathbf{dr} + \mu_0 J.$$

If we take P as datum and try to define a 'magnetic potential' at points external to

R as equal to $\int \mathbf{B} \cdot d\mathbf{r}$ from P to the point in question, we can obviously get several alternative values which differ by $\mu_0 J$ (or multiples thereof if the route goes through the ring several times); in other words a *multi-valued potential* results.

To make the situation more orderly we therefore rule that the routes used to evaluate the potential must not go through the hole in R. This point needs to be made more precise: as a criterion for identifying routes 'through the hole' we need some 'cut' surface S spanning the hole (shown in part in the figure) and then the potential is defined uniquely by going along any route which does not penetrate R and S. (Mathematically we say that S has transformed the region external to R from being *multiply connected* to being *simply connected.*) The potential does not exist inside R.

Consider adjoining points A and B on the two sides of S. We have

$$\int_P^A \mathbf{B} \cdot d\mathbf{r} + \int_B^P \mathbf{B} \cdot d\mathbf{r} = \mu_0 J \quad \text{or} \quad \int_P^A \mathbf{B} \cdot d\mathbf{r} - \int_P^B \mathbf{B} \cdot d\mathbf{r} = \mu_0 J.$$

This states that the potentials at A and B differ by $\mu_0 J$, i.e. S is a surface of discontinuity in the potential field. But realise that this is a purely mathematical, not a physical thing, introduced merely to make the potential single-valued. The exact location of S is arbitrary.

It is standard practice in electromagnetic theory to choose a *negative* sign convention for magnetic potential, consistently with the choice for electric potential, and also to divide by μ_0, so that we actually define magnetic potential U as

$$U = -\frac{1}{\mu_0} \int_P \mathbf{B} \cdot d\mathbf{r},$$

the integral being taken from the datum P by any route not cutting R and S. The convention is such that:

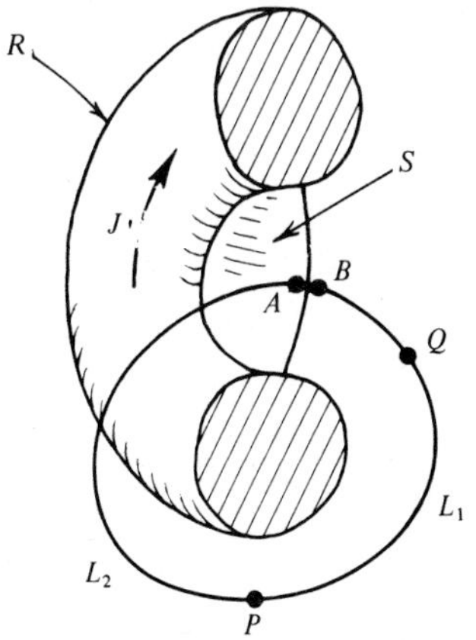

Fig. 4.2

(i) if a point moves in a direction in which **B** has a positive component, U decreases, because

$$dU = -\mathbf{B}\cdot d\mathbf{r}/\mu_0,$$

and

(ii) for points A and B in the figure,

$$U_B - U_A = J.$$

So the surface S acts like a source of magnetic potential, which restores the potential after it has declined while the point circulated from B to A via P. Electrical engineers refer to J (the ampere-turns) as *magneto-motive force* (m.m.f.) because of the obvious analogy with electromotive force (e.m.f.).

Example 4.2 Express the magnetic potential for the field of a long straight wire carrying a current J (as discussed in example 4.1) in polar coordinates.

This is evidently a two-dimensional problem. Figure 4.3 shows a plane transverse to the wire, containing the circular field lines. We may apply

$$\mathbf{B}\cdot d\mathbf{r} = -\mu_0\, dU$$

along either of the two excursions PQ, PR shown, where P is the point (r, θ). Along PR, $\mathbf{B}\cdot d\mathbf{r}$ and dU vanish, for **B** is perpendicular to $\overrightarrow{PR}$, i.e. U is independent of r. Along PQ (of length $r\,d\theta$)

$$-\mu_0 dU = \mathbf{B}\cdot d\mathbf{r} = Br\,d\theta = \mu_0 J\,d\theta/2\pi,$$

for $B = \mu_0 J/2\pi r$. Evidently $U = -J\theta/2\pi$ if we choose the datum for U where $\theta = 0$.

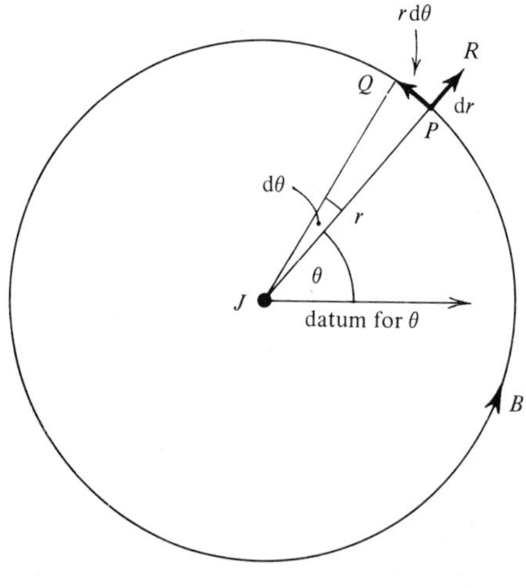

Fig. 4.3. Field of line current J (directed out of page).

Note that θ (and therefore also U) is multi-valued unless a restriction is applied such as $|\theta| < \pi$. (This then makes the 'cut' surface the half-plane $\theta = \pi$.)

If the region R is a wire loop such as appeared in figure 4.1, the magnetic potential has a simple geometrical interpretation in terms of the solid angle Ω subtended at a point by the loop. Comparing the result

$$\mathbf{B} \cdot \mathrm{d}\mathbf{r} = (\mu_0 J/4\pi)\mathrm{d}\Omega$$

in section 4.2 with the equation for $\mathrm{d}U$ we see that

$$\mathrm{d}U = -(J/4\pi)\mathrm{d}\Omega \quad \text{and so} \quad U = -J\Omega/4\pi$$

if we choose the datum at some point where $\Omega = 0$. S could conveniently be chosen as the surface composed of points at which $\Omega = \pm 2\pi$ (half a complete solid angle).

Problem 4.3 explores the important case of a small current loop which turns out to behave like a magnetic dipole. It forms the basis for the study of material magnetisation, which we have been excluding from consideration.

Just as for electric potential, we can superpose the magnetic potential due to several causes. The following example shows how, as a consequence, the use of magnetic potential can be more efficient than calculating magnetic field components direct from the Biot–Savart law. It is a case where Ampère's law is quite useless.

Example 4.3* A long circular cylinder bears axial current distributed at an intensity $J \sin\theta$ per unit perimeter, θ being angular position measured at the centre of the cross-section. Show that the field inside the cylinder is of uniform intensity $\frac{1}{2}\mu_0 J$ and in the direction $\theta = 0$.

We must first establish two purely mathematical results. Figure 4.4 defines the lengths R, l, m and n and the angles θ, α, λ, ϕ and ψ.

Evidently

$$OM = l \sin\psi = R \sin\phi$$

and

$$m = R\cos(\psi - \phi) = R\sin\phi\sin\psi + R\cos\phi\cos\psi = l\sin^2\psi + R\cos\phi\cos\psi.$$

P is a fixed point and R, l and α are constant. Consider integrals in which N moves round the circle while ψ and θ increase by 2π, but ϕ undergoes no net change, and m and n also vary:

$$\oint m\mathrm{d}\psi = \oint l\sin^2\psi\,\mathrm{d}\psi + \oint R\cos\phi\cos\psi\,\mathrm{d}\psi = \pi l,$$

for we may put $\cos\psi\,\mathrm{d}\psi = (R/l)\cos\phi\,\mathrm{d}\phi$ in the second integral, which then vanishes. However, $\oint n\,\mathrm{d}\psi = 0$ for n is an odd function of ψ.

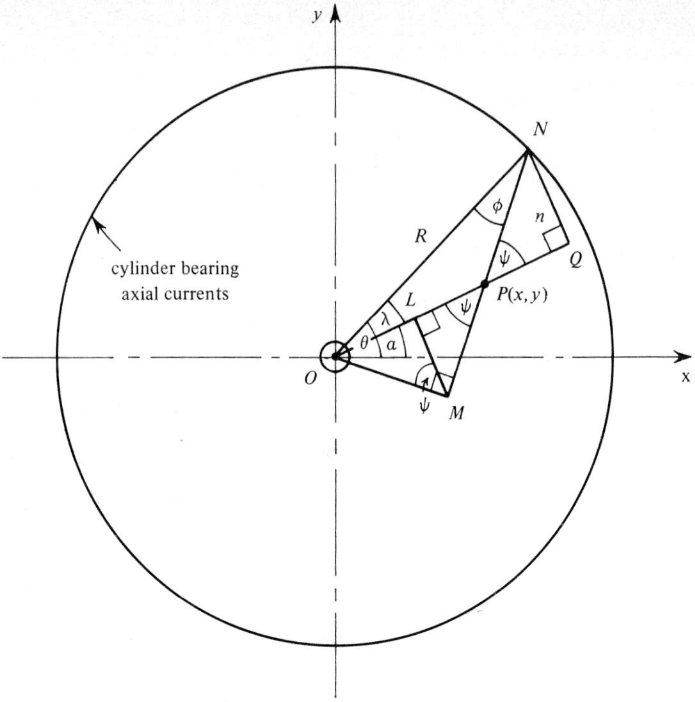

Fig. 4.4. Cross-section of cylinder. l, m are the lengths OP, OQ respectively.

Example 4.2 indicates that the magnetic potential at the point $P(x, y)$ inside the cylinder due to an axial line current $JR \sin\theta\, d\theta$ in an element of perimeter of width $R\, d\theta$ at N is

$$-\phi JR \sin\theta\, d\theta/2\pi,$$

if we measure ϕ, the angle of rotation about N, from the datum ON, in order that the magnetic potential shall be zero at O. Integrating with respect to θ gives U, the total magnetic potential at P, as

$$U = -\frac{JR}{2\pi}\oint \phi \sin\theta\, d\theta.$$

If we integrate by parts and note that $[-\phi\cos\theta]$ does not change we find that

$$U = \frac{JR}{2\pi}\oint \cos\theta\, d\phi = -\frac{JR}{2\pi}\oint \cos\theta(d\psi - d\theta) \quad (\text{for } \phi = \psi - \theta - \alpha,\ \alpha \text{ const.})$$

$$= -\frac{JR}{2\pi}\oint \cos(\alpha+\lambda)d\psi \quad (\text{for } \oint \cos\theta\, d\theta = 0 \text{ and } \theta = \alpha + \lambda)$$

$$= -\frac{J}{2\pi}\left\{\cos\alpha \oint R\cos\lambda\, d\psi - \sin\alpha \oint R \sin\lambda\, d\psi\right\}$$

$$= -\frac{J}{2\pi}\left\{\cos\alpha \oint m\,\mathrm{d}\psi - \sin\alpha \oint n\,\mathrm{d}\psi\right\}$$

$$= -\frac{J\cos\alpha}{2\pi}(\pi l) = -\tfrac{1}{2}Jx \quad (\text{for } x = l\cos\alpha).$$

But $\mathbf{B}\cdot\mathrm{d}\mathbf{r} = B_x\mathrm{d}x + B_y\mathrm{d}y + B_z\mathrm{d}z = -\mu_0\mathrm{d}U = \frac{1}{2}\mu_0 J\mathrm{d}x$, for all $\mathrm{d}x, \mathrm{d}y, \mathrm{d}z$.

Hence

$$B_x = \tfrac{1}{2}\mu_0 J, \quad B_y = 0, \quad B_z = 0,$$

i.e. a uniform field in the direction $\theta = 0$.

This problem is reconsidered by other more efficient methods later.

4.4 Ampère's law and Kirchhoff's first law

If we look at Ampère's law

$$\oint_L \mathbf{B}\cdot\mathrm{d}\mathbf{r} = \mu_0 \iint_S \mathbf{j}\cdot\mathrm{d}\mathbf{a},$$

we may note that the left-hand side is unique, given L and $\mathbf{B}$, and so the surface integral on the right must take the same value for all surfaces S spanning L. If Φ_1 and Φ_2 in figure 3.10 apply to current flux, $\Phi_1 = \Phi_2$ and we infer that $\mathbf{j}$ must be solenoidal. We can then legitimately talk about 'current flux linked' or 'ampere-turns linked' without ambiguity. Problem 4.4(*b*) offers an alternative argument. The statement that $\mathbf{j}$ is solenoidal, a corollary of Ampère's law, is known as *Kirchhoff's first law*. It simply states that as much current must flow into a closed region as flows out of it.

Example 4.4 A node in an electric circuit is a junction between three wires, bearing currents J_1, J_2, J_3 respectively towards the node. Deduce from Ampère's law that

$$J_1 + J_2 + J_3 = 0,$$

the 'circuit form' of Kirchhoff's first law.

Figure 4.5 shows part of the wires near the node. Consider the application of Ampère's law to the loop *ABCDEFGHIJKLA*. The portions *CD* and *LA*, etc. actually coincide but are shown separated for clarity. The contributions to $\oint \mathbf{B}\cdot\mathrm{d}\mathbf{r}$ from such paired portions cancel out because each is traversed in both directions leaving only the contributions from the loops L_1, L_2 and L_3. As the complete loop clearly links no current, Ampère's law gives

$$0 = \oint \mathbf{B}\cdot\mathrm{d}\mathbf{r} = \oint_{L_1} \mathbf{B}\cdot\mathrm{d}\mathbf{r} + \oint_{L_2} \mathbf{B}\cdot\mathrm{d}\mathbf{r} + \oint_{L_3} \mathbf{B}\cdot\mathrm{d}\mathbf{r}.$$

But Ampère's law also implies that

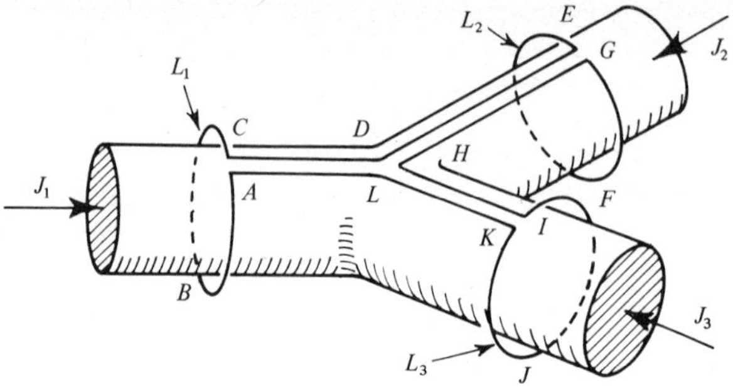

Fig. 4.5. The junction between three wires.

$$\oint_{L_1} \mathbf{B} \cdot \mathbf{dr} = \mu_0 J_1, \text{ etc.}$$

and so

$$J_1 + J_2 + J_3 = 0.$$

We shall reexamine many of these ideas in chapter 6. Why?

4.5 Faraday's law

We have stated, by means of the Coulomb and Biot–Savart laws, how $\mathbf{E}_s$, the electrostatic part of $\mathbf{E}$, and the magnetic field $\mathbf{B}$ are related to their sources (stationary and moving charges) and we have inferred from these laws that $\mathbf{E}_s$ is conservative and also solenoidal (except in the regions containing its sources) while $\mathbf{B}$ is solenoidal but not in general conservative. We have still to state how $\mathbf{E}_i$, the magnetic-induction part of $\mathbf{E}$, is related to its source, which is any changing magnetic field, with a time-derivative $\partial\mathbf{B}/\partial t$ (the rate of change of $\mathbf{B}$ in time at a fixed point in space).

It would be possible to state the experimental facts in Biot–Savart form, giving $\mathbf{E}_i$ as the sum of contributions due to little 'elements' of $\partial\mathbf{B}/\partial t$ distributed in space, but it is more usual to express the relation in terms of the two separate implications of a Biot–Savart-type relationship:

(a) $\mathbf{E}_i$ is solenoidal, and

(*b*) $\mathbf{E}_i$ is not in general conservative. Instead its circulation $\oint \mathbf{E}_i \cdot \mathbf{dr}$ round a closed loop L is given by a law, analogous to Ampère's law, which relates it to a surface integral taken over any surface S spanning the loop. This equation is *Faraday's law* of magnetic induction,

$$\oint_L \mathbf{E}_i \cdot \mathbf{dr} = -\iint_S \frac{\partial \mathbf{B}}{\partial t} \cdot \mathbf{da}, \quad \text{a } \textit{flux} \text{ integral.}$$

Note that the vector $\partial\mathbf{B}/\partial t$ is not in general parallel to $\mathbf{B}$. As

$$\mathbf{E} = \mathbf{E}_s + \mathbf{E}_i \quad \text{and} \quad \oint \mathbf{E}_s \cdot d\mathbf{r} = 0,$$

the equation

$$\oint_L \mathbf{E} \cdot d\mathbf{r} = -\iint_S \frac{\partial \mathbf{B}}{\partial t} \cdot d\mathbf{a}$$

is true even in the presence of an electrostatic contribution to **E**. L can be any loop and need not follow an **E**-field line. As L and S are stationary it is legitimate to alter the order of space integration and time differentiation on the right-hand side. Then

$$\oint_L \mathbf{E} \cdot d\mathbf{r} = -\frac{\partial}{\partial t} \iint_S \mathbf{B} \cdot d\mathbf{a} = -\frac{\partial \Phi}{\partial t},$$

if $\Phi = {}_S\!\iint \mathbf{B} \cdot d\mathbf{a}$, the magnetic flux through S. The negative sign on the right-hand side is necessary to be consistent with the normal right-hand screw convention: looking at S in the positive $d\mathbf{a}$ direction, positive $d\mathbf{r}$ gives a clockwise circuit of L. *Partial* $\partial/\partial t$ is used in $\partial\Phi/\partial t$ to emphasise the fact that it refers to a loop fixed in space.

Faraday's law bears the encouraging news that **E** is not conservative, which opens the way to continuous or repeated exchange of power by electrical means. The quantity $\oint \mathbf{E} \cdot d\mathbf{r}$ at first sight seems physically appealing as it could refer to the work done by the electric field on a unit charge taken round a closed path L, but there is a concealed subtlety: $\oint \mathbf{E} \cdot d\mathbf{r}$ refers to the *instantaneous* state of **E** and so the unit charge would have to make its circuit in a time short enough for **E** not to have changed significantly meanwhile. An electron in a transformer winding does not meet this requirement, not even remotely. (See example 2.3.)

Faraday's law also reveals the importance of the concept of magnetic flux in electromagnetism, even though **B** is apparently an abstraction and does not represent a real flow in any sense. Faraday's law in fact leads to the way in which magnetic flux and magnetic flux intensity **B** are measured, for $\oint \mathbf{E} \cdot d\mathbf{r}$ can be detected as a voltage with a search coil and integrated in time to yield magnetic flux or flux changes (which are therefore measured in volt-seconds, also known as webers). Faraday's law involves no constant factor such as μ_0 and so is the starting point for the erection of the system of magnetic units. Once the unit of charge (the coulomb) has been chosen, the volt is then determined because potential relates to work done on unit charge. (See problem 4.5.)

Just as Ampère's law leads to Kirchhoff's first law, the familiar arguments about unambiguous flux-linkage cause Faraday's law to imply that $\partial\mathbf{B}/\partial t$ must be solenoidal. However, as **B** is solenoidal in any case,

$$\oiint \mathbf{B} \cdot d\mathbf{a} = 0 \quad \text{and so} \quad \oiint \frac{\partial \mathbf{B}}{\partial t} \cdot d\mathbf{a} = \frac{\partial}{\partial t} \oiint \mathbf{B} \cdot d\mathbf{a} = 0$$

and the condition is satisfied.

Example 4.5 A long straight wire of conductivity σ and radius R carries an axial alternating current of intensity $j = k \sin \omega t$ (k, ω const.). The resulting oscillatory magnetic fields must induce perturbations of the electric field. Investigate how they perturb the current density on the centre-line of the wire.

Though the uniformity of the current at each instant must be disrupted to some extent, we shall ignore this in calculating the magnetic field and its effect on the electric field. This is acceptable provided the perturbations of the electric field and current are small. Ohm's law $\mathbf{j} = \sigma\mathbf{E}$ indicates that an axial electric field (k/σ) sin $\omega t = E_0$ has to be applied to drive the current. We shall take this as still being the axial electric field at the surface of the wire even when the electric field inside the wire is perturbed by magnetic induction. To find the axial electric field E_a at the axis we apply Faraday's law round the loop *PQRS* in figure 4.6, *PQ* being of unit length. As there is obviously no radial current flow, there can be no radial electric field components. Therefore

$$\oint E \cdot dr = E_0 - E_a, \quad \text{for} \quad PQ = RS = \text{unity.}$$

For the loop *PQRS* $\iint \mathbf{B} \cdot \mathrm{d}\mathbf{a} = -\int_0^R \frac{1}{2}\mu_0 jr\mathrm{d}r = -\frac{1}{4}\mu_0 R^2 j$ in which the area element, shown shaded, is a strip of unit length and width dr and the value of B has been inserted from example 4.1. The minus sign comes from a proper observance of the right-hand screw rule; in the figure, *PQRS* is a clockwise circuit but B comes out of the paper. Faraday's law therefore gives

$$E_0 - E_a = \tfrac{1}{4}\mu_0 R^2 \mathrm{d}j/\mathrm{d}t = \tfrac{1}{4}\omega\mu_0 R^2 k \cos \omega t,$$

and j_a, the current density on the axis, is given by

$$j_a = \sigma E_a = k(\sin \omega t - \tfrac{1}{4}\sigma\omega\mu_0 R^2 \cos \omega t).$$

This approach is only valid if the perturbation term is small, i.e. if

$$\tfrac{1}{4}\sigma\omega\mu_0 R^2 \ll 1 \quad \text{or} \quad R \ll 2(\sigma\omega\mu_0)^{-1/2}.$$

(The question as to what happens when this is not satisfied is explored in chapter 11.)

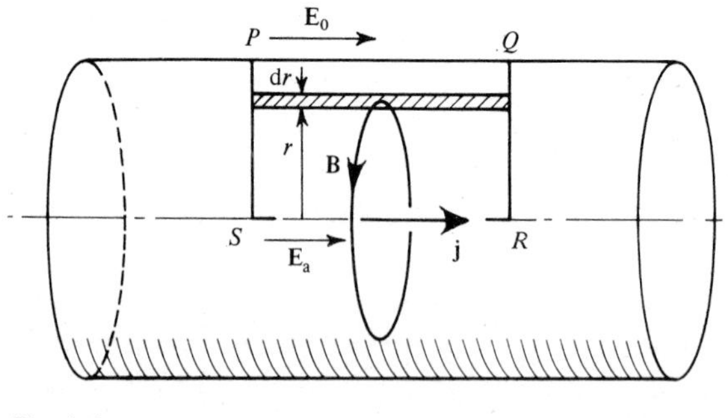

Fig. 4.6

Problem 4.6 is a simple study of charged-particle motion in a changing magnetic field under the influence of the resulting $\mathbf{E}_i$ field.

4.6 The lumped-component approximation

Electrical engineers tend to be preoccupied with *circuits*, in which information or power in the form of currents and fields is channelled by slender conductors (wires) connecting discrete components, each designed to perform specific functions.

A familiar example of such a component is a *capacitor*. This is a configuration of conductors and insulators designed geometrically so as to allow the accumulation of separate, large, equal and opposite charges without the development of excessively large differences of potential. Elementary electrostatics tells us to achieve this by having plates of large area, separated by a small gap, preferably containing a polarisable dielectric, whose bound charges migrate in an endeavour to 'short circuit' the field, thereby reducing the electric potential. (The limited charge migrations result in net charge distributions on the two surfaces of the dielectric adjoining the plates, opposite in sign to the charges on the nearby plate.) It is important to appreciate that when a capacitor is changing its state of charge, currents must flow, causing magnetic fields and inducing an electric field $\mathbf{E}_i$ which is not conservative. Then the idea of electric potential is no longer strictly valid.

Another familiar circuit component is the *inductor*, which is a device designed to maximise the value of $\oint \mathbf{E} \cdot \mathbf{dr}$ induced round a circuit by the changing magnetic field due to the changing currents in the circuit. To achieve this we use the idea of multiple turns not only to make a given current contribute repeatedly to the current flux through a given surface in Ampère's law, so enhancing the resulting magnetic field, but also to capture repeated magnetically-induced contributions to $\oint \mathbf{E} \cdot \mathbf{dr}$, evaluated along the wire in accordance with Faraday's law. Often the magnitude of the magnetic-flux-linked is enhanced by use of a ferromagnetic material as well. This also serves another purpose, that of restricting the magnetic field almost entirely to the interior of the device. But notice that the electric field $\mathbf{E}_i$ due to magnetic induction will not in general be parallel or equal to the total electric field demanded by Ohm's law. There must therefore be an electrostatic part to the field, attributable to charge distributions on the wires.

The conclusion from the two previous paragraphs, then, is that capacitors and inductors are not entirely distinguishable, as capacitors involve inductive behaviour and inductors involve electrostatic or capacitive behaviour to some extent. There are kinds of electrical engineering, particularly in the microwave area, where it is not possible to distinguish capacitors and inductors completely and the full, complex electromagnetic field problem has to be faced. Fortunately, however, most electrical and electronic engineering is amenable to an adequately accurate, approximate treatment which simplifies the situation enormously. This is the *lumped-component approximation*, which becomes valid whenever the frequency or rate of change of events is low enough. The epithet 'lumped' emphasises that it allows a 'black-box' or 'systems' approach to electrical circuits; each component is modelled as having a single, simplified function and simple input/output characteristics. All the inductance is 'lumped' into one component, all the 'capacitance' into another, and so on. The same idea should be familiar in mechanics where, for instance, a mass oscillating on a spring is treated by neglecting the elasticity of the mass and the inertia of the spring, all elastic or inertial effects being lumped into one or other component.

We therefore treat a capacitor as if governed purely by electrostatics, assuming that any $\partial \mathbf{B}/\partial t$ term is small enough for $\mathbf{E}_i$ to be negligible compared with $\mathbf{E}_s$, so that the concept of potential is acceptable. If the plates are also virtually perfect conductors (usually a good approximation), $\mathbf{E} = 0$ in them and $-\int \mathbf{E} \cdot d\mathbf{r}$ taken from any point on one plate to any point on the other will yield the same value, ΔV say, the unique potential difference across the capacitor.

Similarly we can ignore electrostatic phenomena in an inductor. The current all along the wire can be treated as uniform, as only a negligible fraction of it gets side-tracked into changing charge distributions. The potential difference across the terminals AB of the inductor needs care in definition; it is taken as $-\int_A^B \mathbf{E} \cdot d\mathbf{r}$ by any route which keeps sufficiently clear of the magnetic turmoil in and near the inductor, so that a potential can be defined. Treating the resistance of the inductor as a separate resistor in series with a perfectly conducting inductor is acceptable because the errors only concern the neglected phenomena.

Finally, for the resistors and other connectors (assumed perfectly conducting) which complete the circuit, we neglect both electrostatic and Faraday's law effects. The conductors may be moving, as for instance in an electrical machine. Then Ohm's law gains the usual $\mathbf{v} \times \mathbf{B}$ term. If the conductors involve thermoelectric or electrochemical effects or the various phenomena of solid-state electronics that control transistors, etc., Ohm's law has to acquire a further term $\mathfrak{E}$, to model these further causes of charge migration, additional to $\mathbf{E}$ and $\mathbf{v} \times \mathbf{B}$. Inside a discharging battery, for instance, $\mathbf{j}$ flows contrary to $\mathbf{E}$ because of the effects modelled by the $\mathfrak{E}$ term.

4.7 Kirchhoff's second law; electromotive force

A simplified overall statement of circuit behaviour known as Kirchhoff's second law is permitted by the lumped-component approximation. Together with his first law, in circuit form, it provides a complete apparatus for the routine analysis of circuit problems, given the necessary information about the overall characteristics of the various lumped circuit components, e.g. the familiar relations between potential difference (p.d.) and charge content for a capacitor, or between p.d. and rate of change of current for an inductor. To get the second law, consider $\oint \mathbf{E} \cdot d\mathbf{r}$ taken round an electric circuit, crossing the capacitor gaps (if any) by any route. The contributions to the integral are of the form either:

(i) $-\Delta V$ for each capacitor gap
or (ii) $\int \mathbf{E} \cdot d\mathbf{r}$ along conducting components (including inductors)

in which

$$\mathbf{E} = \tau\mathbf{j} - \mathbf{v} \times \mathbf{B} - \mathcal{E},$$

if we include all the effects already referred to. For the circuit as a whole, $\oint \mathbf{E} \cdot \mathrm{d}\mathbf{r}$ is not zero because of the changing magnetic flux Φ linked inside the inductor components, with

$$\oint \mathbf{E} \cdot \mathrm{d}\mathbf{r} = -\partial\Phi/\partial t.$$

Adding together all the contributions to $\oint \mathbf{E} \cdot \mathrm{d}\mathbf{r}$ gives

$$-\sum \Delta V + \sum \int \tau\mathbf{j} \cdot \mathrm{d}\mathbf{r} - \sum \int \mathbf{v} \times \mathbf{B} \cdot \mathrm{d}\mathbf{r} - \sum \int \mathcal{E} \cdot \mathrm{d}\mathbf{r} = -\sum \frac{\partial\Phi}{\partial t}$$

where each Σ denotes a sum over all circuit components of a relevant type. The idealised connectors between components contribute negligibly to the equation. Rearranging this equation and giving each term its normal description renders it more recognisable:

$$\sum \int \tau\mathbf{j} \cdot \mathrm{d}\mathbf{r} + \sum(-\Delta V) = \sum \mathcal{E} \cdot \mathrm{d}\mathbf{r} - \sum \frac{\partial\Phi}{\partial t} + \int \mathbf{v} \times \mathbf{B} \cdot \mathrm{d}\mathbf{r}$$

$\sum \int \tau\mathbf{j} \cdot \mathrm{d}\mathbf{r}$	$\sum(-\Delta V)$	$\sum \mathcal{E} \cdot \mathrm{d}\mathbf{r}$	$\sum \frac{\partial\Phi}{\partial t}$	$\int \mathbf{v} \times \mathbf{B} \cdot \mathrm{d}\mathbf{r}$
potential drops across resistors (e.g. JR)	potential drops across capacitors (e.g. Q/C)	thermoelectric, electrochemical or solid state e.m.f.s	e.m.f.s due to self or mutual inductance (e.g. $L\mathrm{d}J/\mathrm{d}t$)	e.m.f.s due to motion induction

This is *Kirchhoff's second law*. On the right-hand side are gathered all the terms which are described by the conventional generic term *electromotive force* (e.m.f.), quantifying a motley range of physical phenomena, all of which are capable of driving a current against electric resistance and the 'back-pressure' of capacitors. As the reader will no doubt be familiar with circuit theory, a worked example is unnecessary.

The interesting thing about Kirchhoff's circuit laws is that they take the geometry out of that very geometrical subject, electromagnetic theory, and reduce it to algebra. All the geometry has been confined to the working out of the individual component characteristics. This is in the best tradition of engineering progress, whereby a complicated system is broken down into distinct, simple sub-systems which are capable of analysis on their own. Their collective behaviour as a complete system then constitutes a separate problem.

All goes well until an inexperienced circuit-builder puts two connecting wires too close together or makes the frequency too high and gets 'cross-talk'. The inductance and capacitance are then rebelling against their unnatural confinement in the various 'black boxes', for there is really inductance and capacitance everywhere. The lumped-component approximation is then hastily repaired by references to 'stray capacitance', etc.

The lumped-component model has its dangers, therefore. The serious student must remember that it is only a good approximation and may fail him. Problem 4.7 is intended to encourage a critical attitude to lumped-circuit theory.

4.8 Faraday's law for a moving loop

The last two e.m.f. terms in Kirchhoff's second law are both of magnetic origin. Moreover both change if we refer the situation to a moving reference frame, while all other terms in the equation obviously stay the same. These reflections strongly suggest that the two terms represent a single effect, as Faraday's genius, unimpeded by mathematics, readily perceived. We first consider a particular case.

Example 4.6 Figure 4.7 shows a metal ring L of radius R moving along at speed v outside a coil C, out of which magnetic flux is escaping. The current and field of the coil are constant and the configuration is axisymmetric. Ignoring any perturbation of the field by currents induced in L, demonstrate the equivalence of the two magnetic e.m.f. terms referred to a frame fixed (*a*) to the coil, (*b*) to the ring.

Relative to a frame fixed to the coil, $\partial\Phi/\partial t$ is zero, because $\partial\Phi/\partial t$ refers to the flux through a *stationary* surface, spanning the *instantaneous* position of the ring, as $\partial/\partial t$ refers to rate of change at a fixed location. The other magnetic e.m.f. term, $\oint \mathbf{v} \times \mathbf{B} \cdot \mathbf{dr}$ equals $2\pi R v B_r$, if B_r is the radial field component at L.

In contrast, relative to a frame moving with the ring, $\mathbf{v}$ and $\oint \mathbf{v} \times \mathbf{B} \cdot \mathbf{dr}$ vanish but $\partial\Phi/\partial t$ is obviously not zero. The magnetic flux through a plane surface attached to the ring is increasing. We can find $\partial\Phi/\partial t$ by considering the ring in position L' at a time $\mathrm{d}t$ later, when the flux through it has risen to $\Phi + \mathrm{d}\Phi$. Since $\mathbf{B}$ is solenoidal, $\oiint \mathbf{B} \cdot \mathbf{da} = 0$ over the closed surface composed of two planes spanning L and L' and the curved strip joining L and L', which has width $v\mathrm{d}t$ and area $2\pi R v \mathrm{d}t$. For this surface,

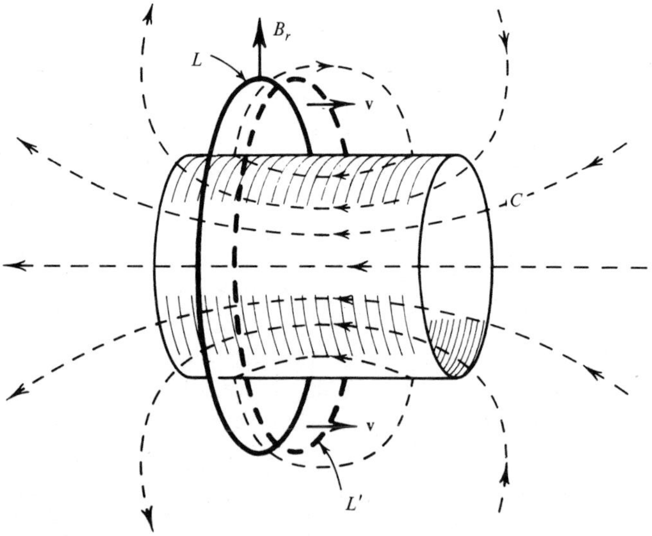

Fig. 4.7. A ring moving axially outside a coil.

$$\text{flux in} = \Phi + \mathrm{d}\Phi = \text{flux out} = \Phi + (2\pi R v \mathrm{d}t)B_r$$

and so

$$\partial\Phi/\partial t = 2\pi R v B_r.$$

Thus the total magnetic e.m.f. is indeed the same relative to either frame. In one case it all comes from $\mathbf{v} \times \mathbf{B}$, in the other from $\partial\Phi/\partial t$.

Problem 4.8 exploits related ideas.

More generally, consider a loop L which may or may not be a material wire circuit, moving and perhaps deforming in space. We must now distinguish clearly between $\partial\Phi/\partial t$, the rate of change of flux linked relative to a *stationary* loop, instantaneously coterminous with L, and $\mathrm{d}\Phi/\mathrm{d}t$, the rate of change of flux linked by L *as it moves.*

Consider a short time interval $\mathrm{d}t$, and let L_i be the initial position of L, and L_f its final position, as shown in figure 4.8. During the interval $\mathrm{d}t$, a point in L travels a distance $\mathbf{v}\mathrm{d}t$ and an element $\mathbf{dr}$ of L traverses a parallelogram area element $\mathbf{da}$ equal to $\mathbf{v}\mathrm{d}t \times \mathbf{dr}$. Consider a closed surface consisting of the endless strip joining L_i and L_f, composed entirely of elements such as $\mathbf{da}$, together with surfaces S_i and S_f spanning L_i and L_f, respectively. We shall apply the condition $\oint\!\!\oint \mathbf{B} \cdot \mathbf{da} = 0$ to this closed surface at the *end* of the time interval $\mathrm{d}t$.

The contribution to the flux integral from the strip is

$$\oint_L \mathbf{B} \cdot \mathbf{v}\mathrm{d}t \times \mathbf{dr} = \oint_L (\mathbf{B} \cdot \mathbf{v} \times \mathbf{dr})\mathrm{d}t = -\oint_L (\mathbf{v} \times \mathbf{B} \cdot \mathbf{dr})\mathrm{d}t,$$

from the scalar triple product transformation property. If Φ_i is the flux through S_i at the start of the time interval $\mathrm{d}t$, the flux through S_i at the end of the interval is

$$\Phi_\mathrm{i} + (\partial\Phi/\partial t)\mathrm{d}t$$

whereas the flux through S_f at the end of the interval is

$$\Phi_\mathrm{i} + (\mathrm{d}\Phi/\mathrm{d}t)\mathrm{d}t.$$

With due attention to the right-hand screw conventions relating Φ, $\mathbf{dr}$ and $\mathbf{da}$, the statement $\oint\!\!\oint \mathbf{B} \cdot \mathbf{da} = 0$ becomes

$$-\oint_L \mathbf{v} \times \mathbf{B} \cdot \mathbf{dr}\mathrm{d}t + \left(\Phi_\mathrm{i} + \frac{\partial\Phi}{\partial t}\mathrm{d}t\right) - \left(\Phi_\mathrm{i} + \frac{\mathrm{d}\Phi}{\mathrm{d}t}\mathrm{d}t\right) = 0.$$

It follows that

$$\frac{\mathrm{d}\Phi}{\mathrm{d}t} = \frac{\partial\Phi}{\partial t} - \int_L \mathbf{v} \times \mathbf{B} \cdot \mathbf{dr}.$$

This is *Faraday's law for a moving loop*, which states that the total e.m.f. due to both kinds of magnetic induction equals the rate of change of flux linked by the loop *as it moves*. This formally confirms our suspicions that choice of frame of reference is irrelevant and that $\mathbf{E}_\mathrm{i}$ and $\mathbf{v} \times \mathbf{B}$ are physically interchangeable.

Notice that the need for the two kinds of magnetic e.m.f. to be interchangeable

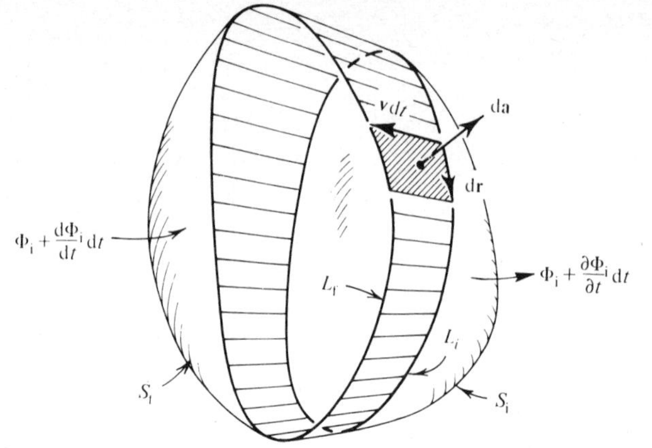

Fig. 4.8

explains why no constant (such as μ_0) can occur in Faraday's law, in contrast to Ampère's law.

When the loop is a *wire* loop, it is common for Φ to be proportional to J, the current in the wire. Then $\Phi = LJ$, where L is a geometrical property of the loop, called the self-inductance, and then the e.m.f. equals $\mathrm{d}/\mathrm{d}t(LJ)$. L may itself vary, as in problem 4.9.

The most spectacular applications of these ideas occur in superconducting technology. In a circuit made of superconductor, which has zero resistance, the e.m.f. must be zero if currents are to stay finite. Therefore a moving or deforming wire loop must always link the same magnetic flux, with the electric currents changing spontaneously as demanded by the geometrical relations between current and magnetic flux. All kinds of intriguing superconducting flux-manipulators, suspension systems, etc. immediately become possible in principle. Problem 4.10(*b*) provides some food for thought in this general area.

It is important to notice that $\mathrm{d}\Phi/\mathrm{d}t$ equals the total magnetically induced e.m.f. only when it arises either by the motion of conductors or by the change in time of the magnetic field. Other ways of changing the flux linked by a circuit, such as switching or commutation, do not produce an e.m.f. For example, figure 4.9 shows a cross-section of a long cylindrical solenoid S within which there is a uniform axial magnetic field. There is no field outside such a solenoid. $ABCD$ is a thin sheet of anisotropic conductor, penetrating the solenoid as shown. It conducts freely in the direction parallel to sides AB and CD but is impervious to current components parallel to BC or DA. AD is a busbar, connected through a voltmeter V to a sliding contact E touching the edge BC. When E moves, does V record the rate of change of flux linked by the circuit? The answer is: 'No'; there is no e.m.f.

This point is crucial in connection with electrical machinery. A d.c. dynamo generates a continuous e.m.f. by constantly raising the flux linked by the armature circuit (by motion of the conductors). But such a process obviously cannot be continued indefinitely. The conflict is resolved by repeated switching by means of a

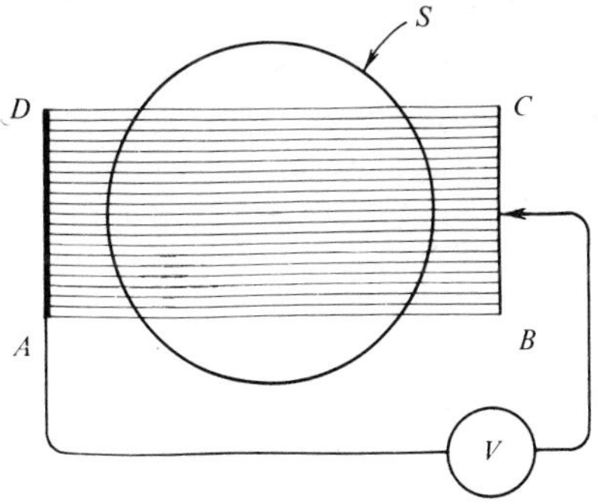

Fig. 4.9

commutator. This turns the rising graph of flux-linked into a saw-tooth, in which the flux-linked is repeatedly dropped back to its original starting value in a way that generates no reverse e.m.f.

Problems 4

4.1 The electric current J in a long, straight, insulated wire PO, which abruptly terminates, spreads radially outwards equally in all directions from O, the bare end of the wire, through a surrounding conducting medium. The resulting magnetic field lines are circles centred on the line of OP. Show that the magnetic field strength B at a point Q at a distance r from O equals $\mu_0 J(\tan \frac{1}{2}\theta)/4\pi r$, where $\angle POQ = \pi - \theta$. Can a magnetic potential be defined?

4.2 A fluid of electrical conductivity σ flows steadily at a volumetric flow rate Q along a circular pipe of radius a. The velocity profile is axisymmetric, i.e. $v = f(r)$, where r is distance from the axis. An electric current of uniform intensity j is passed axially downstream through the fluid, producing a non-uniform magnetic field B with concentric, circular field lines. There are radially-induced e.m.f.s but no radial currents because the pipe is non-conducting.

Show that if the potential difference ΔV between a point electrode on the axis and one at the pipe wall in the same cross-section is measured, then $\Delta V = \mu_0 Qj/4\pi$, whatever the shape of the velocity profile. If there is an axial location error of δ in installing the central and peripheral electrodes show that, if the system is used as a flowmeter, the indicated value of Q is in error by $4\pi\delta/\sigma\mu_0$.

*4.3 A small plane area element $\mathbf{da}$ is located at the origin. Find an expression for the solid angle that it subtends at a point P, defined by position vector $\mathbf{r}$. If a current J circulates round the rim of the element, clockwise as seen in the $\mathbf{da}$ direction, show that the resulting magnetic field has a magnetic potential equal to

$(J\mathbf{da}) \cdot \mathbf{r}/4\pi r^3$.

Compare this with the conclusion of problem 3.10(*d*). This shows that a small current loop has a *dipole* magnetic field, i.e. a field with the same general configuration as an electric dipole. (Knowledge of a potential distribution is sufficient to determine the corresponding vector field.) The *strength* of the dipole is $J\mathbf{da}$. Note that though ϵ_0 appears in the expression for the electric potential, the corresponding μ_0 was suppressed in the conventional choice of magnetic potential.

This result, together with the conclusions of problems 2.3 and 2.4 (which revealed that there is no force on an electric dipole in a uniform field **E** or on a magnetic 'dipole' in a uniform field **B**, but there are couples $\mathbf{p} \times \mathbf{E}$ and $J\mathbf{da} \times \mathbf{B}$ respectively), shows that there is a complete analogy between the two kinds of dipoles.

4.4 (*a*) Two vector fields **u** and **v** are such that $\oint \mathbf{u} \cdot \mathbf{dr}$ round any loop equals $\iint \mathbf{v} \cdot \mathbf{da}$ over any surface spanning the loop. Deduce that **v** is solenoidal.

(*b*) By considering Ampère's law applied to a loop L spanned by a bulbous surface S, and allowing L to shrink to a point and vanish, leaving S a closed surface, deduce that **j** is solenoidal.

4.5 Faraday's law shows that magnetic flux may be measured in volt-seconds, where volts are joules (newton-metres) per coulomb. Express the dimensions of magnetic field **B** and current **J** in terms of mass, length, time and charge and verify that $\mathbf{J} \times \mathbf{B}$ has the dimensions of force per unit length. From the Biot–Savart and Coulomb laws find the dimensions of μ_0 and ϵ_0 and show that $(\mu_0\epsilon_0)^{-1/2}$ has the dimensions of velocity.

4.6 A magnetic system produces a field which on the plane $z = 0$ is in the z-direction and whose magnitude B varies with r, the distance from the origin, according to the relation $B = \alpha/r$ where α is a function of time. The symmetry is such that the induced electric field lines in the plane $z = 0$ are circles centred on the origin. Find how E, the magnitude of the electric field, depends on α and r. There is no electrostatic field.

A charged particle of mass M and charge Q is released with negligible velocity at radius r in the plane $z = 0$ at the instant when $\alpha = 0$ and accelerates away along a circle centred at the origin as B rises. By resolving tangentially find the relation between α and v, the speed of the particle. Then verify that resolving radially gives a consistent equation, however B varies in time, confirming that the circular motion is indeed correct.

4.7 A circular, plane wire loop of inductance L and resistance R carries a decaying current J. In what time does J decay in the ratio e : 1? [N.B. e is 2.71828 . . . , *not* the electronic charge.]

In what sense, if any, does the lumped-component model shown in fig. 4.10 represent the phenomenon?

If the voltage between two ends of a diameter of the loop is measured

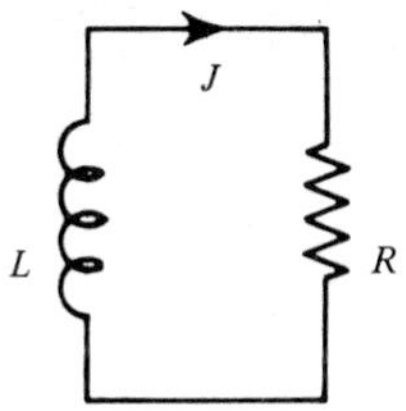

Fig. 4.10. Equivalent circuit of short-circuited ring.

by an ideal voltmeter (which draws negligible current) what signal is recorded? Does it depend on the location of the leads to the meter? How would you measure the e.m.f. round the loop?

*4.8 A circular loop of radius a bearing an alternating current $J(t)$ is fixed in a horizontal plane. A small horizontal circular ring of radius R is magnetically levitated at a height z above the loop and on its vertical axis because the currents induced in it are repelled by the currents in the loop. The ring is such that, when it is isolated, unit current round it produces a flux L through it. During levitation, the e.m.f. round the ring is zero because its resistance is negligible, and therefore the oscillatory fluxes through it (due to the currents in both the loop and the ring) must cancel out. The axial magnetic field component through the ring due to the loop may be treated as uniform over the plane of the ring and equal to its value on the axis, as given by problem 3.12(a). Show that the radial field component B_r due to the loop current J at the rim of the ring equals $(\mathrm{d}\Phi/\mathrm{d}z)/(2\pi R)$ where Φ equals the flux through the ring due to the loop current, and hence show that the mean levitation force reaches its greatest value when $z = a/\sqrt{7}$.

Note that there is no net force on the ring due to its own magnetic field.

4.9 The magnetically-induced e.m.f. $\mathcal{E}$ in a coil of wire is the rate of change of flux Φ linked where $\Phi = LJ$, J being the current in the coil and L the self-inductance. Electrical energy is exchanged with the rest of the circuit at a rate $\mathcal{E}J$, resistance being neglected. Consider the following cycle:

(a) J is raised from zero to a value J_0, with $L = L_0$ (const.),
(b) with $J = J_0$, the coil is deformed until L is zero,
(c) J is lowered to zero, and
(d) with $J = 0$, the coil is restored to its original shape.

Calculate the electrical energy exchanged with the rest of the circuit during the cycle and hence deduce the mechanical work done in overcoming magnetic forces during process (b).

4.10 (a) A long, straight solenoid of arbitrary cross-section bearing a constant current J has N turns per unit length. There is no field external to the solenoid, and the field inside is axial. Show by using Ampère's law that it is uniform and equal to $\mu_0 JN$.

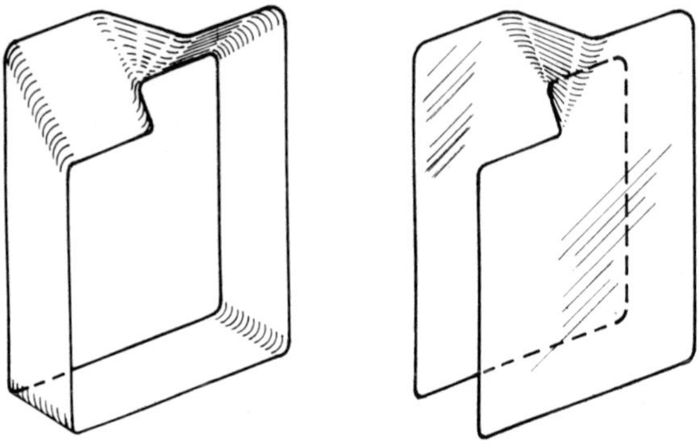

Fig. 4.11. Two alternative surfaces spanning a two-turn loop.

(*b*) Within the solenoid, a long , thin-walled, flexible, highly conducting tube is increasing its cross-sectional area A at a rate $\mathrm{d}A/\mathrm{d}t$. Induced circumferential currents in the tube are sufficient to exclude all magnetic fields from the tube's interior, because the e.m.f. round its circumference must be virtually zero. Find the induced e.m.f. per turn in the solenoid if J is still constant and show that a power equal to $\mu_0 N^2 J^2 \mathrm{d}A/\mathrm{d}t$ is generated in it per unit length. Where has the power come from? (*N.B.* There are two sources.)

*4.11 Figure 4.11 shows a two-turn short-circuited coil. To calculate the e.m.f. round it in a changing magnetic field it is necessary to find the magnetic flux through a surface 'spanning' the loop. But the figure shows that two alternative surfaces are possible. Which is the appropriate one?

(The answer is (*b*), the so-called two-sided or orientable surface. One test is to check which surface could be traversed if the loop were shrunk to a point like a lassoo. The surface (*a*) is a one-sided or Möbius surface. This dilemma is not just a characteristic of two-turn loops. All loops give rise to the same problem, as they can be spanned by either a one-sided or a two-sided surface.)

If the loop carried a current J in a uniform magnetic field, which surface should be used to evaluate $\iint \mathrm{d}\mathbf{a}$ in the expression for the torque on the loop? (See problem 3.19(*b*).)

5

Volume integrals

5.1 Triple integrals

Chapter 3 revealed that generalisations of single and double integrals which could apply to vectors in three dimensions were very useful. There was particular interest in *closed* integrals, whether round a closed loop L or over a closed surface S. It is reasonable to enquire whether similar generalisations are possible for triple integrals.

An ordinary triple integral takes the form $\iiint f \, \mathrm{d}x \, \mathrm{d}y \, \mathrm{d}z$ if referred to cartesian coordinates. It is a sum over some domain in three dimensions of contributions from volume elements $\mathrm{d}x \, \mathrm{d}y \, \mathrm{d}z$, weighted by a factor f (the integrand) which is some specified function of x, y and z. An obvious example is the case where $f = \rho$, the mass density, where the integral represents the total mass, for $\rho \, \mathrm{d}x \, \mathrm{d}y \, \mathrm{d}z$ is the mass of the individual element.

There are several, related points which reveal why the scope for generalisation in three-dimensional, vector terms is narrower for the triple than for the single or double integrals. The first point is that although f can be replaced by $\mathbf{v}$, a vector, to yield a vector volume integral $\iiint \mathbf{v} \, \mathrm{d}x \, \mathrm{d}y \, \mathrm{d}z$ (e.g. the momentum of the contents of the domain of integration, if $\mathbf{v}$ is replaced by $(\rho\mathbf{v})$, where $\mathbf{v}$ is a velocity field), the volume element $\mathrm{d}x \, \mathrm{d}y \, \mathrm{d}z$ is intrinsically a *scalar* and cannot be replaced by a vector. In contrast, with line and surface integrals there was a choice between $\mathbf{dr}$ and $\mathrm{d}s$, $\mathbf{da}$ and $\mathrm{d}a$ respectively. A volume element does not have a direction or orientation, in any sense.

Associated with the directional properties of $\mathbf{dr}$ and $\mathbf{da}$ is the fact that there is a choice of several routes for a line 'integral' between two points A and B, and a choice of several surfaces for a surface integral between the limits defined by the closed loop forming the rim of these surfaces. However, there is only *one* volume that fills a domain whose limits are defined by a closed surface. There can therefore be no analogues in terms of volume integrals of the interesting results we have seen concerning a conservative field (a concept which depends on the multiplicity of alternative routes between two points) or a solenoidal field and flux-linkage (concepts which revolve round the multiplicity of alternative surfaces spanning a given loop). There is no place for a symbol $∰$; i.e. there is no triple analogue of $\oint$ indicating integration round a closed loop made from two alternative routes between two points, or of $\oiint$ indicating integration over a closed surface made from two alternative surfaces spanning a single loop.

The upshot of all this is that there are only two kinds of volume integral:

$$\iiint f \, \mathrm{d}\tau \quad \text{and} \quad \iiint \mathbf{v} \, \mathrm{d}\tau,$$

of scalar and vector form respectively. Notice that, to cover the possibility of using coordinates other than x, y and z, the scalar volume element is generally written as $\mathrm{d}\tau$, which is normally a third-order small quantity. It is *not* the increment of any variable τ. To help the student we shall retain the triple integral sign, although for economy many texts use a single sign for double and triple integrals.

5.2 The evaluation of volume integrals

Their evaluation is easier in some ways than for surface integrals, because $\mathrm{d}\tau$ involves no geometrical problems of orientation, such as occur with $\mathbf{da}$. The vector volume integral can be evaluated in terms of its cartesian components:

$$\iiint \mathbf{v}\,\mathrm{d}\tau = \mathbf{i}\iiint v_x\,\mathrm{d}\tau + \mathbf{j}\iiint v_y\,\mathrm{d}\tau + \mathbf{k}\iiint v_z\,\mathrm{d}\tau.$$

There may be scope for skilful selection of coordinates for defining position. The same considerations as for surface integrals (see section 3.8) largely apply and so these will not be repeated. Care is needed in expressing $\mathrm{d}\tau$ in terms of increments of the coordinates, once cartesians are abandoned. Another difficulty concerns the specification of the inner limits. Consider a scalar integration over the domain shown in figure 5.1, using cartesian coordinates taken in the order x, y, z for the three stages of integration. For clarity we write the triple integral of $f(x, y, z)$ as:

$$\int_{z_1}^{z_2}\underbrace{\left\{\int_{y_1}^{y_2}\underbrace{\left\{\int_{x_1}^{x_2} f\,\mathrm{d}x\right\}}_{y,\,z\ \text{const.}}\mathrm{d}y\right\}}_{z\ \text{const.}}\mathrm{d}z$$

which must be tackled from the inside, working outwards. Evaluation involves the following ideas:

(i) Curve C on the surface of the domain is composed of all the points which constitute the outermost profile of the domain when viewed in the x-direction. C is non-planar in general. It divides the closed surface into a 'back' surface $x = x_1(y, z)$ and a 'front' surface $x = x_2(y, z)$, x_1 and x_2 being two specified functions. The first stage of integration is the accumulation of contributions along the prism AB with $y, z, \mathrm{d}y, \mathrm{d}z$ kept constant, f being treated as a function of x only. x then ranges from x_1 to x_2 in AB.

(ii) The second stage of integration has an integrand which is a function of y and z, whose y, z-dependence stems from the y, z-dependence of f, x_1 and x_2. This integration, with z and $\mathrm{d}z$ constant, represents the accumulation of contributions from prisms such as AB over the thin slab S, which is that part of the domain lying between the two values z and $z + \mathrm{d}z$. The outline of S is shown projected on to the plane $z = 0$. In covering it, y ranges between the values $y = y_1(z)$ and $y = y_2(z)$ which are specified relations characterising the two limbs (low y and high y) of the curve C, separated by the points P_1 and P_2, the 'top' and 'bottom' points of C. The relations are the equations of the two limbs of C', the projection of C on to the plane $x = 0$.

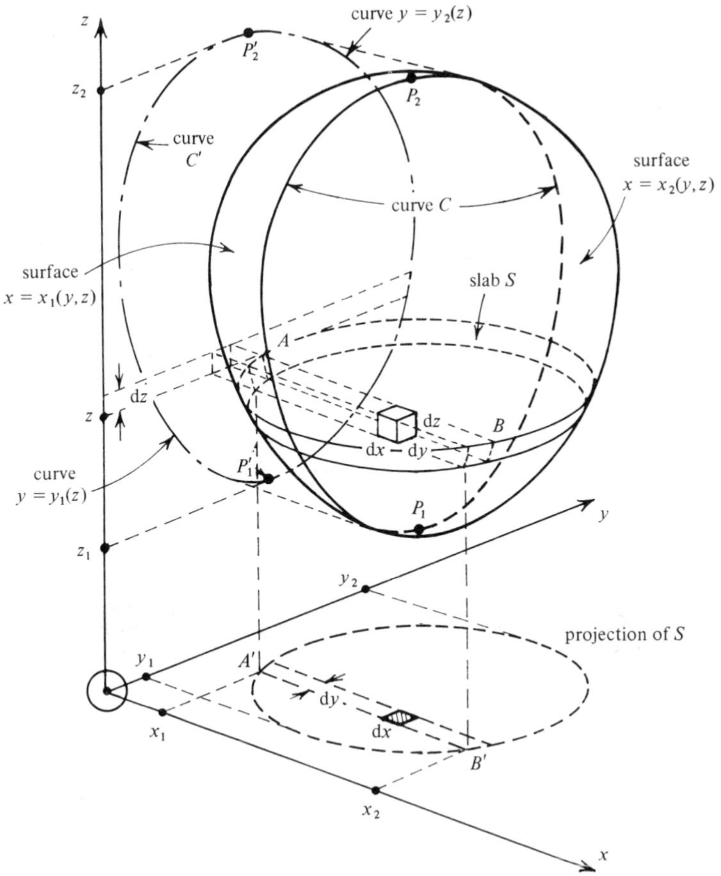

Fig. 5.1. Triple integration over a three-dimensional domain.

(iii) In the final stage of integration, the integrand is a function of z only, with z-dependence stemming jointly from the z-content of f, x_1, x_2, y_1 and y_2, in general. This integration represents the accumulation of contributions from all slabs such as S until the whole domain is covered. The limits on z, namely z_1 and z_2, are constants, the z-coordinates of P_1 and P_2.

Example 5.1 A solid object occupies that part of the octant in which x, y and z are positive which lies within a sphere of unit radius centred at the origin. Its mass density is equal to $kxyz$ where $k = \text{const}$. Find its mass by using (i) cartesian coordinates, and (ii) coordinates comprising radial distance R from the origin, latitude λ and longitude ϕ. (See figures 3.9 and 5.2.)

(i) *Cartesians.* At the sphere, $x^2 + y^2 + z^2 = 1$. We shall integrate in the order x, y, z.

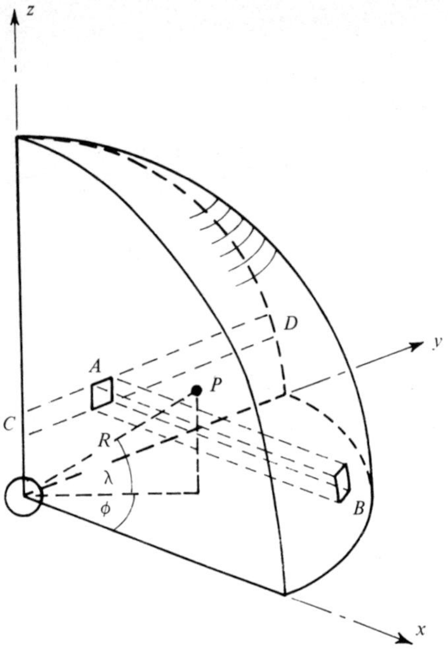

Fig. 5.2. Triple integration over an octant.

The x limits are 0 and $\sqrt{(1-y^2-z^2)}$ (i.e. A to B in figure 5.2).
The y limits are 0 and $\sqrt{(1-z^2)}$ (i.e. C to D in the figure).
The z limits are 0 and 1.
The mass is therefore

$$\int_0^1 \mathrm{d}z \int_0^{\sqrt{(1-z^2)}} \mathrm{d}y \int_0^{\sqrt{(1-y^2-z^2)}} kxyz \, \mathrm{d}x.$$

(Note the use of an alternative disposition of the differentials $\mathrm{d}x$, $\mathrm{d}y$, $\mathrm{d}z$. The evaluation proceeds from right to left, as follows.)

$$\begin{aligned}
\text{Mass} &= \int_0^1 \mathrm{d}z \int_0^{\sqrt{(1-z^2)}} \mathrm{d}y \, [\tfrac{1}{2}kx^2yz]_0^{\sqrt{(1-y^2-z^2)}} \\
&= \int_0^1 \mathrm{d}z \int_0^{\sqrt{(1-z^2)}} \mathrm{d}y \, \{\tfrac{1}{2}kyz(1-y^2-z^2)\} \\
&= \int_0^1 \mathrm{d}z \int_0^{\sqrt{(1-z^2)}} \tfrac{1}{2}kz(y-y^3-z^2y)\mathrm{d}y \\
&= \int_0^1 \mathrm{d}z \left\{\tfrac{1}{2}kz\left[\frac{y^2}{2}-\frac{y^4}{4}-\frac{z^2y^2}{2}\right]_0^{\sqrt{(1-z^2)}}\right. \\
&= \int_0^1 \mathrm{d}z \, \{\tfrac{1}{8}kz(1-2z^2+z^4)\} = \tfrac{1}{8}k\left[\frac{z^2}{2}-2\frac{z^4}{4}+\frac{z^6}{6}\right]_0^1 = \frac{k}{48}.
\end{aligned}$$

(ii) Coordinates of this kind are usually called *spherical polars* (except that the complementary angle $\theta = (\pi/2) - \lambda$ is normally chosen instead of λ). The appropriate volume element $d\tau$ can be inferred from figure 3.9, by erecting a rectangular parallelepiped of radial height dR on the shaded area element $d\mathbf{a}$. Thus

$$d\tau = (R\cos\lambda\, d\phi)(R\, d\lambda)\, dR = R^2 \cos\lambda\, d\lambda\, d\phi\, dR,$$

a third-order small quantity. Although identifying the volume element is more complicated than in cartesians, identifying the limits is much easier (see figure 5.2). They are $\lambda = 0$ to $\pi/2$, $\phi = 0$ to $\pi/2$, $R = 0$ to 1, i.e. the bounding surfaces are all such that one coordinate is constant over each one. Also

$$x = R\cos\lambda\cos\phi, \quad y = R\cos\lambda\sin\phi, \quad z = R\sin\lambda$$

and the density is

$$kR^3\cos^2\lambda\sin\lambda\cos\phi\sin\phi.$$

Hence

$$\text{mass} = \int_0^1 \int_0^{\pi/2} \int_0^{\pi/2} (kR^3\cos^2\lambda\sin\lambda\cos\phi\sin\phi)R^2\cos\lambda\, d\lambda\, d\phi\, dR.$$

Because the limits are all constants, there is no interaction between the stages of integration and the mass is the product of three independent integrals:

$$k\int_0^1 R^5 dR = \tfrac{1}{6}k, \quad \int_0^{\pi/2} \cos\phi\sin\phi\, d\phi = \tfrac{1}{2}, \quad \int_0^{\pi/2} \cos^3\lambda\sin\lambda\, d\lambda = \tfrac{1}{4},$$

and so the mass equals

$$\frac{k}{48}.$$

In many problems there is some symmetry which can be exploited by a suitable choice of coordinates so as to save one or more stages of integration. If in example 5.1 the density had been kx, we could have chosen a volume element $d\tau$ consisting of a quadrant-shaped slab lying between two x-const. planes, dx apart. The radius of this element is $\sqrt{(1-x^2)}$ and its volume is $\frac{1}{4}\pi(1-x^2)dx$. Hence

$$\text{total mass} = \int_0^1 \tfrac{1}{4}\pi kx(1-x^2)dx = \tfrac{1}{16}\pi k.$$

Here two integrations have been avoided. The reader should check that the use of R, λ, ϕ would have been much more cumbersome.

Problems 5.1 to 5.6 give some further exercises on scalar and vector volume integrals and their applications to geometry, mechanics, fluid mechanics, gravitation and electrostatics.

5.3 An important application of volume and surface integrals

In chapter 4 we saw that certain important physical statements took the form:

(closed line integral round loop) = (integral over surface spanning loop).

There are also many physical statements which take the form:

(closed surface integral) = (integral over volume contained).

First notice a general point: in both cases above, the left-hand side is an integral which involves less information than the right-hand side. In the first case one needs only the value of the integrand on the loop instead of over the surface, in the second case one needs only its value on the surface instead of throughout the volume. This cutting-down of the information necessary to make progress is one of the most useful stratagems in physical and engineering science. A fine example is the steady flow energy equation of thermodynamics, expressed relative to a control surface. (See problem 5.7.) All the complex energy exchanges within the volume enclosed can be encompassed purely in terms of various energy fluxes at the surface. Other examples abound: one may cite the 'method of sections' used in the statics of structures or the steady flow momentum equation in fluid mechanics. More abstract examples, which we shall encounter later, concern Maxwell's stress and the Poynting vector in electromagnetism. Another important example of a statement relating a surface integral and a volume integral now follows.

5.4 Gauss's law

Gauss's law concerns the flux of the electric field **E** through a closed surface, i.e. the question whether **E** is solenoidal or not. It is a consequence of three facts which have already been discussed in chapter 3:

(*a*) Coulomb's law which gives the electrostatic field $\mathbf{E}_s$ at a point P due to a point charge Q at the origin in the form

$$\mathbf{E}_s = \frac{Q}{4\pi\epsilon_0}\frac{\mathbf{r}}{r^3},$$

(*b*) the laws of magnetic induction, which include the fact that $\mathbf{E}_i$, the magnetically induced part of **E**, is solenoidal, and

(*c*) the fact that the flux of the inverse square law field $\mathbf{r}/r^3$ through a closed surface is zero if the origin lies outside, but is 4π if the origin lies inside.

It follows from (*a*) and (*c*) that ${}_S\oiint \mathbf{E}_s \cdot \mathbf{da} = Q/\epsilon_0$ if the point charge Q lies inside the surface S, but zero otherwise. If there are many point charges Q_1, Q_2 at various points, each producing its own Coulomb field $\mathbf{E}_1$, etc., then $\mathbf{E}_s = \Sigma\mathbf{E}_1$ summed over the charges, and

$${}_S\oiint \mathbf{E}_s \cdot \mathbf{da} = {}_S\oiint \sum \mathbf{E}_1 \cdot \mathbf{da} = \sum {}_S\oiint \mathbf{E}_1 \cdot \mathbf{da} = \sum Q/\epsilon_0,$$

in which the term ΣQ must clearly be summed *only* over those charges which lie within S. In view of (b) above, together with $\mathbf{E} = \mathbf{E}_i + \mathbf{E}_s$, we have

$$\epsilon_0 \oiint \mathbf{E} \cdot \mathbf{da} = \sum Q.$$

If the charges form a continuous distribution, with net density q per unit volume, $q\mathrm{d}\tau$ replaces Q and integration replaces summation, with the result that

$$\epsilon_0 \oiint \mathbf{E} \cdot \mathbf{da} = \iiint q\mathrm{d}\tau,$$

which is *Gauss's law*. It shows that, as with **B**, the flux of the abstraction **E** is also significant. Unlike **B**, however, **E** is not in general solenoidal except in regions devoid of charges, e.g. *in vacuo*.

Example 5.2 Find the electric field due to charges distributed uniformly along an infinite straight line at a density q per unit length.

Symmetry arguments indicate that **E** must be purely radial, with a magnitude **E** that is independent of axial or angular position. Consider applying Gauss's law to a flat-ended circular cylinder of radius r and of unit length, concentric with the line. There is no electric flux through its ends, while over its curved surface **E** is uniform in magnitude and everywhere normal. Thus Gauss's law gives

$$2\pi r\epsilon_0 E = q, \quad \text{the charge enclosed,}$$

and so $E = q/(2\pi\epsilon_0 r)$, which is the result of problem 3.10(b), achieved with considerably less effort.

Problems 5.8 to 5.10 give further exercises on the use of Gauss's law.

5.5 Gauss's law and dielectric media

A dielectric is a material which allows restricted migration of bound charges, i.e. polarisation. As q in Gauss's law includes bound charge, it is necessary to explore its implications for dielectrics.

In chapter 2 it was seen that polarisation can be regarded as the net amount of bound charge per unit area which has migrated through a small area element whose plane is oriented normal to the direction of migration. One significance of the word *net* is that the same effect is produced by positive charge migrating one way as negative charge moving oppositely. Non-uniform polarisation can imply that, for each small volume element, the net charge migrating inwards may differ from that migrating outwards and so a net bound charge imbalance can arise, even though in the unpolarised state the bound charges all neutralise each other. In this way volume distributions of net bound charge can occur. Similarly, at the boundary of a dielectric, intense surface layers of net bound charge can occur as charges of one sign move to the boundary from within. If another dielectric or conductor adjoins the boundary, a rival charge layer may accumulate in it, also, and in general the two layers will not exactly neutralise each other.

Consider a closed surface composed of elements **da** drawn in space in the presence of dielectric materials. Where an element **da** coincides with material having polarisation **P** (which in general will be inclined to **da**) the net amount of bound charge which has migrated through the element equals $\mathbf{P}\cdot\mathbf{da}$. Summed over the closed surface, $\oiint \mathbf{P}\cdot\mathbf{da}$ represents the total bound charge which has migrated out. Since bound charge is normally indestructible, and an unpolarised dielectric is electrically neutral, this integral must equal $-\iiint q_\mathrm{b}\mathrm{d}\tau$ over the volume enclosed, if q_b is the net density of bound charge per unit volume, associated with the non-uniform polarisation. We thus have

$$\oiint \mathbf{P}\cdot\mathbf{da} = -\iiint q_\mathrm{b}\mathrm{d}\tau$$

which, although it looks reminiscent of Gauss's law, is a quite different kind of statement; a conservation-of-charge statement, in fact. In general q in Gauss's law consists of $q_\mathrm{b} + q_\mathrm{f}$, where q_f is the net density of *free* charge, if any, so Gauss's law implies that

$$\oiint \epsilon_0\mathbf{E}\cdot\mathbf{da} = \iiint q_\mathrm{b}\mathrm{d}\tau + \iiint q_\mathrm{f}\mathrm{d}\tau.$$

It follows that

$$\iiint q_\mathrm{f}\mathrm{d}\tau = \oiint (\epsilon_0\mathbf{E}+\mathbf{P})\cdot\mathbf{da} = \oiint \mathbf{D}\cdot\mathbf{da},$$

if we introduce a new, compound vector

$$\mathbf{D} = \epsilon_0\mathbf{E}+\mathbf{P}.$$

This is conventionally given the unhelpful title '*displacement*'. Its importance lies in the fact that its flux depends only on the *free* charge enclosed, and free charge can often be measured directly (e.g. by integrating ammeter readings) whereas bound charge can only be inferred. For this reason the properties of dielectrics tend to be measured and stated in terms of relationships between **E** and **D** rather than **E** and **P**, even though **P** appears to be the more basic quantity. Where **P** (and therefore also **D**) is proportional to **E** it is usual to write

$$\mathbf{D} = k\epsilon_0\mathbf{E},$$

k being called '*dielectric constant*' of the medium.

In a region composed purely of dielectric, devoid of free charge, $\oiint \mathbf{D}\cdot\mathrm{da} = 0$ and **D** is seen to be solenoidal, having some affinities with magnetic field **B** in this case. The units of **D** are coulombs per square metre, as for **P**. Problems 5.11, 5.12 and 5.13 take these ideas further, as does the example below.

Example 5.3 A polarisable medium with the property that $\mathbf{P} = \alpha\mathbf{E}$ occupies the space between two long, concentric, highly-conducting cylinders of radii R and $2R$ at different, uniform potentials. The medium also contains free charges which provide conductivity σ and carry a total current J per unit length of the cylinders. Is $\mathbf{D}$ solenoidal in the medium?

The problem is obviously axisymmetric, with the vectors $\mathbf{j}$, $\mathbf{E}$, $\mathbf{P}$ and $\mathbf{D}$ all purely radial. Consider a concentric cylinder of unit length and radius $r(R < r < 2R)$. The current intensity across it is $j = J/2\pi r$ and so

$$E = \frac{j}{\sigma} = \frac{J}{2\pi\sigma}\frac{1}{r}, \quad P = \frac{J\alpha}{2\pi\sigma}\frac{1}{r}, \quad \text{and} \quad D = \frac{J(\epsilon_0 + \alpha)}{2\pi\sigma}\frac{1}{r}.$$

The flux of $\mathbf{D}$ through the curved surface of the cylinder is $2\pi rD$ per unit length, i.e. $J(\epsilon_0 + \alpha)/\sigma$, which is independent of r if α and σ are uniform. In this case, $\mathbf{D}$ is solenoidal, for the flux of $\mathbf{D}$ through successive concentric cylinders is the same. Similarly $\mathbf{P}$ is solenoidal and q_b is zero, which indicates that $\mathbf{P}$ is non-uniform in a way that still does not result in net bound charge appearing anywhere.

Note, however, that if α and σ vary the three vectors $\mathbf{E}$, $\mathbf{P}$, $\mathbf{D}$ are not solenoidal, i.e. the net charges q responsible for $\mathbf{E}$ (a combination of free and bound charges) have to be distributed appropriately throughout the medium, and not merely at its surfaces.

5.6 Plasma oscillation

Gauss's law is the key to understanding this important phenomenon. *Plasma* is another name for ionised gas, composed of a neutral or near-neutral mixture of negative, positive and neutral particles. A form of it can also occur in condensed matter. Gas can become ionised because of high temperatures (molecular collisions being then so vigorous as to knock electrons off molecules and atoms, producing ions and free electrons) as in stars or in the hot air round a spacecraft re-entering the Earth's atmosphere; or because of imposed electric fields, as in lightning, welding arcs, or fluorescent tubes; or because of radiation, as in the ionosphere.

Plasma oscillation consists essentially of a motion of all the particles of one sign in a region relative to those of the opposite sign. The large electric fields which result from these charge separations imply restoring forces that cause very high frequency oscillation, as the plasma tries to 'spring' back to neutrality, but keeps overshooting because of the inertia of the particles.

We adopt the simplest possible model of the phenomenon, making the following reasonable simplifications:

(i) The plasma is tenuous enough for collisions between the particles to occur with negligible frequency (compared with the oscillation frequency). This means that the neutral particles can be totally ignored.

(ii) The oscillation frequency is so high that, by comparison, thermal motion of all particles can be ignored, even that of the light electrons, whose thermal speeds are the highest.

(iii) The only negative particles are electrons, all the positive particles being ions of mass very large compared with the electrons. This means that the positive ions can be treated as virtually stationary. (Think of an oscillatory system consisting of a ping-pong ball and a cannon ball, joined by an elastic link – the cannon ball hardly moves.)

(iv) The ions are all singly ionised, with charge e, equal and opposite to that of the electron, and distributed uniformly at a number density n per unit volume throughout the plasma.

(v) The oscillation is one-dimensional: all motions are in the x-direction and all quantities vary only with x and time t. All the electrons in each plane perpendicular to the x-axis move collectively at the same velocity. We also assume that no plane of electrons overtakes a neighbour. (See problem 5.14(*b*).) Subject to this condition, the amplitudes and phases of oscillation of different planes need not be the same.

(vi) The oscillation is localised, i.e. outside the oscillatory region there is *quiescent* plasma, in an undisturbed, neutral state devoid of electric fields.

Note that here, just as in the discussion of media in chapter 2, the fields referred to are blurred-out averages over many particles, not the detail micro-fields round each charged particle.

In figure 5.3, *ABGF* represents a tubular region running in the x-direction with unit cross-section. *AB* is part of a plane of electrons whose equilibrium position would be *CD* if all planes of electrons were pushed back and the plasma restored to neutrality. Let x be the distance between *AB* and *CD*. *FG* is another cross-section of the tube, located outside the oscillatory region in quiescent plasma, where $\mathbf{E} = \mathbf{0}$. The electric fields $\mathbf{E}$ in the oscillatory region are in the x-direction, from reasons of symmetry.

We can easily find the value of $\mathbf{E}$ at *AB* from Gauss's law applied to the volume *ABGF*. It follows from the definition of *CD* that the total charge on all the electrons in *ABGF* is equal and opposite to the charge on the ions in *CDGF*. Thus the net charge in *ABGF* is merely the charge on the ions in *ABDC*, all the other charges in *ABGF* cancelling, as was just remarked. The volume of *ABDC* is simply x, and the ion charge density is ne. So the total net charge in *ABGF* is nex. The electric flux out of *ABGF* is all contributed by face *AB*, since on all other faces $\mathbf{E}$ is either zero or parallel to the surface. Face *AB* is of unit area and the electric field there, of magnitude E, is uniform and normal to *AB*. Gauss's law for *ABGF* therefore yields the very simple result

$$\epsilon_0 E = nex,$$

the desired relation between charge separation distance and electric field. Problem 5.14(*a*) considers this result in the case where x is negative. The analysis is completed by expressing Newton's law for an electron of mass m and charge $-e$ in the plane *AB*, viz.

$$-eE = m\mathrm{d}^2x/\mathrm{d}t^2,$$

so that

$$\frac{\mathrm{d}^2x}{\mathrm{d}t^2} + \frac{ne^2}{m\epsilon_0}x = 0.$$

This equation reveals that the basic plasma oscillation is a simple harmonic motion of frequency

$$\frac{1}{2\pi}\sqrt{\left(\frac{ne^2}{m\epsilon_0}\right)}$$

for all electron layers, irrespective of their amplitudes of oscillation. Typical values in dense plasmas are of the order 10^{12} Hz. These very high frequency charge motions make an interesting contrast with Ohm's law behaviour. They represent a dynamic balance between electric force and electron inertia instead of collisonal drag. It becomes clearer why it is only at these very high frequencies that electron inertia becomes a significant factor, and why it is normally quite negligible in Ohm's law.

The *plasma frequency* $(1/2\pi)(ne^2/m\epsilon_0)^{1/2}$ is an important parameter in connection with radio transmission through or reflection off the ionosphere or the thermally ionised gas round a re-entering spacecraft, because it acts as a cut-off frequency at which transmission ceases. This question is explored further in the problems following chapter 11.

5.7 Conservation statements

In section 5.5 we expressed the conservation of bound charge in the form (surface integral) = (volume integral). There are many other physical statements which take that same mathematical form, usually expressing the notion that, for some indestructible physical entity,

(rate of flow out of a closed surface) = (rate of depletion of contents).

Such an equation is called a *conservation statement*. Important examples include the following:

(*a*) Heat flow intensity **Q**. Heat is released by the cooling of a medium of

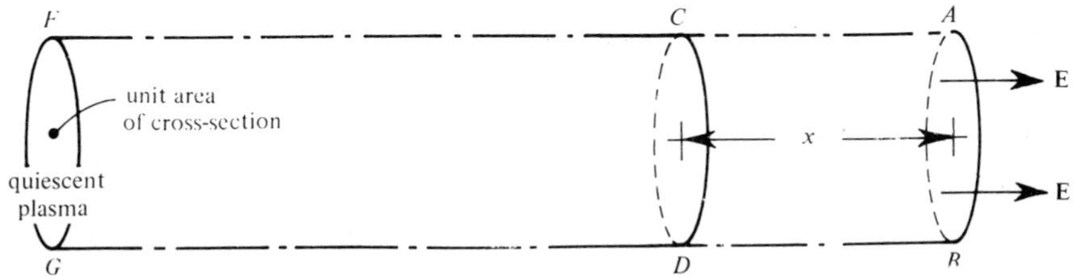

Fig. 5.3

volumetric heat capacity C at a rate $-C\partial T/\partial t$ per unit volume and time, T being temperature. The heat release rate from a volume element $d\tau$ is $-C(\partial T/\partial t)d\tau$ and overall conservation indicates that

$$\oiint \mathbf{Q}\cdot\mathbf{da} = -\iiint C\frac{\partial T}{\partial t}\,d\tau.$$

Alternatively the heat release might be from chemical or nuclear sources, in which case it would be specified mathematically in some other way.

(*b*) Mass flow intensity $\rho\mathbf{v}$. The rate of depletion of the mass $\rho d\tau$ in a fixed volume element $d\tau$ when there is a variable fluid density ρ is $-(\partial\rho/\partial t)d\tau$. The overall conservation of mass then requires that

$$\oiint \rho\mathbf{v}\cdot\mathbf{da} = -\iiint \frac{\partial\rho}{\partial t}\,d\tau = -\frac{\partial}{\partial t}\iiint \rho\,d\tau$$

(in which $\partial/\partial t$ is used because the integration is over a volume fixed in space), the rate of fall of the total mass of the contents. As usual the order of time differentiation and space integration can be interchanged. In the case of fluid of uniform, constant density, the equation degenerates to

$$\oiint \mathbf{v}\cdot\mathbf{da} = 0$$

and the velocity and mass flow fields are solenoidal.

(*c*) Charge flow intensity $\mathbf{j}$, which may be either free charge flow or bound charge flow $\partial\mathbf{P}/\partial t$ or both together. The charge depletion rate in a volume element $d\tau$, where the variable net charge density (free, bound or combined as appropriate) is q, is $-(\partial q/\partial t)d\tau$, and the equation of charge conservation is therefore

$$\oiint \mathbf{j}\cdot\mathbf{da} = -\iiint \frac{\partial q}{\partial t}\,d\tau = -\frac{\partial}{\partial t}\iiint q\,d\tau,$$

a result comparable with that for mass conservation. Problems 5.15 and 5.16 give some exercises on the use of surface and volume integrals in relation to conservation statements. An example follows.

Example 5.4 Heat is released uniformly throughout a spherical nuclear reactor of radius a at a rate R per unit volume and time. The relation between Q, the magnitude of the radial heat flux intensity vector and the temperature T is

$$Q = -K\,dT/dr,$$

where K = thermal conductivity and r = distance in the direction of heat flow. If conditions are steady, calculate the temperature difference between the centre and the outer surface.

Consider a concentric sphere of radius $r(<a)$ in which heat is released at a total

rate $\frac{4}{3}\pi r^3 R$ and from which heat escapes at an equal rate $-4\pi r^2 K(\mathrm{d}T/\mathrm{d}r)$. It follows that $\mathrm{d}T/\mathrm{d}r = rR/3K$. Integrating from $r = 0$ to a gives the temperature difference as $a^2R/6K$.

5.8 Two-dimensional problems

Before we leave the subject of vector integration, some comment on the way that it applies to two-dimensional problems is called for.

Some problems are literally two-dimensional because the phenomena involved occur solely in a plane (e.g. many problems in elementary mechanics). However, the epithet 'two-dimensional' is also applicable to certain field problems which extend into three dimensions. This usage is important and deserves clarification. It applies in situations where the physical quantities involved depend upon only two of the cartesian coordinates, provided that the third one, z say, is chosen correctly, i.e. the z-axis is oriented in 'the direction of no variation'. The quantities may also vary in time, of course.

An example is a fluid motion in which all three of the velocity components v_x, v_y and v_z are independent of z. Note that it is not necessary for v_z to be zero, although this may often be the case. If $v_z = 0$ the motion then occurs in planes $z =$ const., also referred to as 'x, y planes'.

Two-dimensional phenomena may occur for one of two contrasting reasons: first, the region concerned may extend so far in the $\pm z$-directions that any end-effects that might have caused z-variation are negligible over most of the region (a good example is the magnetic field round a long, straight wire, so long that the influence of the end-connections can be ignored); secondly the extent of the region in the z-direction is limited so as to lie between two boundary planes which constrain the phenomenon to be two-dimensional (as for instance with electric current flow in a plane metal sheet of uniform thickness).

Closely analogous to two-dimensional problems are those of high symmetry in which the quantities involved are independent of one of three suitably chosen coordinates other than cartesians. If we refer a problem which is axisymmetric about the z-axis, say, to *cylindrical polar coordinates* (r, θ, z) – where r, θ are plane polar coordinates in x, y planes – then all quantities become independent of θ. The problem could then be described as two-dimensional in terms of r and z (e.g. swirling flow along a nozzle).

As regards cartesian two-dimensional problems, independent of z, it is important for the reader to appreciate how line, surface and volume integrals may degenerate in such cases. Figure 5.4 shows a loop L in an x, y plane, which when moved through unit displacement $\mathbf{k}$ in the z-direction delineates a cylindrical surface S. The plane areas A and B spanning L in its extreme positions, together with S, constitute a closed surface which encloses a volume V.

Here line integrals, whether closed or open, are normally taken along a loop such as L (or a portion thereof) lying in an x, y plane. The line element $\mathrm{d}\mathbf{r}$ then has zero z-component and so $\int \mathbf{v} \cdot \mathrm{d}\mathbf{r}$ becomes $\int (v_x \mathrm{d}x + v_y \mathrm{d}y)$ whereas $\int \mathbf{v} \times \mathrm{d}\mathbf{r}$ becomes $\mathbf{k} \int (v_x \mathrm{d}y - v_y \mathrm{d}x)$ if $v_z = 0$, as is often the case. Surface integrals may

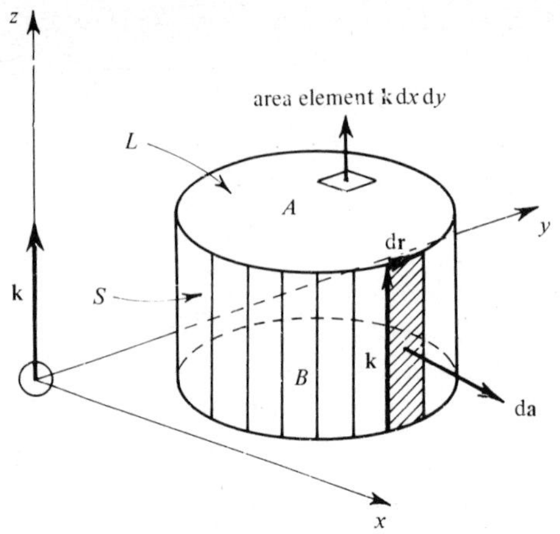

Fig. 5.4. A prism of unit length in a two-dimensional problem.

usefully be taken over surfaces such as A, in which case the apprôpriate area element is $\mathbf{k}\,\mathrm{d}x\mathrm{d}y$; or over surfaces such as S, in which case the area element $\mathrm{d}\mathbf{a}$ may be taken as shown in the figure, where it is seen to equal $\mathrm{d}\mathbf{r} \times \mathbf{k}$; or over the complete closed surface $S + A + B$.

The surface integrals over S bear an interesting relationship to the line integrals round L. For instance, the surface integral of *flux* form,

$$_S\iint \mathbf{v} \cdot \mathrm{d}\mathbf{a} = {}_L\int \mathbf{v} \cdot \mathrm{d}\mathbf{r} \times \mathbf{k} = \mathbf{k} \cdot {}_L\int \mathbf{v} \times \mathrm{d}\mathbf{r} = {}_L\int (v_x \mathrm{d}y - v_y \mathrm{d}x),$$

a line integral. Problem 5.17 gives a simple exercise while problem 5.18 concerns a case of the type $\iint \mathbf{v} \times \mathrm{d}\mathbf{a}$. Problem 3.21 involved the comparable axisymmetric degeneracy.

In integrals over a volume such as V, the obvious volume element $\mathrm{d}\tau$ is a prism of unit length and cross-section $\mathrm{d}x\mathrm{d}y$. The integral therefore degenerates to a surface integral over area A.

Example 5.5 A two-dimensional net charge distribution $q(x, y)$ per unit volume and a changing magnetic field in the z-direction together give rise to an electric field with components E_x and E_y. Express Gauss's and Faraday's laws in terms of integrals referred to a domain in the x, y plane.

Here Gauss's law $\epsilon_0 \oiint \mathbf{E} \cdot \mathrm{d}\mathbf{a} = \iiint q\mathrm{d}\tau$ degenerates to

$$\epsilon_0 \;{}_S\iint \mathbf{E} \cdot \mathrm{d}\mathbf{a} = {}_A\iint q\mathrm{d}x\mathrm{d}y,$$

where the surfaces S and A are defined in figure 5.4. As was seen earlier the left-hand side can be replaced so as to give

$$\epsilon_0 \oint_L (E_x \mathrm{d}y - E_y \mathrm{d}x) = \iint_A q \mathrm{d}x\mathrm{d}y.$$

Meanwhile Faraday's law

$$\oint \mathbf{E} \cdot \mathbf{dr} = -\frac{\partial}{\partial t} \iint \mathbf{B} \cdot \mathbf{da}$$

degenerates to

$$\oint_L (E_x \mathrm{d}x + E_y \mathrm{d}y) = -\frac{\partial}{\partial t} \iint_A B_z \mathrm{d}x\mathrm{d}y.$$

The direction taken round the loop L must appear clockwise when viewed in the z-direction.

The ideas of multiple-connectedness carry over into two-dimensional phenomena. The two-dimensional magnetic field round a conducting bar carrying currents in the z-direction is a case in point. The plane region external to the bar's cross-section R is multiply-connected, and so the magnetic potential is multi-valued unless a 'cut' line S is made in the x, y plane (going from the bar to infinity, in this case) so as to render the x, y plane external to R and S simply-connected. (See example 4.2.) These themes will reappear when we discuss fluid mechanics in chapter 12.

Problems 5

5.1 (*a*) Find the radii of gyration of a solid sphere of unit radius about the x and y axes if its density is given by (i) $k|x|$ or (ii) $k|xyz|$, k being a constant. The origin is at the centre of the sphere.
(*b*) A metal cube with unit edge length is drilled with three holes of unit diameter, each centred on a line through the centre of the cube and parallel to a set of edges. Calculate the volume of metal remaining.

5.2 Fluid of constant density ρ flows radially and symmetrically between two concentric spheres of radius a and $2a$ at a total volumetric flow rate Q. Show that the total kinetic energy of the fluid between the spheres is $\rho Q^2/16\pi a$.

*5.3 A deep cylindrical tank of radius R is pivoted freely about its axis, which is vertical. Initially it is rotating at angular velocity Ω_1 and contains frozen liquid to a uniform depth R. The liquid then melts, without change of volume, until finally viscosity causes it to be again rotating at a uniform angular velocity, Ω_2, with the free surface of the liquid forming a paraboloid of revolution. The depth of liquid just reaches zero at the centre.

If the moment of inertia of the tank is negligible, find the ratio Ω_2/Ω_1. What fraction of the kinetic energy has been dissipated? Has the potential energy changed? (N.B. Equilibrium requires that $R\Omega_2^2 = 4g$.)

5.4 (*a*) Show that the position vector $\bar{\mathbf{r}}$ of the centroid of a body of density ρ is given by $\iiint \rho\mathbf{r}\,d\tau / \iiint \rho\,d\tau$.

*(*b*) A non-uniform solid object is formed from a cone of semi-angle α with a plane base inclined at an angle β to the axis of the cone, such that $\tan\beta = 2\tan\alpha$ and the distance from the apex to the intersection of axis and base is h. A cylindrical hole of radius $\frac{1}{2}h\tan\alpha$, centred on the axis, is drilled through it. In terms of (r, θ, z), with the apex as origin and z along the cone axis, the polar coordinates r, θ being in x, y planes with $\theta = \frac{1}{2}\pi$ the longest generator of the cone, the density is $kz(2 - \sin\theta)^2/r$ (k const.)

Find the r, θ, z equation of the base and the r, θ relation for the curve which is the edge of the base and verify that $d\tau$ is $r\,dr\,d\theta\,dz$. Find the mass and the z-coordinate of the centroid. Integrate in the order: z (with limits of form $f(r, \theta)$), then r (with limits of form $f(\theta)$) and finally θ. Also investigate the feasibility of integrating in other orders or using different coordinates altogether. (r, θ, z) are called *cylindrical polars.*

*5.5 It is shown in books on dynamics that a particle moving at velocity $\mathbf{v}$ relative to a frame of reference that is rotating at an angular velocity Ω suffers an extra acceleration $2\Omega \times \mathbf{v}$ called the Coriolis acceleration. Consider a thin, rigid, uniform, circular ring of radius R and mass dm rotating at angular velocity $\boldsymbol{\omega}$ about an axis through its centre C normal to its plane, and let this axis be itself rotating about another axis at an angular velocity $\boldsymbol{\Omega}$. Show that a couple $R^2\boldsymbol{\Omega} \times \boldsymbol{\omega}\,dm$ must be applied to the ring to account for the Coriolis accelerations of all its parts.

Hence show that for a *gyroscope*, which is an axisymmetric rigid body with an arbitrary density distribution (also axisymmetric), rotating at angular velocity $\boldsymbol{\omega}$ about its axis, where this axis is itself rotating ('*precessing*') at an angular velocity $\boldsymbol{\Omega}$, a torque must be applied equal to $I\boldsymbol{\Omega} \times \boldsymbol{\omega}$, where I is the moment of inertia of the gyroscope about its axis. (This problem reveals new outlets for the vector product concept.)

*5.6 The law of gravitation states that $\mathbf{g}$, the force on a small body per unit of its mass due to another concentrated mass m at a vector distance $\mathbf{r}$ from the body, is equal to $Gm\mathbf{r}/r^3$ (where G is the universal gravitational constant).

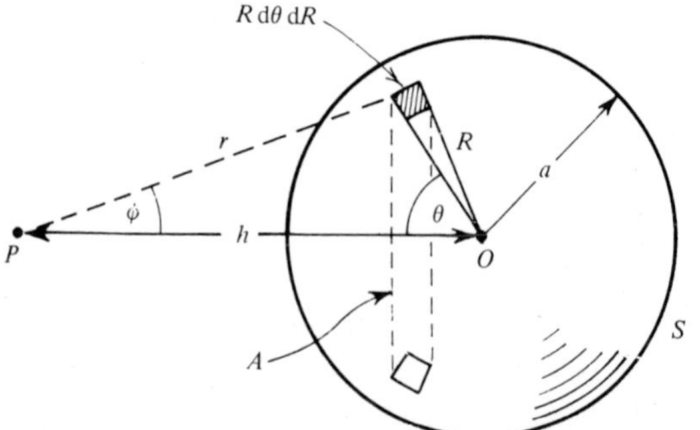

Fig. 5.5

Figure 5.5 shows a massive, spherical star S of radius a and total mass M in which the mass density ρ is a function of distance R from its centre O. Express the total gravity field $\mathbf{g}$ at a point P at a distance $h(>a)$ from O as a double integral in terms of the coordinates R and θ, noting that it will be directed along PO by symmetry. (Proceed by considering the contributions to $\mathbf{g}$ from each annular ring A, of cross-section $R\mathrm{d}\theta\mathrm{d}R$, and first prove that

$$\int_0^{\pi} \frac{\cos\phi \sin\theta \, \mathrm{d}\theta}{r^2} = \frac{2}{h^2} \quad \text{if} \quad h > R.)$$

Deduce that $\mathbf{g}$ is the same as it would be if all the mass M were concentrated at O.

How are the calculation and result modified if $h < a$? Write down, by analogy, the similar results which apply to the electrostatic field due to a spherically symmetric electric charge distribution. Show that there is no electric field inside a uniformly charged spherical shell.

5.7 A stationary, closed *control surface* is drawn in a steady fluid flow field. Write down the *mass conservation* and *steady flow energy equations* in terms of integrals over the surface of the following quantities: $\mathbf{v}$ velocity, ρ mass density, h enthalpy per unit mass, $\mathbf{Q}$ heat flux intensity. Gravity should be ignored. The two terms h and $\frac{1}{2}\mathbf{v}^2$ (which should appear together) are often combined into h_0, the so-called *stagnation enthalpy*.

If the control surface contained some device such as a fan which was supplying work to the fluid stream at a total rate P, how would the energy equation be modified?

5.8 Develop a law for gravitation (see problem 5.6) which is analogous to Gauss's law for electric flux.

By applying it to situations that are spherically symmetric, deduce the conclusions of problem 5.6 in a very much easier manner.

5.9 A sphere of copper, 1 m in diameter, has all its free electrons removed. From Gauss's law find the electric field at a radius r within the sphere and deduce the potential difference between rim and centre. Is it a large or small voltage, by normal standards?

($\epsilon_0 = 8.854 \times 10^{-12}$, charge on electron $= -1.6 \times 10^{-19}$ coulombs, number density of free electrons in copper $= 1.14 \times 10^{29}$ per cubic metre.)

5.10 If two of the equipotential cylinders (of equal radius) in problem 3.10(*c*) are replaced by suitably charged conducting cylinders, each at a uniform potential, the electric field and potential external to them will be unchanged. By applying Gauss's law to unit length of each cylinder, show that the total charge per unit length on a cylinder is $\pm q$. (It will not be uniformly distributed.)

Deduce that the capacity per unit length between two equal parallel conducting cylinders in vacuo is

$$\pi\epsilon_0/\cosh^{-1}(D/2R),$$

in which D = distance between cylinder axes and R = cylinder radius.

5.11 A charged, parallel-plate capacitor has its gap entirely filled with uniform material of dielectric constant k. Show that **D**, which is everywhere normal to the plates, is constant across the surface layer of bound charge in the polarised dielectric and also that its magnitude equals Q, the charge on the front of the plates per unit area, if there is no electric field inside the plates.

Show that k is the natural empirical measure of dielectric quality because it is the ratio by which the capacity of the capacitor is enhanced by the introduction of the dielectric.

Calculate the energy (per unit volume of gap) needed to charge the capacitor with and without the dielectric present, for a given value of **E**. Show that the extra energy to polarise the linear dielectric is $\frac{1}{2}\mathbf{E}\cdot\mathbf{P}$ per unit volume.

5.12 A plane, thin layer of charge (e.g. near the surface of a dielectric) has a density q per unit volume which varies only with x, distance measured normal to the plane layer. There is an electric field E in the x-direction which varies from a uniform value E_1 on one side of the layer to E_2 on the other. By applying Gauss's law to a thinner sub-layer of thickness $\mathrm{d}x$ within the layer, deduce that $\epsilon_0\,\mathrm{d}E/\mathrm{d}x = q$ (which need not be constant). By integrating the force qE per unit volume, show that the x-wise electric force per unit area on the layer is equal to $\sigma\bar{E}$, where $\bar{E}$ is the mean of E_1 and E_2, and σ is the total charge per unit area of layer, i.e. $\int q\mathrm{d}x$ taken across the layer.

Where a normal electric field E emerges from the plane surface of a dielectric of constant k, show that there is a tensile force $\frac{1}{2}\epsilon_0 E^2(1-k^{-2})$ per unit area exerted on the surface layer of the dielectric.

*5.13 A large plane slab of dielectric constant k and uniform thickness moves in its own plane at constant velocity v in the presence of a uniform magnetic field B, parallel to the slab's surface but perpendicular to v. There is no electric field in the laboratory frame of reference outside the dielectric. If the dielectric polarises in proportion to the electric field as seen by an observer *moving with it*, calculate the resulting fractional perturbation of the magnetic field due to convection of the surface charge layers.

5.14 (*a*) In relation to section 5.6, demonstrate the relation $\epsilon_0 E = nex$ for the case where x is negative.

*(*b*) In a one-dimensional plasma oscillation the position x of each plane layer of electrons can be given by

$$x = x_0 + a\sin(\omega t + \phi) \quad (\omega^2 = ne^2/m\epsilon_0),$$

which is a function of x_0, the equilibrium position of the layer, and of an amplitude a and phase angle ϕ, both of which may vary with x_0. If no layer is ever to overtake a neighbour, $(\partial x/\partial x_0)_t$ must never fall to zero. Show that this is secured if

$$\left(\frac{\mathrm{d}a}{\mathrm{d}x_0}\right)^2 + \left(a\frac{\mathrm{d}\phi}{\mathrm{d}x_0}\right)^2 < 1.$$

5.15 A long, uniform cylinder of radius a, thermal conductivity K (see example 5.4) and heat capacity C per unit volume is initially at ambient temperature. An electric current is then passed along it so as to release heat at a uniform rate R per unit volume, while the exterior of the cylinder is maintained at ambient temperature.

By considering a cylinder of radius $r(<a)$ and unit length, find (i) the temperature distribution in the cylinder when a steady state has been reached, and (ii) the amount of energy which was used originally to raise the temperature of the cylinder per unit length. There is no axial heat flow.

5.16 Gas is flowing radially in a spherically symmetric manner from an unsteady point source. By considering the conservation of mass for the space enclosed between two stationary concentric spheres of radius r and $r + \mathrm{d}r$, demonstrate the relationship

$$\frac{\partial}{\partial r}(r^2\rho v) + r^2\frac{\partial \rho}{\partial t} = 0$$

between radial velocity v and density ρ.

5.17 A two-dimensional fluid motion has velocity components

$$v_x = x + y, \quad v_y = x - y, \quad v_z = 0.$$

Show that the integral $\oint \mathbf{v}\cdot \mathrm{d}\mathbf{r}$ round any loop in the x, y plane vanishes. Show also that the net volumetric flow out of a prismatic region with arbitrary cross-section, unit length in the z-direction and plane, normal, ends is zero.

5.18 A prismatic body with arbitrary cross-section and plane, normal ends is held at rest relative to the uniformly rotating fluid in a centrifuge, with its generators parallel to the axis of the centrifuge. The pressure in the (possibly non-uniform) fluid is a single-valued function $p(r)$ of distance r from the axis. By the substitution $\mathrm{d}\mathbf{a} = \mathrm{d}\mathbf{r} \times \mathbf{k}$ ($\mathbf{k}$ a unit vector, parallel to the axis), express the moment about the axis of the pressure forces on the body's surface $-\iint \mathbf{r} \times p\,\mathrm{d}\mathbf{a}$ (in which $\mathbf{r}$ is the two-dimensional radius vector) as a two-dimensional closed line integral round the body's cross-section. Deduce the result that the resultant of these pressure forces passes through the axis.

6

A crisis in electromagnetic theory

6.1 Another look at the Kirchhoff and Ampère laws

When we explore more carefully the implications for electromagnetic theory of the mathematical problems of expressing charge conservation and current flux-linkage, we discover some serious inconsistencies. Resolving them turns out to be a remarkably fruitful undertaking.

Consider the charge conservation law stated in the previous chapter

$$\oint\!\!\oint \mathbf{j} \cdot \mathbf{da} = -\iiint \frac{\partial q}{\partial t}\, \mathrm{d}\tau.$$

In general, the right-hand side is non-zero, for at any instant the state of an electrical system is characterised by electric fields that will usually be in part attributable to electric charge distributions q in the vicinity, which themselves must be changing at some rate $\partial q/\partial t$. Thus, whether $\mathbf{j}$ and q are associated with free or bound charge, or both, it will normally be the case that

$$\iiint \frac{\partial q}{\partial t}\, \mathrm{d}\tau \quad \text{and} \quad \oint\!\!\oint \mathbf{j} \cdot \mathbf{da}$$

do not vanish. So much for Kirchhoff's first law! It now transpires that $\mathbf{j}$ is not in general solenoidal, after all.

More alarming still is the reflection that Kirchhoff's first law is a corollary of Ampère's law and so its failure brings down Ampère's law, too.

We have now reached the famous crisis in the history of electromagnetic theory whose resolution by Maxwell in the 1860s was so elegant and later so remarkably productive in technological terms. What has gone wrong, then, and how did Maxwell put it right?

This is such an important part of electromagnetic theory that it deserves careful treatment. Let us first ponder the role of electrostatics in relation to circuits and other electrical systems. Consider, for instance, the simplest possible d.c. circuit shown in figure 6.1. Kirchhoff's first law asserts that the current J is the same at all points in the circuit, in this case. This is true in the steady d.c. state.

What agency forces the current to flow through the resistor R? There must be a vertical electric field $\mathbf{E}$ in (and around) R which can only be of electrostatic origin, in the absence of changing magnetic fields. Where are the positive and negative charges that account for $\mathbf{E}$? Students have been heard to opine that they are at the battery, perhaps at A and B, but charges at A and B as drawn could never produce a vertical field at R. The battery and resistor might be on opposite sides of a continent, moreover!

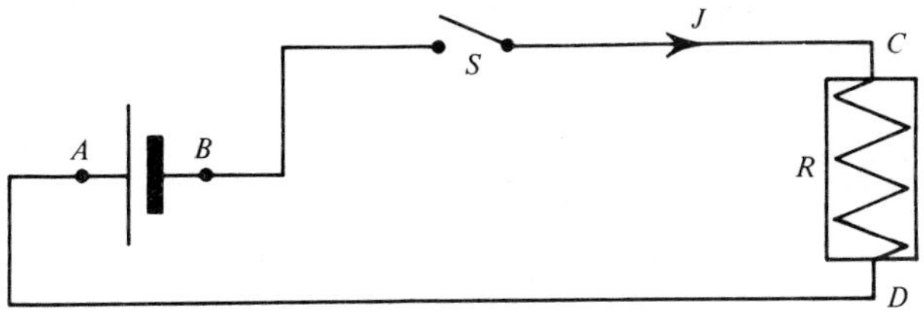

Fig. 6.1. A d.c. circuit.

The charges responsible for **E** are in fact dotted about the system in groups of opposite sign, particularly in the general vicinity of *C* and *D*. The amount of charge needed is determined by the electrical capacity of the particular geometrical configuration taken together with the potential distribution demanded by Ohm's law. Problem 6.1 invites the reader to explore a case where this calculation is easily accomplished.

Now consider the problem of starting the current flow in the first place. When the switch *S* is closed, the battery first has to feed all this capacitance before the electric fields and ultimate steady direct current can build up. For the first few nanoseconds, not all the current leaving the battery reaches *R*; much of it is being sidetracked into laying down the necessary charges. So Kirchhoff's law fails very severely during the starting process.

The same general ideas apply to more complicated situations. Whenever the times in which changes take place are sufficiently short, or whenever frequencies are sufficiently high, a significant part of the currents which flow goes into feeding changing charge distributions. But at lower frequencies, the fraction of the currents thus employed becomes less and is finally negligible. Then simple circuit theory, using Kirchhoff's first law, becomes an acceptable model. We then only allow for charge accumulation specifically in *capacitors*, where it has been deliberately enhanced to the point where, even at low frequencies, their charging currents can be comparable with the other currents in the circuit.

At high frequencies, however, and especially in microwave systems, the changing charges are important. The circuit theorist says: '*Stray* capacitance occurs at high frequencies.' What he is then ignoring is that this same 'stray' capacitance accommodates the charge necessary to keep even a d.c. circuit operating, although its influence on the book-keeping of circuit theory can be neglected.

The untruth of Kirchhoff's first law should be apparent to any owner of a radio or TV set. One of its implications would be that no current could flow into a wire that goes nowhere, i.e. is a cul-de-sac with no connections at its end. But the lead to a TV aerial comes into this category; it is the currents in this lead which are the basis of the whole broadcast reception process.

The failure of Kirchhoff's first law destroys Ampère's law. Ampère's law cannot possibly be correct in the presence of changing charge distributions because (as section 3.17 revealed) the concept of flux-linkage on which Ampère's law depends fails if the vector in question is not solenoidal. A good example is provided by the charging of a parallel-plate capacitor, as shown in figure 6.2.

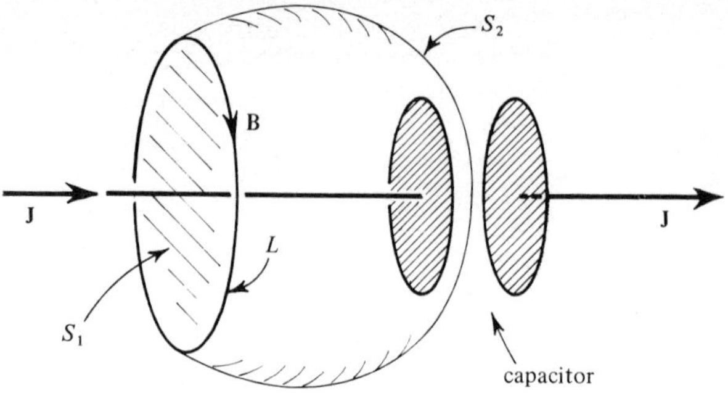

Fig. 6.2. A charging capacitor.

The circular magnetic field **B** around the wire is presumably calculable in the usual way from

$$\oint_L \mathbf{B} \cdot \mathrm{d}\mathbf{r} = \mu_0 J$$

since J is the current flux through S_1, a surface spanning the loop L. But if instead we span L by the surface S_2 which goes through the capacitor gap and intercepts no current flux, Ampère's law gives

$$\oint_L \mathbf{B} \cdot \mathrm{d}\mathbf{r} = 0,$$

a clear contradiction. Other worrying questions now arise as well. Is there a magnetic field in the capacitor gap, and if so, why? But if there is not, where does the magnetic field round the wire peter out as the capacitor is approached?

6.2 Maxwell's resolution of the difficulty

Maxwell perceived how Ampère's law could be repaired and made consistent: it must be written in terms of the flux of some vector which *is* always solenoidal and concerning which it is therefore valid to refer to flux-linkage, because the flux intercepted by two surfaces such as S_1 and S_2 will always be the same. Because Kirchhoff's first law must be replaced by the more general equation of charge conservation

$$\oiint \mathbf{j} \cdot \mathrm{d}\mathbf{a} = -\iiint \frac{\partial q}{\partial t}\,\mathrm{d}\tau = -\frac{\partial}{\partial t}\iiint q\,\mathrm{d}\tau,$$

where

$$\iiint q\,\mathrm{d}\tau = \epsilon_0 \oiint \mathbf{E} \cdot \mathrm{d}\mathbf{a}, \quad \text{by Gauss's law,}$$

it follows that

$$\oiint \mathbf{j} \cdot \mathrm{d}\mathbf{a} = -\epsilon_0 \frac{\partial}{\partial t} \oiint \mathbf{E} \cdot \mathrm{d}\mathbf{a} = -\oiint \epsilon_0 \frac{\partial \mathbf{E}}{\partial t} \cdot \mathrm{d}\mathbf{a}$$

and hence that

$$\oiint \left(\mathbf{j} + \epsilon_0 \frac{\partial \mathbf{E}}{\partial t}\right) \cdot \mathbf{da} = 0.$$

Thus the compound vector $\mathbf{j} + \epsilon_0\, \partial \mathbf{E}/\partial t$ is always a solenoidal one. It would therefore be mathematically consistent to modify Ampère's law into

$$\oint \mathbf{B} \cdot \mathbf{dr} = \mu_0 \iint \left(\mathbf{j} + \epsilon_0 \frac{\partial \mathbf{E}}{\partial t}\right) \cdot \mathbf{da},$$

and this was in effect *Maxwell's conjecture*, greeted with scepticism by many of his contemporaries. He was not finally vindicated for a further twenty years, when Hertz demonstrated the existence of the electromagnetic waves that are a theoretical consequence of Maxwell's insertion of the new term. Once accepted, the conjecture became the *Maxwell–Ampère law*. (The theory of the waves we leave to chapter 11; but see problem 6.2.)

The law states that a changing electric field **E** is just as effective a source of magnetic field as is a charge flow **j**, a not unreasonable idea since Faraday's law states that a changing magnetic field acts as a source of electric field.

Notice that we could have chosen the equally good solution (from a mathematical standpoint) of adding just $\epsilon_0\, \partial \mathbf{E}_s/\partial t$ instead of the full $\epsilon_0\, \partial \mathbf{E}/\partial t$ to **j**, for $\iiint q\, d\tau$ also equals $\epsilon_0 \oiint \mathbf{E}_s \cdot \mathbf{da}$ as $\mathbf{E}_i$ is solenoidal. Since the whole difficulty arose from the changing charge distributions responsible for $\mathbf{E}_s$, this might have been the natural choice. But physically, this alternative conjecture is much less productive; had it been correct there would never have been a telecommunications industry, for it does not predict electromagnetic waves.

Maxwell was probably motivated in his choice by the elegance and symmetry of the equations

$$\oint \mathbf{E} \cdot \mathbf{dr} = -\iint \frac{\partial \mathbf{B}}{\partial t} \cdot \mathbf{da}, \quad \oint \mathbf{B} \cdot \mathbf{dr} = \mu_0 \epsilon_0 \iint \frac{\partial \mathbf{E}}{\partial t} \cdot \mathbf{da},$$

which describe the reciprocal arrangements that prevail between **E** and **B** when charge flow **j** is absent. Philosophically speaking, the correctness of Maxwell's choice is a splendid example of the maxim: 'Beauty is truth'.

At a point out in empty space, far from any material or charges, these equations describe a situation where the two abstractions **E** and **B**, invented to systematise observations on material behaviour, appear to be driving each other without any material intervention. The true boldness of Maxwell's conjecture is then exposed. The lack of material involvement led those scientists of the time who wanted to believe Maxwell to postulate an 'ether' for **E** and **B** to ride upon.

We have still to verify that Maxwell's conjecture does resolve the problem of the charging capacitor, shown in figure 6.2. The fluxes of $\mathbf{j} + \epsilon_0\, \partial \mathbf{E}/\partial t$ across the surfaces S_1 and S_2 are equal; in the case of S_1 it would be due purely to J, while for S_2 it would come from $\epsilon_0\, \partial \mathbf{E}/\partial t$ in the gap, where the electric field is obviously rising as the capacitor charges up. Moreover there *is* a magnetic field in the capacitor gap, as is explored further in problem 6.3(*a*). Another interesting case follows.

Example 6.1 Figure 6.3 shows two concentric spheres, equally and oppositely charged. In the gap **E** is radial and its magnitude is uniform over each concentric sphere, from symmetry arguments. The gas in the gap suddenly breaks down electrically and the capacitor discharges symmetrically with a radial current flow. What magnetic field results?

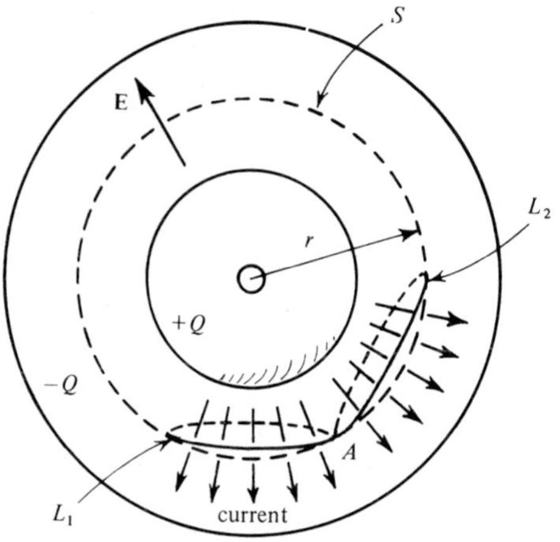

Fig. 6.3. A discharge between spheres.

Gauss's law for the sphere S of radius r reveals that

$$E = Q/(4\pi\epsilon_0 r^2),$$

just as it would be for a point charge Q, and charge conservation for the sphere requires that the radial discharge current intensity **j** is

$$-(\mathrm{d}Q/\mathrm{d}t)/4\pi r^2.$$

Any magnetic field it produced would have to be spherically symmetric, and such a magnetic field cannot exist. If it had a radial component B_r, this would imply a non-zero value for $\oiint \mathbf{B} \cdot \mathrm{d}\mathbf{a}$ over any concentric sphere, which is impossible. But transverse components of **B** are also precluded by the symmetry – any argument for a component in a particular tangential direction can be countered by an equally good argument for a component in exactly the opposite direction. For example loop L_1, which embraces radial current as shown suggests that there might be a circular magnetic field along L_1, coming *out of* the paper at A. The same argument applied to L_2 then implies a field at A directed *into* the paper! The conclusion is that there can be no magnetic field. Thus $\oint \mathbf{B} \cdot \mathrm{d}\mathbf{r} = 0$ round L_1 in flagrant violation of Ampère's law in view of the current 'linked'. The word 'linked' is put in inverted commas because if L_1 is spanned by a different surface, e.g. that consisting of the sphere S of radius r with the cap within L_1 removed, a current flux of different magnitude and sign (relative to L_1) results. Maxwell removes the difficulty because L_1 embraces no net flux of **j** and $\epsilon_0\ \partial\mathbf{E}/\partial t$ combined, for the magnitude of $\epsilon_0\ \partial\mathbf{E}/\partial t$ is $(\mathrm{d}Q/\mathrm{d}t)/4\pi r^2$. (See also problem 6.3(*b*).)

For the same reasons, no magnetic field is produced in the plasma oscillation, discussed in section 5.6; the electron current and the $\epsilon_0\ \partial\mathbf{E}/\partial t$ term have mutually cancelling magnetic effects.

6.3 Dielectrics and 'displacement' current

In the Maxwell–Ampère law,

$$\oint \mathbf{B}\cdot\mathrm{d}\mathbf{r} = \mu_0 \iint \left(\mathbf{j} + \epsilon_0 \frac{\partial\mathbf{E}}{\partial t}\right)\cdot\mathrm{d}\mathbf{a},$$

the term **j** includes all kinds of charge motion whereas $\epsilon_0\ \partial\mathbf{E}/\partial t$ *per se* has nothing to do with charge motion, but is a new physical phenomenon, discovered by Maxwell. In general **j** includes free charge flow (conduction current) denoted by $\mathbf{j}_\mathrm{f}$, say, and bound charge migration $\partial\mathbf{P}/\partial t$. The two very different terms $\partial\mathbf{P}/\partial t$ and $\epsilon_0\ \partial\mathbf{E}/\partial t$ can be written together in one single term $\partial\mathbf{D}/\partial t$, the so-called *displacement current*, if we again invoke the compound quantity $\mathbf{D} = \epsilon_0\mathbf{E} + \mathbf{P}$. This procedure can be very confusing for students because it mixes $\partial\mathbf{P}/\partial t$ together with $\epsilon_0\ \partial\mathbf{E}/\partial t$, which is in no sense a current. It is defensible chiefly for its convenience in connection with problems involving media for which empirical data are likely to be expressed in the form of relations between **E** and **D**. Problems 6.4 and 6.6(*a*) take these ideas further.

*6.4 The Biot–Savart and Maxwell–Ampère laws

The failure of Ampère's law might appear to imperil also the Biot–Savart law from which it was deduced, but this is not so. Ampère's law only follows from the Biot–Savart law in the case where currents flow in closed loops (or elementary field-tubes) each carrying a uniform current, i.e. where the current is solenoidal. This is the situation where Ampère's law is true. But Biot–Savart current elements can be used to describe a current flow pattern which is not solenoidal, i.e. with local charge accretion or depletion.

This is most simply done by regarding a Biot–Savart current element $J\mathrm{d}\mathbf{r}$ as an electric dipole of strength $Q\mathrm{d}\mathbf{r}$ that is changing. The dipole consists of two equal and opposite charges Q and $-Q$ at the ends of d**r** and the current flow J along d**r** causes Q to change at a rate $J = \mathrm{d}Q/\mathrm{d}t$. Consider the case where J is constant in time. The electric field of the dipole is also changing but at a constant rate. The *steady* magnetic field due to the Biot–Savart element can then be seen to be that associated not merely with the current J along d**r** but also with the *steady* $\epsilon_0\ \partial\mathbf{E}/\partial t$ of the whole dipole field. Here the Biot–Savart law can easily be reconciled with Maxwell's view that changing electric fields contribute to magnetic fields. As $\partial\mathbf{B}/\partial t$ is zero in the present case the electric fields can still take their electrostatic form, with $\mathbf{E}_i$ absent. Some of the finer points in this situation are explored in the problems after chapter 11. Obviously events will be more complicated when J and the magnetic field are varying in time.

In the case where many elements are superposed so as to constitute a complete current field which is solenoidal, at each point where one element joins on to the next the 'Q' and '$-Q$' of adjoining dipoles can cancel out, along with their superposed electric fields, and the simple 'Ampère' view of the origin of magnetic fields is seen to apply.

6.5 Material magnetisation

To complete the application of the integral ideas of vector analysis to the description of electromagnetism, we must finally recognise the technologically important fact that a magnetic field can have a source other than moving charges or a changing electric field, namely *material magnetisation*. Though we come to it last, historically it was the first known source of magnetic field because of the natural occurrence of magnetic minerals such as lodestone.

A full study of magnetisation would be out of place here. For present purposes a simple model in terms of magnetic dipoles will suffice. A magnetic dipole is a small loop round which circulates a current J. This produces a magnetic field which is analogous to the electric field of an electric diple (e.g. see problem 4.3). The strength of the dipole is the vector $J\mathbf{da}$, if $\mathbf{da}$ is the loop's vector area pointing in the direction in which J appears to circulate clockwise, as shown in figure 6.4.

Imagine a medium containing N such dipoles per unit volume. If this arrangement is totally at random, their macroscopic magnetic effects will cancel out. But if there is any tendency for the loops to point in the same direction they will produce a new magnetic field inside and outside the medium. Thermal agitation normally tends to randomise such an ordered array.

In order to see how the Maxwell–Ampère law must be modified, consider a section $\mathbf{dr}$ of length ds of a loop L along which $\oint \mathbf{B} \cdot \mathbf{dr}$ is to be evaluated. Consider a cylinder of cross-sectional area A (small in macroscopic terms but large compared

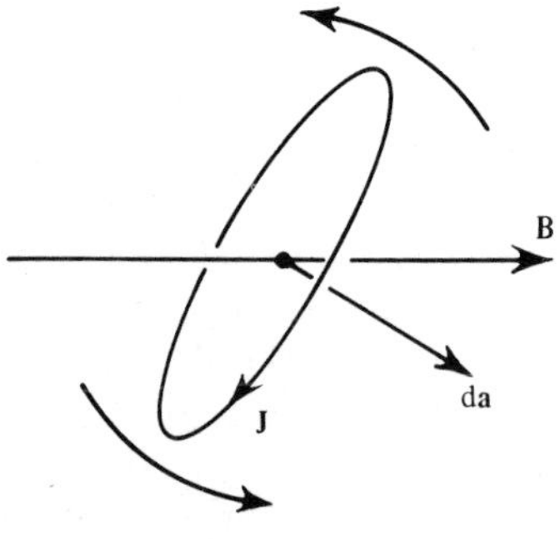

Fig. 6.4. A magnetic dipole $J\mathbf{da}$. The curved arrows indicate the torque due to an imposed field $\mathbf{B}$.

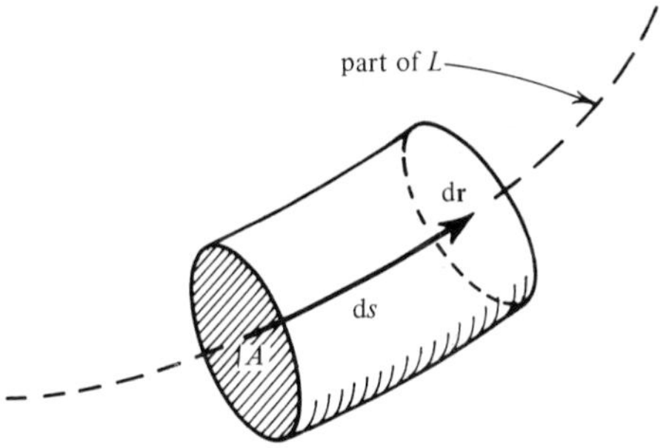

Fig. 6.5

with each of the dipole areas **da**) which has **dr** as its axis. (See figure 6.5.) Its volume A ds contains NA ds dipoles, assumed to be uniformly distributed. The probability that any one of these will be penetrated by the line element **dr** is given by the ratio $\mathrm{d}a_r/A$, where $\mathrm{d}a_r$ is the average projected area of one dipole as seen in the **dr**-direction. Thus $N\,\mathrm{d}a_r\,\mathrm{d}s$ dipoles are penetrated. This part of the loop L therefore links a dipole 'current' equal to $JN\,\mathrm{d}a_r\,\mathrm{d}s$, which equals $JN\,\mathbf{da}\cdot\mathbf{dr}$, or $\mathbf{M}\cdot\mathbf{dr}$ if we write **M** for NJ **da**, the total magnetic dipole strength per unit volume. **M** is also known as the *magnetisation*. Since in general the dipoles will not all be oriented parallel to each other, **da** must be interpreted as an average vector value. If we add all such contributions to the effective current linked by the complete loop L, we get the generalised Maxwell–Ampère law

$$\oint_L \mathbf{B}\cdot\mathbf{dr} = \mu_0\iint\left(\mathbf{j}+\epsilon_0\frac{\partial\mathbf{E}}{\partial t}\right)\cdot\mathbf{da}+\mu_0\oint_L \mathbf{M}\cdot\mathbf{dr},$$

which expresses **B** in terms of all its possible sources. There is no contribution from the final term in regions devoid of magnetic material. If we introduce a new compound vector field

$$\mathbf{H} = \mathbf{B}/\mu_0-\mathbf{M},$$

which equals $\mathbf{B}/\mu_0$ except inside magnetic material, the law simplifies to

$$\oint \mathbf{H}\cdot\mathbf{dr} = \iint\left(\mathbf{j}+\epsilon_0\frac{\partial\mathbf{E}}{\partial t}\right)\cdot\mathbf{da}.$$

Problems 6.5 and 6.6(*b*) are simple exercises on its use, while problem 6.7 generalises magnetic potential to allow for magnetic material.

The statement of the Maxwell–Ampère law in terms of **H** is particularly useful in connection with the testing of materials to establish their magnetic properties. Typically the material is formed into a symmetrical torus (see figure 6.6) and an exciting winding is added, so as to produce a field round the torus. The magnetic flux density **B** can be inferred from an integration in time of the voltages induced in the winding while **H** can be inferred from the known, imposed ampere-turns linking the torus, $\epsilon_0\,\partial\mathbf{E}/\partial t$ being negligible at the low frequencies employed. It is

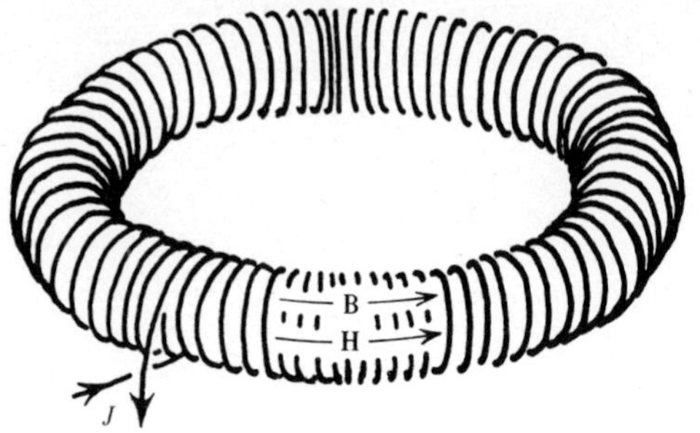

Fig. 6.6. A toroidal solenoid; l = mean circumference of torus.

the convenience of such procedures that leads to magnetic properties of materials being normally recorded as a relation between **B** and **H**, even though **M** is the more fundamental material property. Then, given the data in this form, the $\oint \mathbf{H} \cdot \mathrm{d}\mathbf{r}$ form is the more convenient equation for solving problems concerning electromagnets, etc. because it relates the magnetic quantity **H** to that part of the source of magnetic field which is known *ab initio*, namely the true currents **j** in the windings. The units of **H** and **M** are amperes per metre. These ideas are exploited in problem 6.8 and the example below.

Example 6.2 Figure 6.6 shows a circular torus of magnetic material bearing a uniform thin winding of N_1 turns which carries a variable current $J_1(t)$. The magnetic fields **B**, **H** and **M** all have the same circular field lines and can be treated as uniform in intensity over the cross-section A of the torus. Show that energy is put into the torus via the winding at a rate $H\,\mathrm{d}B/\mathrm{d}t$ per unit volume. (The material is non-conducting and non-polarisable so that no extra currents are induced in it. Also $\epsilon_0\ \partial \mathbf{E}/\partial t$ may be neglected.)

The e.m.f. in the winding is $-A\ \mathrm{d}B/\mathrm{d}t$ per turn or $-N_1A\ \mathrm{d}B/\mathrm{d}t$ *in toto.* If l is the mean length of the field lines, the Maxwell–Ampère law gives

$$Hl = N_1J_1$$

in the absence of other currents and $\epsilon_0\ \partial E/\partial t$.

The electrical input power is e.m.f. times current, i.e.

$$\left(N_1A\,\frac{\mathrm{d}B}{\mathrm{d}t}\right)\left(\frac{Hl}{N_1}\right) = AlH\,\frac{\mathrm{d}B}{\mathrm{d}t}$$

or $H\,\mathrm{d}B/\mathrm{d}t$ per unit volume. If we assume no energy is radiated away, this energy is stored in the torus. (We have also neglected ohmic losses in the winding.)

6.6 Different kinds of magnetisation

Some materials always contain intrinsic magnetic dipoles, which may be ineffective if randomly arranged. In others, the dipoles only appear when a magnetic field is imposed. Some materials are weakly *diamagnetic*, i.e. the materials' response is to oppose any imposed field. A suggestive example is provided by an ionised gas, tenuous enough for collisions to be neglected. In the presence of a magnetic field **B**, imagined to be directed *into* the paper in figure 6.7, a particle with positive charge Q orbits anticlockwise, for the $Q\mathbf{v} \times \mathbf{B}$ force can then produce the requisite centripetal acceleration. The orbiting of such particles is equivalent to dipoles producing a weak magnetic field that is clearly *opposed* to **B**. The reader should check that negative particles contribute to the same effect.

The extreme of diamagnetism is provided by the Meissner effect in a Type I superconductor, which expels *all* magnetic field.

In contrast, a *paramagnetic* material responds by enhancing any imposed field. A simple model is provided by considering the moment on a randomly oriented, intrinsic magnetic dipole of strength $J\mathbf{da}$ when a magnetic field **B** is applied, as in figure 6.4. The moment $J\mathbf{da} \times \mathbf{B}$ (e.g. see problem 2.4) tends to rotate the dipole as shown by the curved arrows until the moment vanishes when **da** becomes parallel to **B**. Notice that the axial field of the dipole then *enhances* **B**.

The extreme case of paramagnetic behaviour is provided by *ferromagnetic* materials, where there is very great enhancement. The relations between **B**, **H** and **M** are then usually non-linear and dependent on the history of the medium. The ratio $|\mathbf{B}|/|\mathbf{H}|$, when **B** and **H** are parallel, is called the permeability μ, whether it is constant or not. As might be expected, there is an upper limit on **M** in a ferromagnetic material which is reached when all available intrinsic magnetic dipoles are more or less fully deployed. This is the phenomenon of *saturation*.

Ferromagnetic materials can be further classified as *soft* or *hard*, depending on the extent to which the dipoles relax back to a more or less random arrangement, producing very little field, as soon as all other sources of magnetic field are removed. The magnetically hard material is the one where a strong field persists, i.e. magnetisation on its own can provide a sustained source of magnetic field. We then have a *permanent magnet*.

It is particularly instructive to consider the nature of the **B**, **M** and **H** fields in

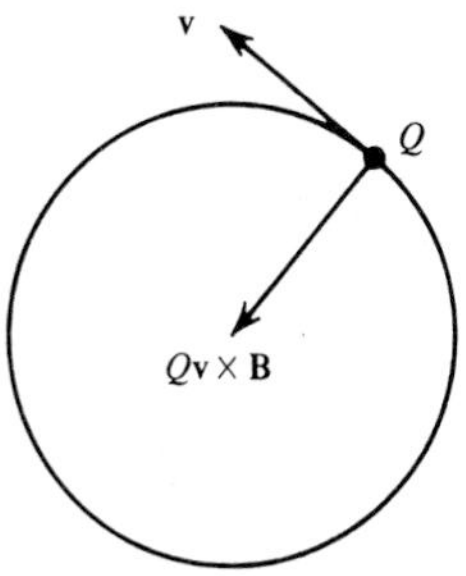

Fig. 6.7. A charged particle gyrating in a magnetic field (directed into the page).

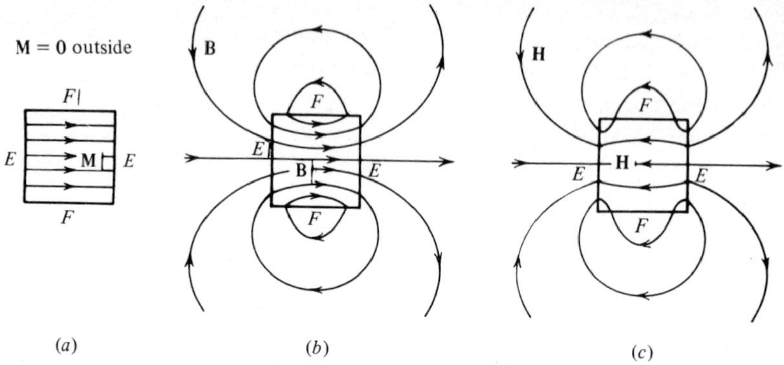

Fig. 6.8. The **M**, **B**, and **H** fields of a uniformly magnetised, cylindrical bar magnet.

and around a permanent magnet. Figure 6.8 presents the fields of a short, cylindrical bar magnet, assumed for simplicity to have uniform permanent magnetisation **M** parallel to its axis. It turns out under this condition that **B**, **M** and **H** cannot be parallel to each other in the medium.

It is noteworthy that **B**, being solenoidal, maintains its direction of advance along its field lines, but **H** does not, because $\oint \mathbf{H} \cdot \mathbf{dr} = 0$ round an **H**-field line when there are no true currents acting as a source of magnetic field, and so $\mathbf{H} \cdot \mathbf{dr}$ has to change sign over part of the loop. Outside the magnet **B** and **H** differ only by the factor μ_0. Inside the magnet **H** is approximately opposed to **B** and to **M**. This is not particularly significant in an intuitive sense because **H** is a rather contrived quantity, justifiable mainly by its convenience for testing and calculation. This particular configuration is discussed further at the end of the chapter and in chapter 11.

Ferromagnetic materials play a big part in electrotechnology. This is not the place for a full discussion, but one important example appears below, while problem 6.9 goes further.

Example 6.3 In order to create a simple *transformer*, a second thin winding of N_2 turns carrying a current $J_2(t)$ is added to the torus shown in figure 6.6. If the material is very highly magnetic, it can be arranged that **B** is large enough to induce significant voltages in the two windings but H is so small that Hl is negligible compared with N_1J_1 and N_2J_2. Show that then energy is exchanged purely between the two windings, in which the e.m.f.s are $\mathcal{E}_1$ and $\mathcal{E}_2$ say, and that

$$|\mathcal{E}_1/\mathcal{E}_2| = |J_2/J_1| = |N_1/N_2|.$$

The e.m.f.s $\mathcal{E}_1$ and $\mathcal{E}_2$ are respectively

$$-N_1A\frac{\mathrm{d}B}{\mathrm{d}t} \quad \text{and} \quad -N_2A\frac{\mathrm{d}B}{\mathrm{d}t},$$

A being the cross-sectional area of the torus. But now

$$N_1J_1 + N_2J_2 = Hl = 0, \quad \text{virtually,}$$

and so

$$\frac{\mathscr{E}_1}{\mathscr{E}_2} = \frac{N_1}{N_2} = -\frac{J_2}{J_1}.$$

Note that the minus sign indicates that the currents must flow in opposing directions. The winding in which current flow is opposed by the e.m.f. and to which energy must therefore be applied from external sources is called the *primary*. The other (the secondary) delivers power to the external circuitry.

The input power $|\mathscr{E}_1 J_1|$ and output power $|\mathscr{E}_2 J_2|$ are evidently equal, energy storage in the material being negligible once Hl is treated as negligible compared with N_1J_1 and N_2J_2.

6.7 The comparison with polarisation

Magnetisation has certain parallels with polarisation. Both are dipole densities. When **P** is not parallel to **E** or **M** is not parallel to **B**, there is a torque per unit volume equal respectively to $\mathbf{P} \times \mathbf{E}$ or $\mathbf{M} \times \mathbf{B}$.

Some texts carry the analogy to the point where magnetisation is treated in terms of dipole pairs of fictitious positive and negative magnetic *poles*, analogous to electric charges. Then each end face of the magnet in figure 6.8 would be regarded as bearing a layer of positive or negative poles, just as a slab of uniformly polarised dielectric acquires surface charges. There is little to be gained from such procedures.

There are also parallels between the compound vectors **D** and **H**. Both are used for similar reasons: they are more readily related to measureable currents and charges than are the basic vectors **P** and **M** which describe the material state. But notice that there is a sign difference when the two relevant equations are set side by side:

$$\mathbf{D} - \mathbf{P} = \epsilon_0 \mathbf{E}, \quad \mathbf{H} + \mathbf{M} = \mathbf{B}/\mu_0.$$

This difference expresses the fact that introducing a magnetisable material into a given configuration of currents generally enhances **B**, whereas introducing polarisable material into a given configuration of free charges reduces **E** on the whole. From the standpoint of technology, both these effects are desirable because they make for better inductors and better capacitors respectively.

6.8 A summary of the equations of electromagnetism in integral form

For reference we now bring together the complete equations that relate the electric and magnetic fields to their sources. These equations generally need to be supplemented by knowledge of the material properties of any media involved, relating **B** to **H** or **M**, **E** to **P** or **D**, or **j** to **E** or e.m.f., in the case of conductors. The suffices 'b' and 'f' refer to bound and free charges or currents respectively.

Flux equations:

$$\epsilon_0 \oiint \mathbf{E} \cdot \mathrm{d}\mathbf{a} = \iiint q \,\mathrm{d}\tau, \quad \text{where} \quad q = q_\mathrm{b} + q_\mathrm{f},$$

$$\oiint \mathbf{B} \cdot \mathrm{d}\mathbf{a} = 0.$$

Circulation equations:

$$\oint \mathbf{E} \cdot \mathrm{d}\mathbf{r} = -\iint \frac{\partial \mathbf{B}}{\partial t} \cdot \mathrm{d}\mathbf{a},$$

$$\oint \mathbf{B} \cdot \mathrm{d}\mathbf{r} = \mu_0 \iint \left(\mathbf{j} + \frac{\partial E}{\partial t}\right) \cdot \mathrm{d}\mathbf{a} + \mu_0 \oint \mathbf{M} \cdot \mathrm{d}\mathbf{r}, \quad \text{where} \quad \mathbf{j} = \mathbf{j}_\mathrm{f} + \partial \mathbf{P}/\partial t.$$

In order to avoid reference to the less accessible quantities q_b, **P** and **M**, we can replace the first equation by

$$\oiint \mathbf{D} \cdot \mathrm{d}\mathbf{a} = \iiint q_\mathrm{f} \mathrm{d}\tau,$$

and the last equation by

$$\oint \mathbf{H} \cdot \mathrm{d}\mathbf{r} = \iint \left(\mathbf{j}_\mathrm{f} + \frac{\partial \mathbf{D}}{\partial t}\right) \cdot \mathrm{d}\mathbf{a}.$$

(We may comment, without proving it, that these laws are correct even from a relativistic point of view.)

These integral statements are complete but not in general very useful. The reader may have noticed in this and the preceding chapters that all the problems that can be solved by means of these integral relations are ones that possess high symmetry, e.g. in which $\oint \mathbf{B} \cdot \mathrm{d}\mathbf{r}$ is easily written down for a circular field line because **B** has constant magnitude along it. But situations in general do not have such symmetry and it is not possible to know in advance how **B**, for instance, varies along a field line, nor what the shape of the field line is.

Notice also that an alternative approach using integration based on the Coulomb or Biot–Savart laws for relating fields to their sources is usually inadequate because we rarely know in advance the complete detailed charge and current distributions, and $\mathbf{E}_\mathrm{i}$ could not be obtained in this manner as a rule because its source is the distribution of $\partial \mathbf{B}/\partial t$, which is rarely known *ab initio*. So, for the majority of problems, some new techniques are needed. These are provided by using the *differential approach* instead of the integral approach which we have followed so far. The rest of this book is devoted to differential vector calculus, until integrals make a reappearance in the last chapter.

The remainder of this chapter deals with a fairly advanced topic and the reader may jump to the next chapter if he so wishes.

*6.9 Electromagnetic boundary conditions

Electromagnetic field problems often involve interfaces between different media, or between a medium and empty space. It is then necessary to know how the various field quantities, **E**, **P**, **D**, **B**, **M** and **H** are related on the two sides of such an interface. Figure 6.9 shows a portion of an interface. V is a small prismatic volume of

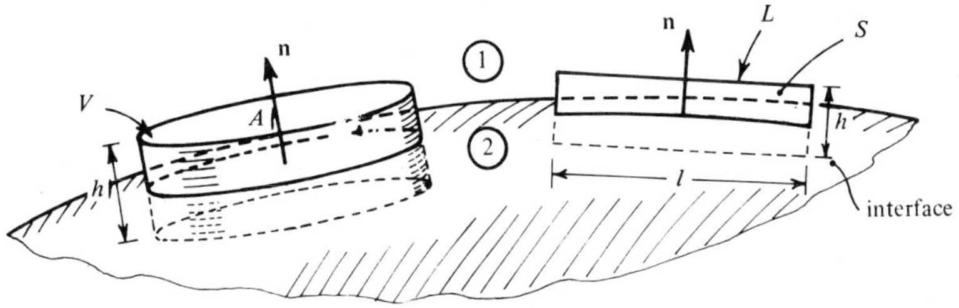

Fig. 6.9. The interface between two regions 1 and 2, pierced by a short prism V and a plane rectangle S, normal to the surface.

cross-sectional area A and short length h, straddling the interface, with the dimension h in the direction of the normal **n**. S is a small rectangular plane area element, also piercing the interface and bounded by the loop L of length l and width h, again with the shorter dimension h in the direction of the normal **n**. h is taken to be so small that any flux through the curved face of V can be neglected, and any contribution from the short sides to a circulation integral round L can be neglected. Let the suffices 1 and 2 denote values on the two sides of the interface, as numbered in the figure, and let suffices 'n' and 't' denote components along the normal **n** and in any tangential direction at the interface.

(a) *Electric boundary conditions*

In general there can be a surface layer of charge in either medium, or possibly both. If one medium is a dielectric, with a component of polarisation P_n normal to its surface it will carry a bound surface charge density equal to P_n. A conductor can also carry a surface layer of free charge. Let q be the *total* charge density per unit area at the interface. Then volume V contains a charge qA and the electric fluxes through its faces are $E_{1n}A$ and $E_{2n}A$, both in the **n**-direction. If we neglect electric flux through the curved sides of V, Gauss's law gives

$$\epsilon_0(E_{1n}A - E_{2n}A) = qA \quad \text{and} \quad \epsilon_0(E_{1n} - E_{2n}) = q,$$

the normal electric boundary condition. If either medium is a dielectric, there can be surface layers of bound charge of density $-P_{1n}$ or P_{2n}. If there is a free charge layer of density q_f, then

$$q_f - P_{1n} + P_{2n} = q = \epsilon_0(E_{1n} - E_{2n})$$

and so

$$D_{1n} - D_{2n} = q_f \quad (\text{for } D = \epsilon_0 E + P).$$

Faraday's law applied to L gives

$$E_{1t}l - E_{2t}l = -\frac{\partial B_t}{\partial t} hl$$

if we neglect the contributions to $\oint \mathbf{E} \cdot \mathbf{dr}$ from the short ends of L and B_t is the component of magnetic field normal to S. As $h \ll l$, we may neglect the terms involving h, with the result that

$$E_{1t} = E_{2t},$$

the tangential electric boundary condition. (At this point the reader may wonder how E_t in two parallel conductors separated by a thin insulator can be different – as can obviously occur, e.g. in two adjoining wires carrying different current intensities. In this case E_n in the insulator is so extremely high that the contributions from the short ends of L are no longer negligible.) Problems 6.10(*a*) and (*b*) test these ideas.

(*b*) *Magnetic boundary conditions*

In contrast to electric field, magnetic field is always solenoidal. Thus the application to $\mathbf{B}$ of the above arguments concerning the volume V simply gives

$$B_{1n} = B_{2n},$$

the normal magnetic boundary condition. An alternative statement of this, when magnetic media are present, is

$$H_{1n} - H_{2n} = M_{2n} - M_{1n} \quad \text{for} \quad B = \mu_0(H + M).$$

If we apply the Maxwell–Ampère law in its $\oint \mathbf{H} \cdot \mathbf{dr}$ form to the loop L, we find that

$$H_{1t}l - H_{2t}l = \left\{ \mathbf{j}_f + \frac{\partial \mathbf{D}}{\partial t} \right\}_{t'} hl$$

if we neglect the contributions to $\oint \mathbf{H} \cdot \mathbf{dr}$ from the short ends of L and suffix 't'' denotes a component normal to S. As $h \ll l$ we may again neglect the terms involving h, with the result that

$$H_{1t} = H_{2t},$$

the tangential magnetic boundary condition. Problem 6.10(*c*) uses these conditions.

Sometimes, however, there is a layer of intense tangential current flow at the surface (a current *sheet*). A layer of parallel conductors (e.g. a thin winding) is one way in which this can occur. Let there be a current J per unit periphery (measured perpendicular to J). If in figure 6.9 we take S perpendicular to J, the current flux through S is Jl, which is not negligible. The Maxwell–Ampère statement then becomes

$$H_{1t}l - H_{2t}l = Jl \quad \text{and} \quad H_{1t} - H_{2t} = J,$$

any other terms still being negligible. H_t is a tangential component *perpendicular* to J; any tangential component of $\mathbf{H}$ *parallel* to J does not change across the interface.

Problems 6.11(*a*) to (*e*) pertain to magnetic boundary conditions. Another instructive example appears below.

***Example 6.4** Compare the **B** and **H** fields of the bar magnet shown in figure 6.8 with those of the short solenoid shown in figure 6.10(*a*) in the case where J (for the solenoid) equals M (for the magnet).

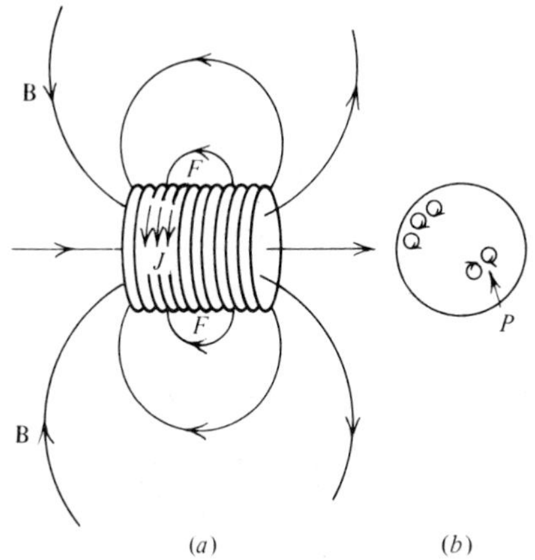

Fig. 6.10. (*a*) A short solenoid bearing J ampere turns per unit length. (*b*) End view of bar magnet showing equivalent current loop dipoles. Internal currents may cancel, e.g. at P, but currents at the rim do not.

The magnet. Let the suffices 'i' and 'o' denote values inside and outside the magnet. To use the equation $\mathbf{B} = \mu_0(\mathbf{H} + \mathbf{M})$ we need the facts that on the ends EE, $M_{\mathrm{it}} = M_{\mathrm{ot}} = 0$, whereas $M_{\mathrm{on}} = 0$ but $M_{\mathrm{in}} = M$. The normal field condition is $B_{\mathrm{in}} = B_{\mathrm{on}}$. The tangential field condition is $H_{\mathrm{it}} = H_{\mathrm{ot}}$ and so $B_{\mathrm{it}} = B_{\mathrm{ot}}$ in this case and we see that the **B**-field is continuous at the faces EE here. However, the **H**-field behaves differently; $H_{\mathrm{on}} - H_{\mathrm{in}} = M$ and so the slope of the **H**-field lines changes at EE in figure 6.8(*c*). In fact H_n changes sign across these faces.

On the flanks FF, $M_{\mathrm{in}} = M_{\mathrm{on}} = 0$, whereas $M_{\mathrm{ot}} = 0$ but $M_{\mathrm{it}} = M$. The normal condition $B_{\mathrm{in}} = B_{\mathrm{on}}$ implies also that $H_{\mathrm{in}} = H_{\mathrm{on}}$. The tangential condition is $H_{\mathrm{it}} = H_{\mathrm{ot}}$ and so this time it is the **H**-field which is continuous at the surface. However, the **B**-field is bent at FF in figure 6.8(*b*) because $B_{\mathrm{it}} - B_{\mathrm{ot}} = \mu_0 M$. In fact B_t changes sign across these faces.

The solenoid. The current is assumed to flow in the thin layer or current sheet FF. There is now no physical surface corresponding to the ends EE of the magnet, and **B** and **H** (which is merely $\mathbf{B}/\mu_0$ here) are continuous everywhere except at the surface FF, which is the source of the fields. The boundary conditions at FF are

$$B_{\mathrm{in}} = B_{\mathrm{on}}$$

and

$$H_{\mathrm{it}} - H_{\mathrm{ot}} = J = M, \quad \text{i.e.} \quad B_{\mathrm{it}} - B_{\mathrm{ot}} = \mu_0 M,$$

if we use the suffices 'i' and 'o' to denote values inside and outside the solenoid. Notice that **B** obeys the same boundary conditions as for the bar magnet. Since it also obeys the same field equations, the **B**-field is in fact identical in the two cases, granted $J = M$, the magnet being uniformly magnetised. Compare figures 6.8(*b*) and 6.10.

One view of the situation is to think of all the virtual dipole currents equivalent to M in the magnet as cancelling out inside it (see figure 6.10(*b*)) but not at its surface, where a net virtual circulating current appears, enabling the magnet to simulate a solenoid.

The **H**-field inside the magnet and solenoid are quite different, however. The magnet's **H**-field can be regarded as 'flowing', both inside and outside the magnet, from its 'sources' on one end E to its 'sinks' on the other end E. The matter is discussed further in chapter 11.

Problems 6

6.1 A d.c. circuit consists of two equal resistances connected in series across a source of constant voltage V_0. If the source is suddenly disconnected, calculate how much charge *subsequently* passes between the two resistances, in the case where the resistances are the inner and outer conductors of a coaxial cable of length d, separated by an annular air gap of inner and outer radii a and $2a$ and short-circuited together at one end, the source being connected at the other end. What happens if the resistances are not equal? What is the effect of a dielectric filling the gap?

(The capacity of coaxial cylinders of radius r_1 and r_2 *in vacuo* is

$$2\pi\epsilon_0/\log_e (r_2/r_1) \quad \text{per unit length.)}$$

6.2 A magnetic field B (in the y-direction) and an electric field E (in the z-direction) are both distributed *in vacuo* in such a way as to depend only upon the x-coordinate and upon time t. Apply Faraday's law to a rectangular loop, lying in the x, z plane, of unit length in the z-direction and small width $\mathrm{d}x$ in the x-direction (so that the electric field strength changes from E to $(E + \mathrm{d}x\, \partial E/\partial x)$ across it). Similarly apply the Maxwell–Ampère law to a loop in the x, y plane. Observe the righthand screw conventions carefully.

Show that the resulting pair of differential equations can be satisfied by a solution of the form

$$\begin{aligned} E &= C \sin (x - ct) \\ B &= D \sin (x - ct) \end{aligned} \qquad (C, D, c \text{ const.}).$$

(This solution represents a wave travelling x-wards at speed c.) Relate c to μ_0 and ϵ_0 and find the ratio C/D.

6.3 (*a*) A parallel plate capacitor has circular plates, normal to and centred on a common axis. Ignoring edge effects, take the electric field in the small air gap as being parallel to the axis. The charges on the plates per unit area are $\pm q$, which is uniform in space but rising at a *constant* rate $\mathrm{d}q/\mathrm{d}t$. There is a magnetic field B in the gap with concentric circular field lines. Find how B varies with radius. Why is the electric field still conservative? (b) What is the effect on example 6.1 if there is a net charge density in the conducting gas? Show that the argument is not invalidated.

6.4 The space between two long, highly conducting, concentric cylinders of radius a and $2a$ is filled with a linear dielectric, with dielectric constant k. The cylinders are connected to an electric supply at the left-hand end, as in figure 6.11. At a certain instant, the charge q per unit length is uniform,

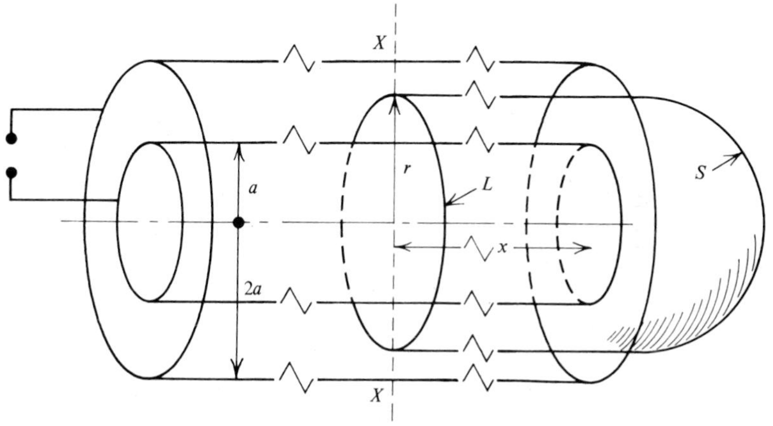

Fig. 6.11. Long, concentric cylinders (only the ends and a short intermediate section are shown).

equal and opposite on the two cylinders and is rising at a uniform rate $\dot{q}$. Show that the total current along the inner cylinder at a cross-section XX (at a distance x from the right-hand end) is equal to $x\dot{q}$ and find the magnetic field $\mathbf{B}$ along the circular field line L, of radius r. (The electric field $\mathbf{E}$ and polarisation $\mathbf{P}$ and their rates of change should be taken as radial vectors having no flux through a plane surface spanning L.) Is $\mathbf{B}$ affected by the presence of the dielectric?

Calculate the magnitudes of $\partial\mathbf{E}/\partial t$, $\partial\mathbf{P}/\partial t$ and the 'displacement current' $\partial\mathbf{D}/\partial t$ at a radius r in the dielectric. Do they depend on x? Note that, as $\mathbf{B}$ is steady, the electric field is purely electrostatic and conservative.

Now consider the surface S which also spans L. Its curved end portion may be taken to be in a region of zero field. The only current crossing S is polarisation current. Which of the following quantities is equal to $_L\oint \mathbf{B}\cdot \mathrm{d}\mathbf{r}/\mu_0$ (through surface S, found by integrating from 0 to x)?

(i) the flux of polarisation current,
(ii) the flux of $\epsilon_0\ \partial\mathbf{E}/\partial t$,
(iii) the sum of (i) and (ii),
(iv) the flux of $\partial\mathbf{D}/\partial t$.

6.5 A linear, loss-free magnetic material is one in which **M** (and so also **H**) is proportional and parallel to **B**, i.e. the permeability μ is constant. A uniform metal rod of radius R bears an axial current of uniform density j. Find how the circular magnetic field **B** depends on radial distance r from the axis, for both $r \lessgtr R$, if μ (const.) $> \mu_0$.

Which of **H**, **B** and **M** are then discontinuous at the surface $r = R$? (Compare example 4.1.)

6.6 Repeat problem 6.2 for the cases where the fields **B** and **E** exist in
(*a*) a linear dielectric such that $\mathbf{D} = k\epsilon_0\mathbf{E}$, or
(*b*) a linear magnetic material such that $\mathbf{B} = \mu\mathbf{H}$,
k and μ being constant. How is the wave speed modified?

6.7 (*a*) In the presence of magnetic material, in regions devoid of current flow or $\partial\mathbf{D}/\partial t$, the definition of magnetic potential U at a point $P(x, y, z)$ can be generalised to be equal to $-\int_D^P \mathbf{H}\cdot\mathrm{d}\mathbf{r}$, D being a datum point. Verify that outside regions of current flow and any 'cut' surface S (as in figure 4.2), the value is uniquely defined and is consistent with the previous definition for the case of non-magnetic material.
(*b*) If part of the region of interest is occupied by highly magnetic material in which **H** can be taken as negligible compared with **H** in the empty space adjoining the material, show that the surface of the material becomes one of constant U, i.e. an equipotential surface, analogous to a highly conducting surface in electrostatic or electric conduction theory.

6.8 The torus shown in figure 6.6 is made of a linear magnetic material for which $\mathbf{B} = \mu\mathbf{H}$ (μ = const.) and has a cross-sectional area of 1 cm^2 and a mean circumference of 30 cm. It carries a winding of negligible resistance with 1000 turns. When an alternating current of 0.1 ampere at 50 Hz is passed through the winding, the voltage across it is 10 volts.

Calculate μ and compare it with μ_0. Does the answer depend on whether r.m.s. values are used? What is the self-inductance?

6.9 A common situation in electrotechnology is the *magnetic circuit*, which can be adequately modelled as a magnetic field tube, forming a closed toroidal loop L, of variable cross-section A over which the magnetic fields **B** and **H** can be treated as uniform and parallel. The magnetic material traversed can be of variable permeability μ and may include airgaps, where $\mu = \mu_0$. Show that the magnetic flux Φ along the field tube is constant and that

$$\Phi \oint \frac{\mathrm{d}s}{\mu A} = \text{ampere-turns linked by } L,$$

ds being an element of the loop L. The integral is called the *reluctance*. Note the crude analogy between

electromotive force = current × resistance

and

'magnetomotive force' (ampere turns) = flux × reluctance.

*6.10 (*a*) Deduce the result relating D_n to q_f in section 6.9(*a*) from the integral statement

$$\iiint q_f \mathrm{d}\tau = \oiint \mathbf{D} \cdot \mathbf{da}.$$

(*b*) The surface of a linear dielectric is exposed to a vacuum. Show that inclined electric field lines are 'refracted' as they cross the interface and find a relation between the dielectric constant k and the angles θ_i and θ_o between the normal to the surface and the field inside and outside the dielectric respectively.

(*c*) A parallel plate capacitor has its gap filled with two layers of non-polarisable insulator between which lies a layer of dielectric for which $k = 5$. All three layers have the same thickness. When the capacitor plates carry charges $\pm Q$ per unit area, what is the bound charge density at the surfaces of the dielectric? By what fraction is the voltage across the capacitor reduced by the presence of these charges?

*6.11 (*a*) Check that the results of problem 6.5 are consistent with the magnetic boundary conditions in section 6.9(*b*).

(*b*) The surface of a linear magnetic material (μ constant and $> \mu_0$) is exposed to a vacuum. Show that inclined magnetic field lines are 'refracted' as they cross the interface and find a relation between μ/μ_0 and the angles θ_i and θ_o between the normal to the surface and the field inside and outside the material respectively. What happens as $\mu \to \infty$? Is the refraction similar to that in problem 6.10(*b*)?

(*c*) A layer of conductors in the plane $z = 0$ carries an x-wise current J per unit length measured in the y-direction. In the region $z < 0$, there is a uniform magnetic field $B = \mu_0 J$ in the z-direction.

Find the inclination of the uniform field in the region $z > 0$ in the cases

(i) no magnetic material present,
(ii) linear magnetic material for $z < 0$ only, and
(iii) linear magnetic material for $z > 0$ only,

taking the permeability μ as equal to $(\sqrt{3})\mu_0$ in (ii) and (iii).

(*d*) Using the same notation as in section 6.9(*b*) show that

$$B_{1t} - B_{2t} = \mu_0(J + M_{1t} - M_{2t}).$$

(*e*) In dealing with configurations containing highly ferromagnetic media, it is often a good approximation to take $\mathbf{H}$ as zero (i.e. $\mu \to \infty$) inside the media. Show that this implies that just *outside* the surface of such a medium:

(i) if there is any field component normal to the surface, it is the *only* component, i.e. the field is normal to the surface, but

(ii) if there is no field component normal to the surface, there is no field at all outside the medium. (The field inside will be parallel to the surface.)

These ideas are crucial to the understanding of iron-cored electromagnets, in which the core is used to 'channel' magnetic flux in much the same way that a conductor channels current. (See also problems 6.7(*b*), 6.9 and 6.11(*b*).)

7

Gradient and divergence, gauges of non-uniformity

7.1 The differential approach

After so much space devoted to the *integral* approach, it is time to enter Zone 2 of figure 3.2 to develop the *differential* approach, necessary for the description of those physical phenomena which feed on non-uniformity, whether in space or time. A body cools down (in time), for instance, because the temperature is non-uniform (in space). In this chapter and the next three we shall seek inspiration particularly from fluid mechanics.

7.2 The influence of fluid pressure

A crucial factor in the statics or dynamics of fluids is the way in which the stresses are distributed. Here we shall restrict attention to cases where viscous stresses are either zero because the fluid is at rest or negligible even though it is in motion. Then the normal stress, the pressure, is isotropic as we saw in chapter 1 and all shear stresses are zero. Since it has no particular direction, the pressure p is then a scalar.

The distribution of the pressure is a scalar field $p(x, y, z)$ at each instant. It is possible to imagine the family of all points sharing a common value of the pressure, defined by $p(x, y, z) = \text{const}$. Such an equation in general defines a *surface* in three dimensions. Different values of the pressure define different surfaces. These cannot intersect, for otherwise the pressure at a point of intersection would have to take two values at once. We describe these constant pressure surfaces as *nesting surfaces*. The best way to picture the scalar field is in terms of these various surfaces of constant pressure, filling the region of interest rather like the layers of an onion. These ideas apply to *any* single-valued scalar field, not just a pressure field.

To express the dynamics or statics of a small fluid element or immersed solid particle, it is necessary to calculate the net force on its surface due to variations in the pressure in the vicinity. This net force, exerted *on* the element *by* the fluid outside, is $-\oiint p\mathbf{da}$, taken over its surface. If we assume the element is small compared with distances over which the pressure changes by a large fraction, it will be sufficient to treat pressure as a linear function of position in the vicinity, i.e. to retain only first-order terms in Taylor's series:

$$p = p_0 + x\left(\frac{\partial p}{\partial x}\right)_0 + y\left(\frac{\partial p}{\partial y}\right)_0 + z\left(\frac{\partial p}{\partial z}\right)_0,$$

where x, y, z are small cartesian distances referred to an origin inside the element

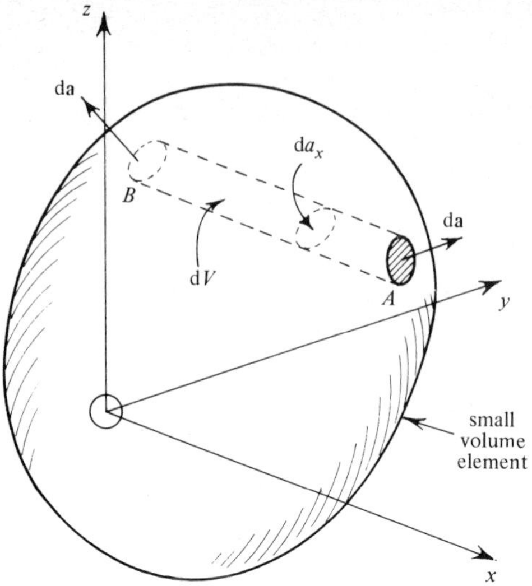

Fig. 7.1

and suffix 0 denotes values at the origin. (See figure 7.1.) Consider the x-component of $-\oiint p\mathbf{da}$:

$$-\oiint p\,da_x = -p_0\oiint da_x - \left(\frac{\partial p}{\partial x}\right)_0 \oiint x\,da_x - \left(\frac{\partial p}{\partial y}\right)_0 \oiint y\,da_x$$
$$- \left(\frac{\partial p}{\partial z}\right)_0 \oiint z\,da_x,$$

in which da_x is the x-wise projection of the surface element $\mathbf{da}$, equal to the cross-section of the tube element dV. The first, third and fourth terms of the right-hand side all vanish because each contribution to the integrals from a point such as A in figure 7.1 is cancelled by the contribution from a point such as B, for which da_x is equal but negative, and $y_A = y_B$ and $z_A = z_B$. However, the contribution to $\oiint x\,da_x$ from A and B is $(x_A - x_B)da_x$ or dV, the volume of the tube AB, if higher-order small quantities are ignored. Thus $\oiint x\,da_x$ equals the volume V of the whole element, and we find that the x-component of $-\oiint p\mathbf{da}$, the net pressure force on the element in the x-direction, is

$$-\left(\frac{\partial p}{\partial x}\right)_0 \text{ times } V.$$

Taking all the components together and dropping the suffices, we conclude that

$$\text{(net pressure force per unit volume)} = -\left(\mathbf{i}\frac{\partial p}{\partial x} + \mathbf{j}\frac{\partial p}{\partial y} + \mathbf{k}\frac{\partial p}{\partial z}\right),$$

whatever the shape of the element, provided it is small. The expression on the right-hand side, obtained by various differentiations of p, is evidently a vector which describes certain interesting aspects of the non-uniformity of the scalar pressure field. It does not matter where the origin is taken, provided $\partial p/\partial x$, etc. are evaluated at the element. Problem 7.1 is a less general version of the above arguments.

7.3 Vector gradient

Generalising, in order to describe the non-uniformity of any scalar field $\phi(x, y, z)$, we can define a *vector gradient* of ϕ, abbreviated to grad ϕ, as

$$\mathbf{i}\frac{\partial\phi}{\partial x} + \mathbf{j}\frac{\partial\phi}{\partial y} + \mathbf{k}\frac{\partial\phi}{\partial z}.$$

Note that grad ϕ is a vector field, deduced from a scalar field by differentiation. From the case of pressure, where grad p was found to have physical significance, we can infer that grad ϕ must be quite independent of the choice of axes, which are after all quite arbitrary, even though grad ϕ is defined in terms of x, y and z-differentiation. A mathematician would say that grad ϕ is *invariant* to change of axes.

The relationship of grad ϕ to the nesting surfaces of constant ϕ should be worth investigating. Consider a small excursion $\mathbf{dr}$ in space (with components $\mathrm{d}x$, $\mathrm{d}y$, $\mathrm{d}z$) along which ϕ changes to $\phi + \mathrm{d}\phi$. Retaining only first-order small quantities, the total differential theorem tells us that

$$\mathrm{d}\phi = \mathrm{d}x\frac{\partial\phi}{\partial x} + \mathrm{d}y\frac{\partial\phi}{\partial y} + \mathrm{d}z\frac{\partial\phi}{\partial z},$$

which can be recognised as the scalar product of $\mathbf{dr}$ and grad ϕ.

Hence $\mathrm{d}\phi = \mathbf{dr}\cdot\text{grad}\,\phi.$

If $\mathbf{dr}$ is any excursion in a surface of constant ϕ, $\mathrm{d}\phi = 0$ and we conclude from the vanishing of the scalar product that grad ϕ is perpendicular to all such excursions, i.e. to the surface itself. This is the first important discovery: *at each point* grad ϕ *is perpendicular to the constant-ϕ surface.*

Figure 7.2 shows a two-dimensional section through two constant-ϕ surfaces, characterised by the values ϕ and $\phi + \mathrm{d}\phi$. Distance n is measured normal to the surfaces, i.e. in the direction of grad ϕ. If we apply the result $\mathrm{d}\phi = \mathbf{dr}\cdot\text{grad}\,\phi$ in the case where $\mathbf{dr}$ is $\overrightarrow{AB}$, parallel to grad ϕ and of magnitude $\mathrm{d}n$, then

$$\mathrm{d}\phi = \mathrm{d}n|\text{grad}\,\phi| \quad\text{and}\quad |\text{grad}\,\phi| = \partial\phi/\partial n.$$

This second important result states that *the magnitude of* grad ϕ *equals the rate of change of ϕ with distance measured normal to the constant-ϕ surfaces.*

If instead $\mathbf{dr}$ is $\overrightarrow{AC}$, inclined at an angle θ to grad ϕ, and of magnitude $\mathrm{d}s$, the scalar product formula for $\mathrm{d}\phi$ gives

$$\mathrm{d}\phi = \mathrm{d}s|\text{grad}\,\phi|\cos\theta \quad\text{and}\quad \frac{\partial\phi}{\partial s} = \frac{\partial\phi}{\partial n}\cos\theta = \text{component of grad}\,\phi\text{ along }AC.$$

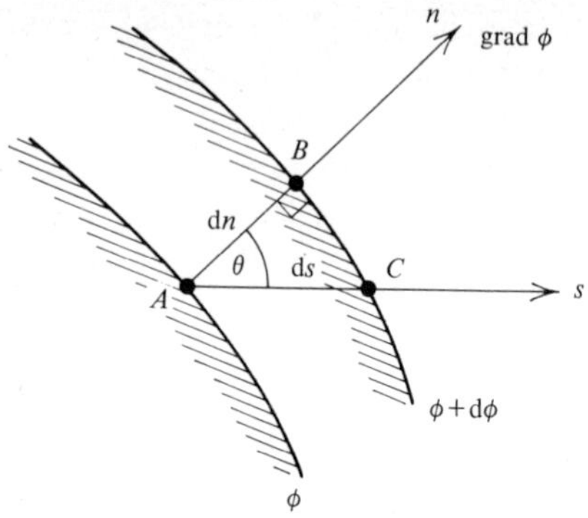

Fig. 7.2. Portion of two adjoining constant-ϕ surfaces.

(This is also obvious because $dn = ds \cos\theta$.) The third important result is therefore that *the component of* grad ϕ *in any direction equals the rate of change of* ϕ *with distance in that direction*. Rate of change with distance is showing the familiar cosine-quality of vectors, discussed in chapter 1: there is one direction (the normal direction) where it reaches its maximum value $|\text{grad}\,\phi|$, while in other directions it is reduced by a factor $\cos\theta$, reaching zero in all directions tangential to the constant-ϕ surface. It is rather obvious that the rate of change with distance will be greatest if the shortest (i.e. normal) route is taken between adjacent constant-ϕ surfaces. Problem 7.2 and the examples below explore the geometrical possibilities of grad.

Example 7.1 Show that the vector $\mathbf{i}a + \mathbf{j}b + \mathbf{k}c$ and the plane $ax + by + cz = A$ are perpendicular, and find the equation of the tangent plane to the surface $\phi(x, y, z) = \text{const.}$ at the point $P(x_0, y_0, z_0)$.

Consider a family of planes with different values of A. A then becomes a scalar field, defined by

$$A(x, y, z) = ax + by + cz.$$

Now grad A is known to be perpendicular to the surface $A = \text{const.}$ and

$$\text{grad}\,A = \mathbf{i}a + \mathbf{j}b + \mathbf{k}c, \quad \text{for} \quad \partial A/\partial x = a, \quad \text{etc.}$$

So the vector and any of the planes are indeed perpendicular. The planes are all parallel.

The relation $\phi(x, y, z) = \text{const.}$ defines one surface belonging to a family defined by taking different values of the constant. So grad ϕ or

$$\mathbf{i}\left(\frac{\partial\phi}{\partial x}\right)_0 + \mathbf{j}\left(\frac{\partial\phi}{\partial y}\right)_0 + \mathbf{k}\left(\frac{\partial\phi}{\partial z}\right)_0$$

(where suffix 0 denotes values at P) is a vector $\mathbf{N}$ normal to the surface at P. If Q, with position vector $\mathbf{r}$, lies in the plane tangential to the surface at P (position vector $\mathbf{r}_0$) then $\vec{PQ}$ is perpendicular to $\mathbf{N}$ and $\vec{PQ} \cdot \mathbf{N} = 0$, i.e.

$$(\mathbf{r} - \mathbf{r}_0) \cdot \mathbf{N} = 0,$$

or

$$(x - x_0)\left(\frac{\partial \phi}{\partial x}\right)_0 + (y - y_0)\left(\frac{\partial \phi}{\partial y}\right)_0 + (z - z_0)\left(\frac{\partial \phi}{\partial z}\right)_0 = 0,$$

the cartesian equation of the tangent plane.

Example 7.2 A point P lies at scalar distances r_1 and r_2 from two fixed points O_1 and O_2 (see figure 7.3). If $r_1 + r_2 = \text{const.}$, P describes an ellipsoidal surface. (See example 3.2.) Show that the normal to the ellipsoid at P bisects $\angle O_1 P O_2$.

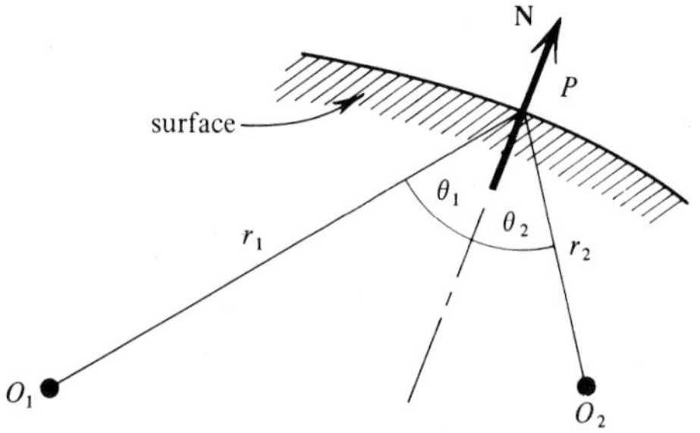

Fig. 7.3

If $\mathbf{r}$ is the position vector relative to an origin O, of magnitude r, the surfaces $r = \text{const.}$ are concentric spheres and therefore grad r is a *radial* vector, normal to the spheres, of magnitude equal to the rate of change of r with distance measured normal to the spheres, which is obviously unity. So grad r is the unit radial vector, $\mathbf{r}/r$.

grad $(r_1 + r_2)$ (= grad r_1 + grad r_2) is a vector $\mathbf{N}$ normal to the surface $r_1 + r_2 = \text{const.}$ But grad r_1 and grad r_2 are equal in magnitude, both being unit vectors, and so their resultant $\mathbf{N}$ bisects the angle between them. This can be formally confirmed by resolving perpendicular to $\mathbf{N}$ and deducing that $\sin\theta_1 = \sin\theta_2$. (See figure 7.3.)

An alternative notation is available for grad ϕ if we invent the idea of a 'vector' differentiation operator, called *del* (or nabla):

$$\nabla \equiv \mathbf{i}\frac{\partial}{\partial x} + \mathbf{j}\frac{\partial}{\partial y} + \mathbf{k}\frac{\partial}{\partial z},$$

which applies to whatever comes next on its right. Then

$$\text{grad}\,\phi \equiv \nabla\phi.$$

An enterprising reader may immediately begin to speculate whether $\nabla \cdot \mathbf{v}$ or $\nabla \times \mathbf{v}$

have any meaning, if $\mathbf{v}$ is a vector field. They do indeed, as we shall see later. In fact this is the reason why in this book we prefer to use the distinct, evocative names such as grad, rather than several notations all involving ∇ which are therefore easily confused. Note that del is not Δ, which is *delta*.

The grad operator has many obvious properties such as

$$\text{grad}\,(\phi + \psi) = \text{grad}\,\phi + \text{grad}\,\psi$$

(used in example 7.2) and

$$\text{grad}\,\phi\psi = \phi\,\text{grad}\,\psi + \psi\,\text{grad}\,\phi.$$

These are easily verified in terms of cartesian components. Problems 7.3 and 7.4 consider other important properties.

7.4 Hydrostatics

In the application to fluid pressure, the net force per unit volume on a small element is seen to be $-\text{grad}\,p$. The minus sign reminds us that this force is *from* high pressure *towards* low, whereas any vector grad ϕ points in the direction in which ϕ *increases.* The fluid element exerts a force on its surroundings equal to $+\text{grad}\,p$ per unit of its volume.

Let us exploit the physical significance of grad p in the context of *hydrostatics*, the theory of fluids at rest, where there is equilibrium between a non-uniform pressure field and whatever body forces prevail. A *body force* is a force field that is distributed throughout a medium instead of acting only on its surface; weight and the magnetic $\mathbf{j} \times \mathbf{B}$ force in a conductor are examples. We postpone study of the role of grad p in fluid *dynamics* to later chapters (but see problem 7.5).

Gravitational hydrostatics describes the balance between weight and the net pressure force per unit volume ($-\text{grad}\,p$). If the gravity force field is $\mathbf{g}$ per unit mass, the weight per unit volume is $\rho\mathbf{g}$ and equilibrium demands that

$$\rho\mathbf{g} - \text{grad}\,p = \mathbf{0}.$$

Gravity hydrostatics is a complicated subject on the astrophysical scale because then the $\mathbf{g}$-field is determined by the distribution of the mass in space and finding $\mathbf{g}$ is part of the problem. Near Earth however we can take $\mathbf{g}$ as known according to the inverse-square law and indeed for many purposes $\mathbf{g}$ can be taken as constant in magnitude and direction. Then hydrostatics becomes a rather easy subject.

If $\mathbf{g}$ is uniform (of magnitude g) and we measure z in the opposite direction to $\mathbf{g}$ (i.e. vertically upwards), and if ρ is uniform, then

$$\text{grad}\,\rho g z = \rho g\,\text{grad}\,z = \rho g\mathbf{k} = -\rho\mathbf{g} = -\text{grad}\,p.$$

Hence

$$\text{grad}\,(p + \rho g z) = \mathbf{0},$$

and

$$p + \rho g z = \text{uniform}.$$

Notice that when grad of some scalar ϕ vanishes, $\partial\phi/\partial x$, $\partial\phi/\partial y$, $\partial\phi/\partial z$ all vanish and ϕ becomes independent of x, y and z, i.e. uniform in space. (It might vary with time t. This does not arise here however for we are dealing with static situations.)

The quantity $p + \rho gz$ also plays an important role in constant-density fluid *dynamics*. We shall denote it by p^*.

Example 7.3 Demonstrate Archimedes' Principle.

In situations where grad p is uniform, as in hydrostatics, the Taylor series for p given in section 7.2 terminates after the first-order terms because the higher derivatives such as $\partial^2 p/\partial x^2$ all vanish. This means that x, y and z in that series need not be small and so the argument applies to large immersed volumes. Thus the pressure force on an immersed object, however large, is upwards (the opposite direction to grad p) and in magnitude equals

$$(|\text{grad}\, p|, \text{i.e. } \rho g) \text{ times (volume)},$$

which equals the *weight of fluid 'displaced'*. This famous result is known as *Archimedes' Principle*. It is intuitively obvious if one reflects that the fluid nearby must behave the same whether it is supporting a fish, let us say, or merely more fluid. It therefore exerts a force on the fish that would have equally well supported the 'displaced' fluid.

Problems 7.6 and 7.7 are further exercises in hydrostatics.

7.5 Magnetohydrostatics

This less familiar area of study is concerned with the static equilibrium of a conducting fluid under the action of pressure gradients and a magnetic body force $\mathbf{j} \times \mathbf{B}$. It finds application in the important research field of controlled thermonuclear fusion (where hot ionised heavy hydrogen gas is to be squashed by $\mathbf{j} \times \mathbf{B}$ forces for periods long enough to turn it into helium and release saleable power). The main application is in astrophysics, for the universe consists almost entirely of electrically conducting gas permeated by magnetic fields. Thus stellar or galactic equilibrium is magnetohydrostatic in character, usually with gravitational overtones.

In pure magnetohydrostatics, the equilibrium relation for a fluid element, expressed per unit volume is

$$\mathbf{j} \times \mathbf{B} - \text{grad}\, p = \mathbf{0}.$$

There is space only for one example.

Example 7.4 ('The linear pinch') This simple scheme for squashing gas by means of $\mathbf{j} \times \mathbf{B}$ forces exploits the tendency of parallel 'like' currents to attract each other. Figure 7.4 shows a cylindrical column of gas of radius R, assumed to be carrying a uniform axial current intensity j. The column is surrounded by vacuum. (This gas has a free surface!) Find what pressure distribution is necessary for equilibrium.

The problem is easily analysed in terms of the currents' magnetic field B, which, from the symmetry, has concentric circular field lines. As in example 4.1, the field

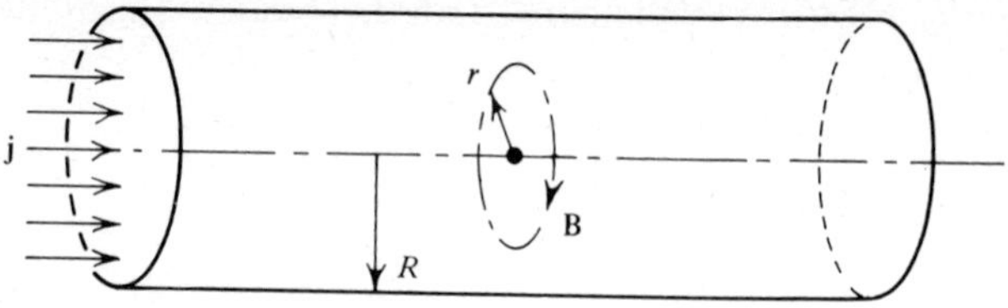

Fig. 7.4. A linear pinch.

at a distance r from the axis is given by $B = \frac{1}{2}\mu_0 jr$. Now $\mathbf{j} \times \mathbf{B}$ acts radially inwards and the radial component of the equilibrium relation is

$$-jB - \mathrm{d}p/\mathrm{d}r = 0,$$

for the radial component of grad p is $\mathrm{d}p/\mathrm{d}r$, the rate of change of p with distance in that direction. (A *partial* differential coefficient is not required because p depends only on r.) Inserting the value of B gives

$$\mathrm{d}p/\mathrm{d}r = -\tfrac{1}{2}\mu_0 j^2 r,$$

which integrates to

$$p = \tfrac{1}{4}\mu_0 j^2(R^2 - r^2) \quad \text{or} \quad p + B^2/\mu_0 = \tfrac{1}{4}\mu_0 j^2 R^2,$$

if we determine the constant of integration from the condition that $p = 0$ when $r = R$. The pressure is distributed parabolically with radius, reaching a maximum on the centre-line. The final thing that must be said about this equilibrium is that it is very unstable indeed. The gas column shows a strong tendency to 'pinch' itself off into a string of sausages and to develop meanders like a varicose vein. Successful controlled fusion depends on much more subtle magnetohydrostatic equilibria. Problem 7.8 takes this example further.

7.6 Grad and potential

Section 7.3 revealed that the change $\mathrm{d}\phi$ in a scalar ϕ along a line element $\mathrm{d}\mathbf{r}$ is given by

$$\mathrm{d}\phi = \operatorname{grad}\phi \cdot \mathrm{d}\mathbf{r}.$$

Such changes may be summed (integrated) along a series of line elements forming a curve joining two points A and B to yield the line integral

$$\int_A^B \operatorname{grad}\phi \cdot \mathrm{d}\mathbf{r} = \phi_B - \phi_A,$$

which takes the same value irrespective of the route from A to B, if ϕ is single-valued. We conclude that *grad ϕ is a conservative vector field and that ϕ is its potential.* (See problem 7.9.) These conclusions are typical of the unexpected but pleasing coming-together of apparently unrelated earlier concepts that we shall constantly encounter from here onwards.

The converse is true: if **v** is conservative then its potential ϕ is such that $\mathbf{v} = \pm\, \text{grad}\, \phi$. (The negative sign applies if the negative convention is adopted in defining potential.) The reason is that

$$\mathbf{v} \cdot \mathrm{d}\mathbf{r} = \mathrm{d}\phi = \text{grad}\, \phi \cdot \mathrm{d}\mathbf{r} \quad \text{for all choices of } \mathrm{d}\mathbf{r}.$$

Taking $\mathrm{d}\mathbf{r}$ parallel to each axis in turn shows that corresponding components of **v** and grad ϕ are equal.

Thus $$\phi = \int_{\text{datum}} \mathbf{v} \cdot \mathrm{d}\mathbf{r}, \quad \mathrm{d}\phi = \mathbf{v} \cdot \mathrm{d}\mathbf{r}, \quad \mathbf{v} = \text{grad}\, \phi$$

are all equivalent ways of stating the same relationship between **v** and ϕ. Notice that knowledge of a potential is sufficient to define the corresponding conservative vector field, which is an important and useful fact.

Example 7.5 A two-dimensional field **v** has a potential $\phi = x + \sqrt{(x^2 + y^2)}$. Find the form of the field lines and equipotentials.

$$v_x = \partial\phi/\partial x = 1 + x/\sqrt{(x^2 + y^2)}, \quad v_y = \partial\phi/\partial y = y/\sqrt{(x^2 + y^2)}.$$

Along a field line,

$$\frac{\mathrm{d}y}{\mathrm{d}x} = \frac{v_y}{v_x} = \frac{y}{\sqrt{(x^2 + y^2)} + x} = \frac{\sqrt{(x^2 + y^2)} - x}{y}$$

(from multiplying top and bottom by $\sqrt{(x^2 + y^2)} - x$).
Hence

$$x + y\, \mathrm{d}y/\mathrm{d}x = \sqrt{(x^2 + y^2)}.$$

Also, along a field line,

$$\frac{\mathrm{d}\phi}{\mathrm{d}x} = 1 + \frac{x + y\mathrm{d}y/\mathrm{d}x}{\sqrt{(x^2 + y^2)}} = 2,$$

and so

$$\phi = 2x + k \text{ (const.)},$$

i.e.

$$\sqrt{(x^2 + y^2)} = x + k \quad \text{or} \quad y^2 = 2kx + k^2.$$

Each field line is therefore a parabola with the origin as focus. On the equipotentials, however,

$$x + \sqrt{(x^2 + y^2)} = \phi \text{ (const.)} \quad \text{or} \quad y^2 = -2\phi x + \phi^2,$$

and these represent orthogonal, confocal parabolas, as shown in figure 7.5. (They are orthogonal because $\mathbf{v} = \text{grad}\, \phi$, which is perpendicular to the equipotentials.)

In electrostatics we have $V = -\int_{\text{datum}} \mathbf{E} \cdot \mathrm{d}\mathbf{r}$, where V is electric potential and so $\mathbf{E} = -\,\text{grad}\, V$. It is now clearer why **E**, being a potential gradient, is measured in volts per metre. Similarly, in regions devoid of current, it is possible to write

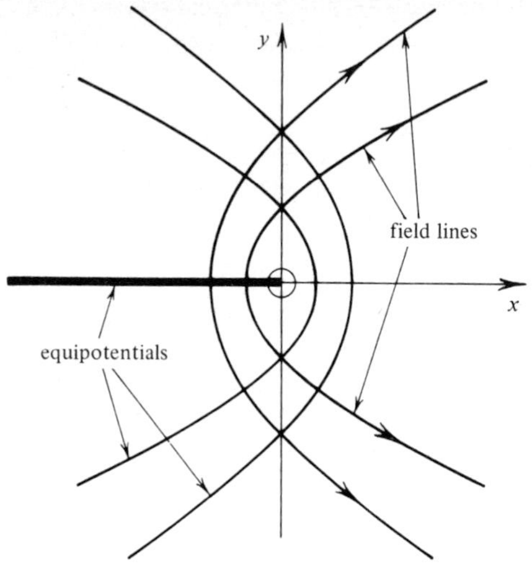

Fig. 7.5

$$\mathbf{H} = -\operatorname{grad} U,$$

where U is magnetic potential, rendered single-valued if necessary by suitable 'cut' surfaces. (See problem 6.7(a).) Problems 7.10 and 7.11 exploit these ideas.

7.7 Flux/gradient laws

It is very common in physical and engineering science to encounter a vector flow field that is equal to the negative gradient of a scalar, which can therefore be described as its potential. These cases differ from the $\mathbf{E}/V$ or $\mathbf{H}/U$ relations in that the vector represents the intensity with which some physical entity is actually flowing and the scalar potential is usually a separately measureable quantity, not an abstraction. The relationships are in fact new physical information, not mathematical deductions. We shall call them *flux/gradient laws*. The grad concept now begins to pay huge dividends.

One of the best examples is Fourier's law of heat conduction, which states that heat flows at a rate (per unit area and time) proportional to the rate of change of temperature T with distance and in a direction normal to the constant temperature surfaces, directed from high temperature to low. The constant of proportionality K is a property of the conducting medium known as the *thermal conductivity*. It is evident that we can sum all this up very tersely by writing

$$\mathbf{Q} = -K \operatorname{grad} T = \operatorname{grad}(-KT), \quad \text{if } K \text{ is constant,}$$

in which case $-KT$ is the potential ϕ for the heat flow intensity field $\mathbf{Q}$. (In this book we shall normally adopt the *plus* convention, $\mathbf{v} = +\operatorname{grad}\phi$.) Problem 7.12 studies the case of variable conductivity.

As might be expected, electric conduction behaves similarly: Ohm's law in the absence of extra e.m.f. states that

$$\mathbf{j} = \sigma\mathbf{E} = -\sigma \operatorname{grad} V, \quad \text{if } \sigma = \text{electric conductivity}$$

$$= \operatorname{grad}(-\sigma V), \quad \text{if } \sigma \text{ is constant,}$$

in which case $-\sigma V$ is the potential for the current intensity field $\mathbf{j}$.

Another important flux/gradient law, of great interest to civil and chemical engineers, is Darcy's law for seepage flow through a porous medium, which is a rigid structure of small solid particles between which fluid can seep. Under appropriate conditions this law governs underground flow of oil or water to wells or springs, seepage under or through dams or flow through filter beds. It involves the slightly subtle concept of *effective velocity* $\mathbf{v}^*$, which is a blurred-out version of the actual complicated detail velocity pattern between and around the solid particles of the porous medium. It represents that velocity field which, in the absence of the solids, would produce the same macroscopic transport of fluid at each point. The actual velocities of the fluid, of course, tend to be higher than $\mathbf{v}^*$ because the solids take up so much of the space. Darcy's law like Ohm's law neglects the inertia of the transported entity. It describes an average dynamic equilibrium between the driving force and the drag, which is viscous and therefore linearly proportional to velocity. The flow occurs at very low Reynolds number (referred to particle size).

When there is no flow, the equilibrium condition is still

$$p^* = p + \rho g z = \text{const.},$$

even in the presence of the fixed solid particles. Flow only comes about when there is non-uniformity of p^*. Darcy's law therefore takes the linear form

$$\mathbf{v}^* = -k \operatorname{grad} p^*,$$

where $-\operatorname{grad} p^*$ is the 'driving force' and k, the *hydraulic permeability*, is a characteristic property of the porous medium and the flowing fluid. If gravity is unimportant we have merely

$$v^* = -k \operatorname{grad} p.$$

If k is constant,

$$v^* = \operatorname{grad}(-kp^*)$$

and $-kp^*$ is the potential of the velocity field. In the literature, the variable *head* ($= p^*/\rho g$) is often used instead of p^*.

A further broad class of flux/gradient laws is important in all kinds of engineering. These are *diffusion laws*, which are an equivalent of Darcy's law at the molecular scale. They relate the non-uniformity of the *concentration* of a foreign substance to the rate at which it spreads, or diffuses, by a series of molecular collisions through the host medium. Familiar examples are sugar in unstirred tea or an odour in a still room. Important technological examples include the interpenetration of reacting species in a diffusion flame (i.e. one where the reactants are not pre-mixed), case-hardening (which involves the controlled diffusion of carbon into steel) or semi-conductor electronics (which involves the controlled diffusion of selected impurities called *dopants*). Civil engineers may find themselves concerned with interdiffusion of fresh water with saltwater or sewage-contaminated water. The spread of pollution in the atmosphere also involves diffusion as well as turbulent mixing.

Diffusion occurs because of gradients in the concentration n of the relevant species, which may be defined as the number of molecules per unit volume (or alternatively, of moles per unit volume). The rate of diffusion is described by a vector field **N**, oriented in the direction of transport and in magnitude equal to the number of molecules (or moles) passing per unit area and time. Then either from kinetic theory or experiment we find that diffusion usually follows the linear relation known as *Fick's law*:

$$\mathbf{N} = -D \operatorname{grad} n,$$

in which D is a characteristic property of the diffusing substance and host medium known as the *diffusivity*. If D is uniform $\mathbf{N} = \operatorname{grad}(-Dn)$ and $-Dn$ is the potential for **N**.

One important application of Fick's law is to nuclear fission power technology. In this case the diffusing particles are neutrons, born in fission and on their way to creating further fission, being absorbed parasitically, dying naturally or escaping uselessly from the reactor. If we ignore some of the finer points, such as the distinction between fast and thermal neutrons, Fick's law is a good model of the way in which neutrons show a net drift from regions of high concentration to low, some thereby escaping from the reactor. The value of D is obviously of great significance in determining how large a reactor must be if too many neutrons are not to escape and prevent a sustained chain reaction.

(We avoid the term *neutron flux* which in reactor physics is used to denote something different from flux as interpreted in this book. Reactor physicists also usually define D slightly differently.)

Finally the reader should note that all these vector intensities **Q**, **j**, **v***, **N** stand in the same geometrical relationship to their corresponding surfaces of constant potential; because of the direction of grad, their field lines are everywhere normal to the constant-potential surfaces. (See figure 7.5, for example.) The same is true of **E** with respect to V and of **H** with respect to U.

As an example of the use of flux-gradient laws we next develop an important result in seepage flow theory.

Example 7.6 Figure 7.6 shows the cross-section of a porous dam with vertical sides and thickness L, resting on impermeable ground, separating two pools of depths H and h. Inside the dam the water has a free surface AB at which the water pressure p is zero (if we measure all pressures above atmospheric and neglect surface tension). p is zero also on BC, that part of the downstream discharge face which is above the downstream pool. In each pool the water is virtually stationary and

$$p^* = p + \rho g z$$

is uniform. From Darcy's law show that Q, the total volumetric flow per unit length of dam, is given by the *Dupuit formula*

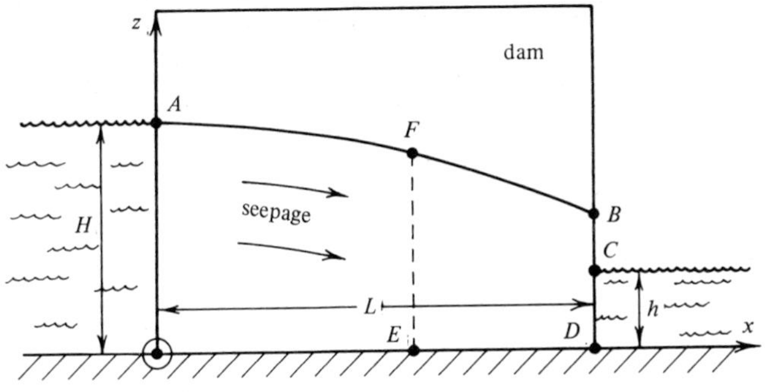

Fig. 7.6. Seepage through a vertical-sided dam.

$$Q = \rho g k(H^2 - h^2)/2L,$$

which relates flow to change in level.

The horizontal component of Darcy's law is

$$v_x^* = -k\, \partial p/\partial x$$

since z is constant in the differentiation. Also,

$$Q = \int v_x^*\, \mathrm{d}z$$

integrated over any vertical section EF. Consider the first equation integrated over the whole domain $ABDO$

$$\int\left\{\int v_x^*\, \mathrm{d}z\right\} \mathrm{d}x = -k\int\left\{\int \frac{\partial p}{\partial x}\, \mathrm{d}x\right\} \mathrm{d}z,$$

in which the two sides use a different order of integration. The left-hand side equals

$$\int Q\, \mathrm{d}x = QL$$

while the right-hand side equals $-k \int \Delta p\, \mathrm{d}z$, in which Δp represents the pressure difference between the two points where a constant-z line intersects the boundary of the domain. Since $p = 0$ along AB and BC, and p^* is constant along OA and CD,

$$-k\int \Delta p\, \mathrm{d}z = k\left\{\int_0^H \rho g(H-z)\mathrm{d}z - \int_0^h \rho g(h-z)\mathrm{d}z\right\} = \tfrac{1}{2}\rho g k(H^2 - h^2).$$

Equating the two sides of the double integral equation gives

$$Q = \rho g k(H^2 - h^2)/2L.$$

7.8 The divergence operator

In order to complete the analysis of any of the above flow or diffusion problems, some account must be taken of the birth, loss or indestructibility of the transported entity, i.e. *conservation statements* are needed. Section 5.7 discussed these in integral form, but we need a *differential* form that can be married with the differential flux/gradient laws.

Consider again the small element of volume shown in figure 7.1, which is now immersed in a vector field such as $\mathbf{Q}$, a heat flow field. The aim is to express

$$d\Phi = \oiint \mathbf{Q} \cdot d\mathbf{a},$$

the net efflux from the small volume, in terms of the non-uniformity (i.e. the gradients) of $\mathbf{Q}$ in the vicinity. Obviously if $\mathbf{Q}$ is uniform there is no net efflux; as much heat enters as leaves. By gradients of $\mathbf{Q}$ we mean such vectors as $\partial\mathbf{Q}/\partial x$, which has components $\partial Q_x/\partial x$, $\partial Q_y/\partial x$, $\partial Q_z/\partial x$. Referred to an origin inside the element, Taylor's series gives

$$\mathbf{Q} = \mathbf{Q}_0 + x\left(\frac{\partial\mathbf{Q}}{\partial x}\right)_0 + y\left(\frac{\partial\mathbf{Q}}{\partial y}\right)_0 + z\left(\frac{\partial\mathbf{Q}}{\partial z}\right)_0,$$

if we retain only first-order terms in the small quantities x, y, z. Suffix 0 denotes values at the origin. Expanding the scalar product $\mathbf{Q} \cdot d\mathbf{a}$ in terms of cartesian components we have

$$\mathbf{Q} \cdot d\mathbf{a} = Q_{x0}\, da_x + x\left(\frac{\partial Q_x}{\partial x}\right)_0 da_x + y\left(\frac{\partial Q_x}{\partial y}\right)_0 da_x + \text{etc.}$$

The reader may find it helpful to write out all 12 terms on the right-hand side. When the contributions to $\oiint \mathbf{Q} \cdot d\mathbf{a}$ are totalled, terms like $Q_{x0} da_x$ do not survive because contributions from A and B in figure 7.1 with equal and opposite da_x cancel. Nor do terms like

$$y(\partial Q_x/\partial_y)_0\, da_x$$

survive, for $y_A = y_B$. On the other hand, from such terms as

$$x(\partial Q_x/\partial x)_0\, da_x$$

come paired contributions at A and B equal jointly to

$$(x_A - x_B)(\partial Q_x/\partial x)_0\, da_x, \quad \text{i.e.} \quad dV(\partial Q_x/\partial x)_0,$$

where dV is the volume of tube AB, if higher-order small quantities are ignored. We conclude that, for the whole volume,

$$d\Phi = \oiint \mathbf{Q} \cdot d\mathbf{a} = \left(\frac{\partial Q_x}{\partial x} + \frac{\partial Q_y}{\partial y} + \frac{\partial Q_z}{\partial z}\right) \times (\text{volume of element}),$$

i.e. $$\textit{net efflux per unit volume} = \frac{\partial Q_x}{\partial x} + \frac{\partial Q_y}{\partial y} + \frac{\partial Q_z}{\partial z},$$

evaluated at the element. The suffix 0 has now been dropped. This simple, scalar, symmetrical, differential quantity is a new way of gauging the non-uniformity of a vector field. It is a scalar field deduced by differentiation of a vector field. The idea is of course applicable to any vector field **v**, not merely **Q**. The quantity

$$(\partial v_x/\partial x + \partial v_y/\partial y + \partial v_z/\partial z)$$

is universally called the *divergence* of **v**, or div **v** for short, but the name *source density* would be more evocative and accurately descriptive.

It is important that the reader should grasp as swiftly as possible the idea that div measures the excess of flux-out over flux-in for any small element, expressed per unit volume. Despite its name, divergence has nothing to do with the question whether the field lines are diverging or converging. Problem 7.13 is a less general version of the derivation of div.

The 'del' notation can be extended to apply to div, by putting

$$\text{div}\,\mathbf{v} \equiv \nabla \cdot \mathbf{v},$$

if we generalise the idea of scalar multiplication in the obvious way. The operator div has the property

$$\text{div}\,(\mathbf{a}+\mathbf{b}) = \text{div}\,\mathbf{a} + \text{div}\,\mathbf{b}.$$

In particular this means that we can 'take the divergence of' an equation such as $\mathbf{a}+\mathbf{b}=\mathbf{0}$ to get div **a** + div **b** = 0. From the facts that the direction of axes in figure 7.1 is arbitrary and that the translation of axes does not affect terms like $\partial Q_x/\partial x$, it follows that div **v** is invariant (independent of axes).

It is rather easy for the beginner to mix div and grad up, as each consists of three superficially similar terms when expressed in cartesians. We repeat, therefore, that div is a *scalar*, describing the non-uniformity of a *vector* field whereas grad is a *vector*, describing the non-uniformity of a *scalar* field. grad expressed in cartesians involves **i**, **j**, **k** but div does not. Each term in div is the derivative of a different quantity – contrast grad!

For a solenoidal vector field the net efflux from any volume vanishes and so div **v** = 0. The converse is also true; if div **v** = 0 then $\oiint \mathbf{v}\cdot d\mathbf{a} = 0$ for any volume, and **v** is solenoidal. This follows from Gauss's theorem, which appears shortly. Problem 7.14 and example 7.7 take up these points.

Example 7.7 Does the field in example 7.5 have the property that 'closeness' of field lines indicates intensity?

The question is, essentially: 'Is **v** solenoidal?' (see section 3.16), i.e. does div **v** vanish? Now

$$\text{div}\,\mathbf{v} = \frac{\partial v_x}{\partial x}+\frac{\partial v_y}{\partial y} = \frac{1}{(x^2+y^2)^{1/2}} - \frac{x^2}{(x^2+y^2)^{3/2}} + \frac{1}{(x^2+y^2)^{1/2}} - \frac{y^2}{(x^2+y^2)^{3/2}} = \frac{1}{(x^2+y^2)^{1/2}} \neq 0.$$

The field intensity is *not* proportional to the 'closeness' of the field lines.

7.9 Flux conservation statements

The properties of the div operator tell us that in electromagnetism div $\mathbf{B} = 0$ always but div $\mathbf{E} \neq 0$ in general. In fact, for a small volume element, Gauss's law states that

$$q \times (\text{volume}) = \epsilon_0 \oiint \mathbf{E} \cdot \mathbf{da} \qquad \text{(if } q \text{ is net charge density)}$$

$$= \epsilon_0 \,\text{div}\, \mathbf{E} \times (\text{volume})$$

in view of the properties of div. We conclude that

$$\epsilon_0 \,\text{div}\, \mathbf{E} = q,$$

which is the differential form of Gauss's law.

For vector fields which represent the actual transport of some physical entity, the mathematical equipment is now available for expressing conservation in differential form. For example, we can translate the statement:

'net efflux of charge from a small volume element, per unit volume
= rate of depletion of net charge inside, per unit volume',

which expresses the indestructibility of charge, into mathematical form as follows:

$$\text{div}\, \mathbf{j} = -\partial q/\partial t.$$

Problem 7.15 applies this result. Kirchhoff's first law describes the case where $\partial q/\partial t$ is negligible or vanishes and then $\mathbf{j}$ is solenoidal, with div $\mathbf{j} = 0$.

As an exercise, before reading on, the reader should similarly write down a statement expressing the indestructibility of mass in terms of $\mathbf{v}$ the fluid velocity (or *volumetric* flux intensity) and ρ the mass density. The answer appears later.

Heat flow is a more involved case because heat is not indestructible but can be converted from and to other forms of energy. An example follows.

Example 7.8 Formulate a conservation statement for a medium of volumetric heat capacity C in the presence of unsteady heat flow.

Here the state of the medium is changing in time and heat is being converted into stored energy E *per unit volume* (or enthalpy if the pressure is constant) in the conducting medium. Now obviously

$$\text{div}\, \mathbf{Q} = -\partial E/\partial t,$$

in which the minus sign reminds us that div measures net *efflux* per unit volume. Of course $\partial E/\partial t$ may be negative, as when a body is cooling down and div $\mathbf{Q}$, the heat release rate, is positive. But $E = CT$, where T is temperature (not necessarily absolute), C being the *volumetric* heat capacity. So

$$\text{div}\,\mathbf{Q} = -C\partial T/\partial t.$$

If C' is 'specific heat' (heat capacity per unit *mass*) then $C = \rho C'$.

In contrast, heat may be being released and be flowing away for reasons other than changing temperature. These include chemical or nuclear reaction, ohmic (Joule) heating or the thermoelectric Thomson effect. The problems of cooling large volumes of setting concrete or fission waste products are cases in point. Microwave heating will cook a loaf from inside. In such cases, information is available to specify R, the heat release rate per unit volume and time.

Then

$$\text{div}\,\mathbf{Q} = R.$$

Students sometimes have trouble in distinguishing $\mathbf{Q}$ from R; R is a scalar field measured in watts per cubic metre, $\mathbf{Q}$ is a vector field measured in watts per square metre. Problem 7.16 exercises these ideas.

7.10 An important vector identity

The operators div and grad come together unexpectedly if we consider the significance of an expression of the form div ($s\mathbf{v}$), where s is a scalar and $\mathbf{v}$ is a vector. In cartesian terms,

$$\text{div}\,s\mathbf{v} = \sum \frac{\partial}{\partial x}(sv_x) = s\sum \frac{\partial v_x}{\partial x} + \sum v_x \frac{\partial s}{\partial x}$$

and so

$$\text{div}\,s\mathbf{v} \equiv s\,\text{div}\,\mathbf{v} + \mathbf{v}\cdot\text{grad}\,s,$$

an identity with a remarkably wide physical repertoire. The second term deserves special comment, for it equals the magnitude of $\mathbf{v}$ times the component of grad s in $\mathbf{v}$'s direction, i.e. the rate of change of s with distance in that direction. If $\mathbf{v}\cdot$ grad s vanishes, this means that grad s has no component in the direction of $\mathbf{v}$ and that s is constant along the field lines of the field $\mathbf{v}$, i.e. they lie in surfaces of constant s. The operator ($\mathbf{v}\cdot$ grad) or $\Sigma v_x(\partial/\partial x)$, applied to a scalar, can be thought of as an operator which extracts the scalar's rate of change with distance in the direction of a vector $\mathbf{v}$ (times the magnitude of v), just as ($\mathbf{dr}\cdot$ grad) extracts the change incurred in going along $\mathbf{dr}$.

One application of the identity is in fluid mechanics. The mass-conservation equation, expressed per unit volume, is

$$\text{div}\,\rho\mathbf{v} = -\partial\rho/\partial t,$$

in which ρ = mass density, $\mathbf{v}$ = velocity and $\rho\mathbf{v}$ = mass flow intensity. (The reader should check whether he generated this correctly for himself at an earlier stage.) This equation is widely known by the rather vacuous name of '*equation of continuity*'. The left-hand side can be explained by means of the identity so that the equation becomes

$$\frac{\partial \rho}{\partial t} + (\mathbf{v} \cdot \text{grad})\rho + \rho \, \text{div} \, \mathbf{v} = 0,$$

an equation which we shall discuss further later. If $\rho = \text{const.}$, we merely have $\text{div} \, \mathbf{v} = 0$. Problem 7.17(*c*) explores another application to fluid mechanics, while problem 7.17(*b*) relates the identity to electric conduction, and the example below relates it to heat conduction.

Example 7.9 Heat conduction is an irreversible process. Express and interpret the resulting creation of entropy in terms of a differential statement, using the concept of entropy flow.

This is a more sophisticated application of the identity. Whenever there is heat flow **Q** present, the entropy produced by irreversibility in one place may reveal itself as the change in state (and in entropy) of matter in some other place.

To account for this quantitatively, the concept of entropy flow **S**, which equals $\mathbf{Q}/T$, has to be introduced. Then

$$\text{div} \, \mathbf{Q} = \text{div} \, T\mathbf{S} = T \, \text{div} \, \mathbf{S} + \mathbf{S} \cdot \text{grad} \, T$$

and

$$\text{div} \, \mathbf{S} + (-\text{div} \, \mathbf{Q})/T = -(\mathbf{Q} \cdot \text{grad} \, T)/T^2,$$

an equation with profound physical significance. First consider the right-hand side. Heat flows from hot to cold. This is true even in anisotropic conductors in which **Q** need not be parallel to $-\text{grad} \, T$. A precise way of saying that heat flows from hot to cold is: $\mathbf{Q} \cdot \text{grad} \, T$ is always negative. It equals the magnitude of **Q** times the rate of change of T with distance in the **Q** direction, which must be negative. Thus the right-hand side in the equation is always positive.

On the left-hand side, the first term is the net efflux of entropy from a volume element (expressed per unit volume). This is entropy on its way to appear in the change of state of matter somewhere else, outside the element. As regards the second term on the left-hand side, $-\text{div} \, \mathbf{Q}$ is the net rate at which heat is deposited inside the element (per unit volume) and so $(-\text{div} \, \mathbf{Q})/T$ is the rate of increase of entropy of the contents of the element itself (per unit volume). The equation therefore states a version of the *Principle of Increase of Entropy* in the form

(Entropy appearing elsewhere) + (Entropy appearing locally) = positive.

7.11 Gauss's theorem

Conservation statements may be written in integral or differential form. In the case of charge conservation we have equivalent forms

$$\oiint \mathbf{j} \cdot d\mathbf{a} = -\iiint \frac{\partial q}{\partial t} \, d\tau \text{ (from section 5.7)}$$

and

$$\operatorname{div} \mathbf{j} = -\frac{\partial q}{\partial t} \text{ (from section 7.9),}$$

and we can deduce that

$$\oiint \mathbf{j} \cdot d\mathbf{a} = \iiint \operatorname{div} \mathbf{j} \, d\tau.$$

This strongly suggests that there might be a general mathematical result that

$$\oiint_S \mathbf{v} \cdot d\mathbf{a} = \iiint_V \operatorname{div} \mathbf{v} \, d\tau,$$

where **v** is any vector for which div **v** is defined throughout the volume V enclosed by the closed surface S. There is indeed such a result, known as *Gauss's theorem* or the *divergence theorem*. Its proof is easy.

Imagine the volume V broken down into small elements $d\tau$ of volume of any shape (provided they include all of V). Note the crucial fact that the surfaces of these elements are mostly composed of common interfaces between adjoining elements, but each element which is at the periphery of V will have as part of its surface a part of the surface S enclosing V. Such parts will *not* be an interface between elements.

Now

$$d\Phi = \operatorname{div} \mathbf{v} \, d\tau$$

is the net efflux of the vector **v** from the element $d\tau$. But at a common interface between elements, flux *out of* one is flux *into* its neighbour. Therefore when we sum all such terms to get

$$\Phi = \iiint \operatorname{div} \mathbf{v} \, d\tau,$$

all the contributions from these common interfaces will cancel; only the contributions from the parts of the outer surface S will survive. These contributions, added together, yield $\oiint \mathbf{v} \cdot d\mathbf{a}$ taken over S, and Gauss's theorem is proved. It is a scalar statement about a vector field.

Put into words as 'the net efflux from a finite volume equals the net efflux from all the elementary volumes comprising that volume', Gauss's theorem seems almost tautologous. It relates the net flux out of a closed surface to the total 'source strength' inside.

Gauss's theorem is pure mathematics and should not be confused with Gauss's law, which is physics. The two are related, however, because the two forms of Gauss's law, integral and differential,

$$\epsilon_0 \oiint \mathbf{E} \cdot d\mathbf{a} = \iiint q \, d\tau \quad \text{and} \quad \epsilon_0 \operatorname{div} \mathbf{E} = q$$

are seen to be equivalent from Gauss's theorem.

Gauss's theorem confirms the earlier assertion that **v** is solenoidal if div **v** = 0.

Example 7.10 Show that, for any surface S enclosing a volume V,

$$\oint\!\!\oint_S \mathbf{r} \cdot \mathrm{d}\mathbf{a} = 3V.$$

Gauss's theorem applied to the vector **r** is

$$\oint\!\!\oint_S \mathbf{r} \cdot \mathrm{d}\mathbf{a} = \iiint \operatorname{div} \mathbf{r}\, \mathrm{d}\tau$$

and

$$\operatorname{div} \mathbf{r} = \partial x/\partial x + \partial y/\partial y + \partial z/\partial z = 3.$$

Thus

$$\oint\!\!\oint_S \mathbf{r} \cdot \mathrm{d}\mathbf{a} = 3 \iiint \mathrm{d}\tau = 3V.$$

It is also possible to demonstrate this result another way. Figure 7.7 shows the portion d**a** of the surface S and the perpendicular of length p from the origin to the tangent plane at the location of d**a**.

$$\mathbf{r} \cdot \mathrm{d}\mathbf{a} = r \cos \theta\, \mathrm{d}a = p\mathrm{d}a = 3\, \mathrm{d}V,$$

where $\mathrm{d}V$ is the volume of the elementary pyramid which has $\mathrm{d}a$ as its base and p as its height. If we integrate over the whole surface

$$\oint\!\!\oint_S \mathbf{r} \cdot \mathrm{d}\mathbf{a} = 3V, \quad \text{again.}$$

Note that if the origin is outside the volume, as in the case shown, some of the contributions to the integrals will be negative.

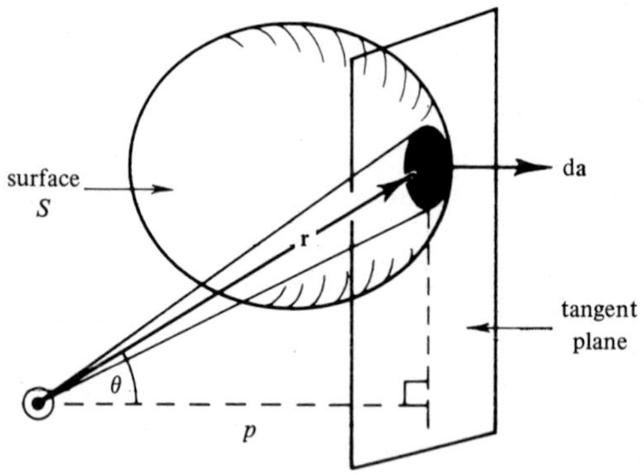

Fig. 7.7

Instead of $\mathbf{r}$, consider the inverse-square law field $\mathbf{v} = \mathbf{r}/r^3$ which we know to be solenoidal and therefore with zero divergence. For this field Gauss's theorem would indicate that $\oiint_S \mathbf{v} \cdot \mathbf{da} = 0$. But this integral equals 4π whenever S encloses the origin, the source of the field, seen to be of total strength 4π. At the origin div $\mathbf{v}$ increases without limit because a finite source strength is packed into a vanishingly small volume. This example illustrates the fact that care is needed in using Gauss's theorem in the presence of singularities.

Problem 7.18 provides further experience with Gauss's theorem.

7.12 A corollary to Gauss's theorem; the gradient theorem

An important general corollary emerges if Gauss's theorem is applied to a vector field $\mathbf{I}s$, in which s is a non-uniform scalar and $\mathbf{I}$ is a uniform unit vector field in some direction. Since div $\mathbf{I} = 0$ we have

$$\text{div}\,\mathbf{I}s = \mathbf{I} \cdot \text{grad}\, s$$

and Gauss's theorem gives

$$\oiint \mathbf{I}s \cdot \mathbf{da} \text{ (i.e. } \oiint \mathbf{I} \cdot s\mathbf{d}a) = \iiint \mathbf{I} \cdot \text{grad}\, s\, \mathrm{d}\tau.$$

Since

$$\mathbf{I} \cdot \mathbf{a} + \mathbf{I} \cdot \mathbf{b} + \text{etc.} = \mathbf{I} \cdot (\mathbf{a} + \mathbf{b} + \text{etc.}),$$

and integration is only a form of addition, we can take out the common factors $\mathbf{I}$ to get

$$\mathbf{I} \cdot \oiint s\mathbf{da} \equiv \mathbf{I} \cdot \iiint \text{grad}\, s\, \mathrm{d}\tau.$$

By taking $\mathbf{I}$ in turn in the x, y and z-directions, we may deduce the equality of the two vector quantities:

$$\oiint s\mathbf{da} = \iiint \text{grad}\, s\, \mathrm{d}\tau.$$

This is the *gradient theorem*, a corollary of Gauss's theorem. It is a vector statement about a scalar field, and should not be confused with Gauss's theorem.

An obvious application is the case where s is the fluid pressure p. Then the corollary merely states that the net pressure force $-\oiint p\mathbf{da}$ on a finite volume of fluid equals the sum of the net pressure forces ($-$grad $p\, \mathrm{d}\tau$) on all the individual elements of the volume. This is self-evident because the equal and opposite forces on the interfaces between adjoining elements cancel mutually, just as the fluxes at these interfaces cancel in the proof of Gauss's theorem.

Problem 7.19 relates the corollary to the case of uniform pressure and to the idea of vector area discussed in section 3.11.

Example 7.11 Show that for a uniform body of volume V the position vector $\bar{\mathbf{r}}$ of the centre of mass relative to some origin is given by the integral

$$\oiint_S r^2\,\mathrm{d}\mathbf{a}/2V$$

evaluated over the body's *surface S*.

We shall need the result grad $r^2 = 2\mathbf{r}$ (for $r^2 = x^2 + y^2 + z^2$). Then Gauss's corollary gives

$$\oiint_S r^2\,\mathrm{d}\mathbf{a} = \iiint (\text{grad } r^2)\mathrm{d}\tau = 2\iiint \mathbf{r}\,\mathrm{d}\tau.$$

But the centre of mass is identified by the relation

$$\bar{\mathbf{r}}V = \iiint \mathbf{r}\,\mathrm{d}\tau \quad \text{(cf. problem 5.4)},$$

and the required result follows.

Problems 7

7.1 Consider a small rectangular parallelepiped with edges $\mathrm{d}x$, $\mathrm{d}y$, $\mathrm{d}z$ immersed in a fluid and show that the net pressure force on it per unit volume equals

$$-\left(\mathbf{i}\frac{\partial p}{\partial x} + \mathbf{j}\frac{\partial p}{\partial y} + \mathbf{k}\frac{\partial p}{\partial z}\right).$$

7.2 (*a*) What is the condition for the planes $ax + by + cz = A$ and $lx + my + nz = B$ to be perpendicular?

How can the equation of the parallel planes $ax + by + cz = A$ (a variable parameter) be put into the form $\Sigma x \cos\alpha = C$, where $\cos\alpha$ etc. are the direction cosines of the planes (i.e. the components of a perpendicular unit vector)? What is the magnitude of grad C? Deduce a method for finding the perpendicular distance of the origin from any plane $ax + by + cz = A$.

(*b*) Show that the tangent plane to the surface $xy = z^2$ at the point $(2, 2, 2)$ passes through the origin.

(*c*) Show that grad $(r^n) = nr^{n-2}\mathbf{r}$, where $r = |\mathbf{r}|$, $\mathbf{r}$ being the position vector. What scalars ϕ have the gradients:

(i) grad $\phi = \mathbf{r}/r^3$ (inverse-square law); (ii) grad $\phi = \mathbf{r}/r^2$?

(*d*) Show that grad $(\mathbf{p}\cdot\mathbf{r}) = \mathbf{p}$ if $\mathbf{p}$ is a constant vector.

7.3 (*a*) Two scalar functions of position, p and q, are such that $p = f(q)$. Show that grad $p = (\mathrm{d}p/\mathrm{d}q)$ grad q. Also show that grad $p \times$ grad $q = \mathbf{0}$, (i) from the previous result, (ii) using cartesian components, and (iii) by considering the constant-p and constant-q surfaces in relation to grad p and grad q.

(*b*) If $p = \tan^{-1}x + \tan^{-1}y$ and $q = (x + y)/(1 - xy)$, evaluate grad $p \times$ grad q. Can you explain your result?

7.4 In thermodynamics, $T\mathrm{d}s = \mathrm{d}u + p\mathrm{d}(1/\rho)$, all variables being scalars. Deduce that
(*a*) T grad s = grad $u + p$ grad $(1/\rho)$ and
(*b*) that

$$T\frac{\mathrm{D}s}{\mathrm{D}t} = \frac{\mathrm{D}u}{\mathrm{D}t} + p\frac{\mathrm{D}}{\mathrm{D}t}\left(\frac{1}{\rho}\right),$$

if $\mathrm{D}/\mathrm{D}t$ refers to rate of change in time as seen by a travelling fluid element.

7.5 A speck of cream of volume ϵ is held at rest relative to the milk in a centrifuge at a radius where the centripetal acceleration is a. In terms of the densities of the two fluids (ρ_c and ρ_m), a and ϵ, find the radial pressure gradient in the milk and the force necessary to hold the speck in position. What is its direction? Neglect gravity.

7.6 Repeat problem 3.17 and example 3.6 by replacing the immersed body with water and considering the equilibrium of the water under its own weight and the pressures that would still prevail.

7.7 A solid spherical planet is surrounded by an atmosphere in which $p \propto \rho$. Using the inverse-square law for gravity, show that the pressure of the atmosphere remains finite at great distances.

7.8 A linear cylindrical pinch (see section 7.5) carries a total axial current J in gas of uniform temperature T, for which $p = \rho RT$ (R = gas constant, p = pressure, ρ = density), and is surrounded by vacuum. Show that the mass of gas per unit length of pinch equals $\mu_0 J^2/8\pi RT$ in the cases: (*a*) current uniformly distributed and *(*b*) arbitrary, axisymmetric distribution of current intensity j. (*Hint*: Integrate for the mass by parts so as to introduce $\mathrm{d}p/\mathrm{d}r$ and use the equation $\mathrm{d}(Br)/\mathrm{d}r = \mu_0 jr$, deduced by applying Ampère's law to the loop shown in figure 7.8, noting that the contributions to the line integral from XX (traversed twice, once in each direction) cancel out.)

7.9 A fluid is in hydrostatic equilibrium under a body force $\mathbf{F}$ per unit volume. Show that the vector field $\mathbf{F}$ must be conservative.

7.10 (*a*) The potential V of an electric dipole of strength p at the origin is given by $4\pi\epsilon_0 V = \mathbf{p}\cdot\mathbf{r}/r^3$. (See problem 3.10(*d*).) Exploiting problems 7.2(*c*) and (*d*), show that the electric field is given by

$$4\pi\epsilon_0\mathbf{E} = 3(\mathbf{p}\cdot\mathbf{r})\mathbf{r}/r^5 - \mathbf{p}/r^3.$$

(*b*) Show similarly that the magnetic field $\mathbf{B}$ of a magnetic loop dipole $J\mathrm{d}\mathbf{a}$ at the origin is given by

$$4\pi\mathbf{B}/\mu_0 J = 3(\mathrm{d}\mathbf{a}\cdot\mathbf{r})\mathbf{r}/r^5 - \mathrm{d}\mathbf{a}/r^3.$$

(Refer to problem 4.3.)

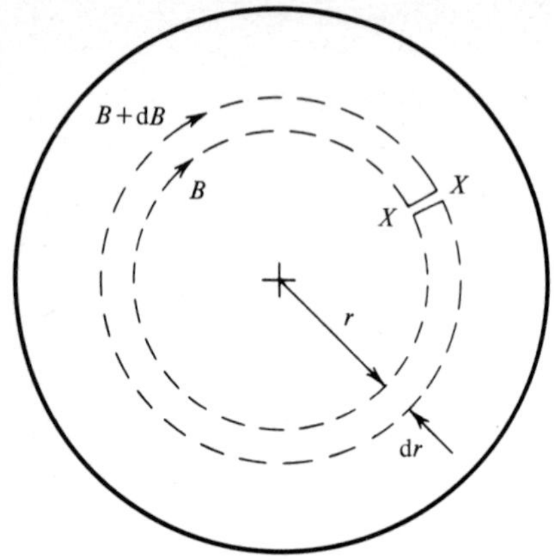

Fig. 7.8. Cross-section of linear pinch.

7.11 A non-magnetic fluid of uniform conductivity σ is subjected to an electric field, given by a potential distribution $V = A(x^2 - y^2)$ and to a magnetic field given by a potential distribution $U = Cxy$, A and C being constants. (The weak, non-conservative magnetic field due to the currents in the fluid may be ignored.) Show that the resulting $\mathbf{j} \times \mathbf{B}$ force on the fluid is in the z-direction and equals $2\mu_0\sigma ACr^2$, where r is distance from the z-axis. (Such an axisymmetric force distribution can be used to produce or modify flow in a pipe.) Sketch the electric and magnetic field lines.

7.12 Show that the potential ϕ for heat flow in a conductor in which the thermal conductivity K is a function of temperature T is given by $\phi = -\int K \mathrm{d}T$.

7.13 Consider a small rectangular parallelepiped with edges $\mathrm{d}x$, $\mathrm{d}y$, $\mathrm{d}z$ and express the net efflux of a vector field $\mathbf{Q}$ from it in terms of Q_x, etc. and their gradients. Verify that for this element the net efflux per unit volume equals div $\mathbf{Q}$.

7.14 (*a*) Would it be proper to refer to 'the flux-linked by a loop' in relation to the vector field

$$\mathbf{v} = x^2\mathbf{i} + xy\mathbf{j} - 3xz\mathbf{k}? \quad \text{(See section 3.17.)}$$

(*b*) Repeat example 7.7 with the potential given by

$$\phi = \{x + (x^2 + y^2)^{1/2}\}^{1/2}.$$

7.15 A non-polarisable medium has a *uniform* electric resistivity τ. Show that any net charge decays in the ratio e:1 in a time $\tau\epsilon_0$. Evaluate this for a metal of resistivity $\tau = 10^{-7}$ ohm-metre. Note that *surface* charges can persist, however. (See also problem 7.17(*b*).) ($\epsilon_0 = 8.854 \times 10^{-12}$ and e = 2.718 28, *not* electronic charge.)

7.16 (*a*) Heat is released uniformly at a rate R per unit volume and time and flows out radially in a uniform spherical body in which temperatures are steady. By expressing energy conservation for a sphere of radius r, show that the heat flow $\mathbf{Q}$ is given by

$$\mathbf{Q} = \tfrac{1}{3}R\mathbf{r}$$

and check that div $\mathbf{Q}$ takes the correct value. (Compare example 5.4.)
(*b*) A long prismatic bar of uniform thermal conductivity K has a cross-section defined by the lines $y = 0, x = y, x = 1$. Ohmic heating produces the temperature distribution

$$T = y(x-y)(1-x).$$

Find the distribution of heat flow $\mathbf{Q}$ and heat release rate R and the total heat release rate per unit length of bar. From the components of $\mathbf{Q}$ normal to the surface calculate the total heat escape rate per unit length, for comparison.

The electric field and current intensity are axial; which is uniform? How must the electric conductivity σ vary with position?

7.17 (*a*) Show that div $(r^n\mathbf{r}) = (3+n)r^n$. (See problem 7.2(*c*).) For what value of n is it zero $(r \neq 0)$?
(*b*) Electric current flows steadily in a *non-uniform* conductor. Relate the net charge density q in the conductor to the rate of change of resistivity τ with distance measured in the current flow direction. (Contrast problem 7.15.)
(*c*) A trawler net with a rigid rim is being drawn through incompressible sea-water, containing n fish per unit volume. In a frame of reference travelling with the rim, the net has a steady shape and the velocity field of the water is $\mathbf{v}$. Write down an integral expression for the rate at which fish impinge on the net and show that the condition for this rate to be independent of the shape of the net, at a given instant, for a given velocity field, is

$$\mathbf{v} \cdot \text{grad } n = 0.$$

Interpret this result in relation to the streamlines. Discuss whether the concept of flux-linkage is relevant.

7.18 The following are exercises on Gauss's divergence theorem;
(*a*) Repeat problem 3.20(*b*).
(*b*) A vector field $\mathbf{v}$ has potential

$$\phi = x^2(1+z) + y^2(1-z).$$

Find its flux out of a unit cube with three edges on the forwards cartesian axes.
(*c*) By considering the vector field $\mathbf{v}$ with components $(P, Q, 0)$ show that

$$\oint_L P\mathrm{d}y - Q\,\mathrm{d}x = \iint_S (\partial P/\partial x + \partial Q/\partial y)\mathrm{d}x\,\mathrm{d}y,$$

where P and Q are any differentiable functions of x and y and S is the area enclosed by a loop L in the x, y plane.

If $\mathbf{v}$ is solenoidal show that a vector field $\mathbf{u}$, with components $(-Q, P, 0)$ is conservative, with a potential ψ, say, such that

$$\mathrm{d}\psi = P\mathrm{d}y - Q\mathrm{d}x.$$

7.19 The following are exercises on the corollary to Gauss's theorem (the gradient theorem):
(*a*) Show that the total force due to a uniform pressure acting on any surface spanning a given closed loop is the same.
(*b*) Prove that the vector area of the loop $\iint \mathrm{d}\mathbf{a}$ (evaluated over any such surface) is unique, so confirming the results of section 3.11.
(*c*) The attractive force between two masses m_1 and m_2, a distance r apart, equals Gm_1m_2/r^2 (G = const.). Show that $\mathbf{g}$, the gravity force on unit mass lying near a large gravitating body of uniform density ρ and arbitrary shape enclosed within a surface S, is given by

$$\mathbf{g} = -G\rho \oiint_S \frac{\mathrm{d}\mathbf{a}}{r},$$

r being distance from the unit mass to the area element $\mathrm{d}\mathbf{a}$. (Refer to problem 7.2(*c*)(i).)

8

Gradient and divergence combined

8.1 The laplacian operator

We have seen that there are many physical vector fields whose properties are best stated in terms of div and grad. In such situations

$$\mathbf{v} = \text{grad}\,\phi$$

and

$$\text{div}\,\mathbf{v} = R,$$

R being some kind of source density about which information is available. There are essentially four simultaneous equations here (because the first, vector equation has three components) and four unknowns (ϕ and the three components of $\mathbf{v}$). By taking the divergence of the first equation we deduce a single (scalar) equation in the one unknown ϕ:

$$\text{div}\,(\text{grad}\,\phi) = R.$$

It is a great simplification to be able to replace four equations by one in this way. In 'del' notation div grad becomes $\nabla \cdot \nabla$ and this is conventionally written as ∇^2, which is spoken as '*del-squared*'. It is also known as the *laplacian operator*. Written out in cartesians

$$\nabla^2\phi = \text{div grad}\,\phi = \sum \frac{\partial}{\partial x}\left(\frac{\partial \phi}{\partial x}\right) = \frac{\partial^2\phi}{\partial x^2} + \frac{\partial^2\phi}{\partial y^2} + \frac{\partial^2\phi}{\partial z^2}.$$

This scalar quantity is invariant to choice of axes. The operator ∇^2 has various properties such as

$$\nabla^2(\phi + \psi) = \nabla^2\phi + \nabla^2\psi.$$

(Contrast the example below.) Problems 8.1 and 8.2 give practice with ∇^2.

Example 8.1 Show that $\nabla^2(\phi\psi) \neq \phi\nabla^2\psi + \psi\nabla^2\phi$ and find the difference between them.

$$\text{grad}\,(\phi\psi) = \phi\,\text{grad}\,\psi + \psi\,\text{grad}\,\phi$$

and

$$\nabla^2(\phi\psi) = \text{div grad}\,(\phi\psi) = \text{div}\,(\phi\,\text{grad}\,\psi) + \text{div}\,(\psi\,\text{grad}\,\phi)$$

$$= \phi\nabla^2\psi + 2\,\text{grad}\,\phi \cdot \text{grad}\,\psi + \psi\nabla^2\phi$$

(from the identity $\text{div}\,s\mathbf{v} \equiv s\,\text{div}\,\mathbf{v} + \mathbf{v} \cdot \text{grad}\,s$).

The required difference is therefore

$$2 \text{ grad } \phi \cdot \text{grad } \psi.$$

The equation

$$\text{div grad } \phi \equiv \nabla^2\phi = R,$$

together with appropriate boundary conditions, forms the basis for solving a very wide range of field problems, and a great variety of mathematical, numerical and analogue methods have been developed for extracting solutions. There is space in this book only for hinting at some of the possibilities, but first we must bring in one other mathematical idea, useful in two-dimensional problems.

8.2 Stream function for two-dimensional, solenoidal fields

When a vector field is conservative and therefore has a scalar potential ϕ, we have already seen that expressing the problem in terms of ϕ can produce convenient simplification. If a vector field is *two-dimensional and solenoidal*, a comparable simplification is possible with the aid of a new scalar field ψ, called the *stream function*. As its name suggests, ψ is useful in fluid mechanics, but it is equally applicable in other subjects.

Figure 8.1 represents an x, y plane in a two-dimensional solenoidal vector field $\mathbf{v}$ which does not vary in the z-direction. As usual we confine attention to a region of the field lying between two planes z = constant, unit distance apart. Then when we speak of the flux 'across a line DB', say, we mean the flux across the surface generated by moving DB unit distance in the z-direction between these two planes.

Choose a particular point D as a fixed datum, and define ψ at a point A as the flux of $\mathbf{v}$ across the line L_1 joining D to A, the point (x, y). Because $\mathbf{v}$ is solenoidal,

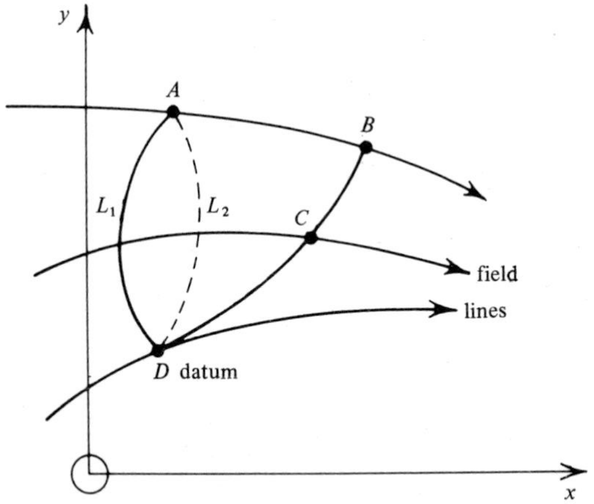

Fig. 8.1. A two-dimensional vector field.

its flux across the line L_2 (also joining D to A) is the same, i.e. ψ depends only on A and not on the route DA and so $\psi = \psi(x, y)$, a function of the position of A.

Note that this argument does not require $v_z = 0$, although this is often the case. If $v_z \neq 0$, the flux for the closed surface defined by L_1 and L_2 includes contributions from its flat ends, lying in the x, y planes between L_1 and L_2, but these cancel in a two-dimensional problem because v_z does not depend on z. Another complication occurs when the otherwise solenoidal field includes a source (or sink), from (or to) which there is a net flux. Then ψ is multi-valued because the curve DA can go round the source any number of times. This ambiguity is normally resolved by introducing a 'cut' across which ψ is discontinuous, in a manner analogous to that used for multi-valued potentials in section 4.3. As the problem is two-dimensional any source is a *line source* (extending in the z-direction) not a point source.

If B is on the same $\mathbf{v}$ field line as A, the fluxes across DA and DB are the same, for $\mathbf{v}$ is solenoidal and there is no flux across AB. Hence $\psi_A = \psi_B$ and we have the result that along any field line $\psi(x, y) = \text{const}$. This is one of the most valuable properties of the stream function concept. Also observe that

$$\psi_B = (\text{flux across } DB) = (\text{flux across } DC)(\psi_C) + (\text{flux across } CB)$$

and so

$$(\text{flux across } CB) = \psi_B - \psi_C.$$

For two field lines close together, characterised by values ψ and $\psi + d\psi$,

$$d\psi = \text{flux passing between them.}$$

Finally we reiterate that ψ always refers to flux *per unit thickness in the z-direction.*

It is instructive to extend the properties of ψ in a tabular format, setting them alongside the comparable results for the potential ϕ in the hope that any confusion between them will be reduced if the properties of ϕ and ψ are seen together and consciously contrasted.

Solenoidal vector field (two-dimensional)	*Conservative vector field (two- or three-dimensional)*
$\psi(x, y)$ exists.	$\phi(x, y, z)$ exists,
Figure 8.2(*a*) shows two closely adjoining field lines between which the flux (per unit z-wise distance) equals $d\psi$. If n is distance measured normal to field lines,	with $\mathbf{v} = +\text{grad}\,\phi$, say. A consequence of this, if s is distance measured along field lines (see figure 8.2(*b*)), is that
$d\psi = v\,dn$ and $v = \partial\psi/\partial n$.	$v = \partial\phi/\partial s$.

These contrasting results have a common aspect, namely, v is larger where field lines or equipotentials are closer together.

Figure 8.2(*c*) shows a line element $d\mathbf{r}$ (of length dl) along which ψ changes to $\psi + d\psi$. To find $d\psi$ we require v_n, the velocity component normal to $d\mathbf{r}$. It may be found by resolving v_x and v_y:

$$v_n = v_x \sin\theta - v_y \cos\theta.$$

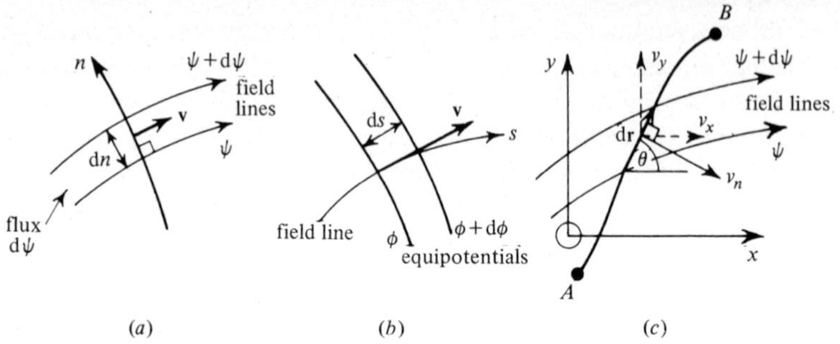

Fig. 8.2

Then

$$d\psi = v_n dl = v_x dl \sin\theta - v_y\, dl \cos\theta$$

or

$$d\psi = v_x dy - v_y dx.$$

We also have

$$d\phi = \mathbf{v} \cdot d\mathbf{r}$$

or

$$d\phi = v_x dx + v_y dy (+\, v_z dz).$$

In these contrasting results, dx, dy (and dz if three-dimensional) are the components of $d\mathbf{r}$. These differential relations can be integrated along the curve AB (figure 8.2(c)):

$$\psi_B - \psi_A = \text{flux across } AB = \int_A^B (v_x dy - v_y dx),$$

which is consistent with the last equation of section 5.8.

From the $d\psi$ equation, it follows that

$$v_x = \frac{\partial \psi}{\partial y}, \quad v_y = -\frac{\partial \psi}{\partial x}.$$

This demands

$$\frac{\partial^2 \psi}{\partial x \partial y} = \frac{\partial^2 \psi}{\partial y \partial x} \quad \text{or} \quad \frac{\partial v_x}{\partial x} = -\frac{\partial v_y}{\partial y},$$

the condition for ψ to exist (div $\mathbf{v} = 0$).

$$\phi_B - \phi_A = \int_A^B v_x dx + v_y dy (+\, v_z dz).$$

(The $v_z dz$ term is necessary in three-dimensional problems.)

In contrast $\mathbf{v} = \text{grad}\, \phi$ gives

$$v_x = \frac{\partial \phi}{\partial x}, \quad v_y = \frac{\partial \phi}{\partial y}, \quad \left(v_z = \frac{\partial \phi}{\partial z}\right).$$

This demands

$$\frac{\partial^2 \phi}{\partial x \partial y} = \frac{\partial^2 \phi}{\partial y \partial x} \quad \text{or} \quad \frac{\partial v_x}{\partial y} = \frac{\partial v_y}{\partial x}, \text{ etc.,}$$

the condition for ϕ to exist.

These equations show how, given ψ or ϕ, $\mathbf{v}$ may be found. In each problem, to determine ψ or ϕ we need boundary conditions in addition to a differential equation. The two commonest kinds of boundary are the impermeable wall and the equipotential wall, each with its characteristic conditions on ψ or ϕ:

The impermeable wall

Here $\mathbf{v}$ must be parallel to the boundary, which is therefore a field line, with $\psi =$ const. In consequence

$$\partial\psi/\partial s = 0$$

if s is measured along the boundary.

As grad ϕ is parallel to the boundary, equipotentials are normal to it. Moving normally we see no change in ϕ, i.e.

$$\partial\phi/\partial n = 0$$

if n is measured normal to the boundary.

The equipotential wall

Here grad ϕ and so **v** also are normal to the boundary. The field lines ψ = const. are normal to the boundary and, moving normally, we see no change in ψ,

i.e. $\partial\psi/\partial n = 0$

if n is measured normal to the boundary.

On the boundary

$$\phi = \text{const.}$$

or $\partial\phi/\partial s = 0$

if s is measured along the boundary.

The reader should notice how the boundary conditions on ϕ and ψ interchange in these two cases..

Though many problems concern two-dimensional fields which are both conservative and solenoidal (so that both columns in the preceding tabulation are applicable) there are many situations where only *one* of the columns is relevant. Problem 8.3(*a*) explores these points, while problem 8.3(*b*) is a harder exercise.

Example 8.2 A vector field **v** has a potential $\phi = x^4 - 6x^2y^2 + y^4$. Find the form of its field lines and stream function, if any.

$\mathbf{v} = \text{grad}\,\phi \doteq \mathbf{i}\,\partial\phi/\partial x + \mathbf{j}\,\partial\phi/\partial y = \mathbf{i}\,(4x^3 - 12xy^2) + \mathbf{j}\,(-12x^2y + 4y^3)$. Along a field line

$$\frac{\mathrm{d}x}{4x^3 - 12xy^2} = \frac{\mathrm{d}y}{-12x^2y + 4y^3},$$

or $E = (-12x^2y + 4y^3)\mathrm{d}x + (4x^3 - 12xy^2)\mathrm{d}y = 0 = -(4y^3\mathrm{d}x + 12xy^2\mathrm{d}y) + (12x^2y\mathrm{d}x + 4x^3\mathrm{d}y)$. This integrates to $4(x^3y - xy^3) = \text{const.}$, the equation of a field line.

For the stream function to exist, **v** must be two-dimensional and solenoidal, with div $\mathbf{v} = 0$, i.e.

$$\frac{\partial}{\partial x}(4x^3 - 12xy^2) + \frac{\partial}{\partial y}(-12x^2y + 4y^3) = 0, \quad \text{which is true.}$$

(Note that this is also the condition which makes E a perfect differential and allows the direct integration above.) So ψ exists and is given by

$$\frac{\partial\psi}{\partial x} = -v_y = 12x^2y - 4y^3, \quad \text{and} \quad \frac{\partial\psi}{\partial y} = v_x = 4x^3 - 12xy^2.$$

Integrating, we have

$$\psi = 4(x^3y - xy^3) + \text{const.}$$

The constant of integration may be set to zero, for the datum for ψ is arbitrary. Keeping ψ = const. gives the same equation for the field lines as before. The ψ-method for finding field lines is very efficient but is of course only applicable to two-dimensional, solenoidal fields.

8.3 Laplace's equation

In this and the next chapter we shall study *five* cases of the equation $\nabla^2\phi = R$. This chapter deals with the case where $\mathbf{v} = \text{grad}\,\phi$ and $\text{div}\,\mathbf{v} = R = 0$ and so $\nabla^2\phi = 0$, a very important equation known as *Laplace's equation*. It occurs whenever the vector field $\mathbf{v}$ is solenoidal because of the absence of sources distributed in the region of interest. The ideas in chapter 7 yield several cases:

(*a*) *Steady heat conduction without heat release* (*uniform conductivity*):

$$\text{div}\,\mathbf{Q} = 0 \quad \text{and} \quad \mathbf{Q} = -K\,\text{grad}\,T \quad \text{so} \quad \nabla^2 T = 0.$$

(*b*) *Constant density flow in a porous medium* (*uniform permeability*):

$$\text{div}\,\mathbf{v}^* = 0 \quad \text{and} \quad \mathbf{v}^* = -k\,\text{grad}\,p^* \quad \text{so} \quad \nabla^2 p^* = 0.$$

There are analogous cases of steady diffusion according to Fick's law.

(*c*) *Steady or 'low frequency' electric conduction* (*uniform conductivity*):

$$\text{div}\,\mathbf{j} = 0 \quad \text{and} \quad \mathbf{j} = -\sigma\,\text{grad}\,V \quad \text{so} \quad \nabla^2 V = 0.$$

(*d*) *Magnetic field in vacuo* (*$\epsilon_0\partial\mathbf{E}/\partial t$ assumed negligible*):

$$\text{div}\,\mathbf{H} = 0 \quad \text{and} \quad \mathbf{H} = -\text{grad}\,U \quad \text{so} \quad \nabla^2 U = 0.$$ (See problem 8.4(*c*).)

(*e*) *Electrostatic field in vacuo*:

$$\text{div}\,\mathbf{E} = 0 \quad \text{and} \quad \mathbf{E} = -\text{grad}\,V \quad \text{so} \quad \nabla^2 V = 0, \text{ again.}$$

Note that the reasons why V obeys Laplace's equation in cases (*c*) and (*e*) are quite different. (Problem 7.17(*b*) showed that $\text{div}\,\mathbf{E} \neq 0$ and so $\nabla^2 V \neq 0$ in a current-bearing electric conductor as soon as the conductivity is non-uniform.) Problem 8.4 discusses V in a dielectric, while problem 8.5 studies the surface charges responsible for a laplacian electrostatic field. There is an important general point to be reiterated here. Most electromagnetic problems are *not* ones where the sources (charges or currents) are known *ab initio* and the solutions can be found by integrations based on the Coulomb or Biot–Savart laws. The charges, etc., spontaneously distribute themselves in some unknown way so as to satisfy some other constraint, e.g. that a particular, charge-bearing boundary shall be at uniform potential. The problem is solved via a *differential* equation such as Laplace's equation subject to boundary conditions that are not expressed in terms of charges, etc. If necessary these sources of the fields can be deduced as a last stage of the calculation, as in problem 8.5.

The prospect of killing so many physical birds with one mathematical stone makes it well worthwhile to develop techniques for solving Laplace's equation: some of these are briefly explored in this chapter. Problems 8.6, 8.7 and 8.8 investigate other laplacian situations.

8.4 Analogue methods

One obvious stratagem is to exploit the analogy between physical systems which Laplace's equation reveals. One merely chooses whichever physical phenomenon is most convenient experimentally and uses it to simulate the system of interest, if this is experimentally less convenient. Usually the most convenient is the electric conduction analogue, using either a tank of uniform electrolyte or, for a two-dimensional problem, a sheet of special conducting paper. With these it is very easy to locate the surfaces (or lines) of constant potential. If it is desired to reveal the field lines instead, the most convenient laplacian experiment is a two-dimensional variant of the porous medium flow known as *Hele–Shaw flow*. In this the porous medium is replaced by two closely-spaced glass plates between which a suitably viscous fluid is passed. The stream lines in steady flow are then easily revealed with dye.

In setting up the analogy one must make the two analogous systems geometrically similar and also model the boundary conditions correctly. To model a constant potential boundary, the electric conduction analogue must use a highly conducting electrode whereas in the Hele–Shaw analogue such a boundary must adjoin a tank of virtually stationary fluid in which $p^* = \text{const}$; in contrast any insulated or impermeable boundary must be modelled in the obvious way.

8.5 Mathematical methods

One mathematical approach is to look for solutions of any special form that will enable the *partial* differential equation to be reduced to one or more *ordinary* differential equations, since the techniques for solving these are so highly developed. The example below exhibits one method, in which ϕ is taken to be a function of a single combination of the independent variables. The subsequent example reveals another way of achieving the change from partial to ordinary calculus.

Example 8.3 If $\phi = f(w)$ where $w = y/x$ and $\nabla^2\phi = 0$, find f.

$$\frac{\partial\phi}{\partial x} = \frac{\mathrm{d}f}{\mathrm{d}w}\frac{\partial w}{\partial x} = -\frac{\mathrm{d}f}{\mathrm{d}w}\frac{y}{x^2}, \quad \frac{\partial\phi}{\partial y} = \frac{\mathrm{d}f}{\mathrm{d}w}\frac{\partial w}{\partial y} = \frac{\mathrm{d}f}{\mathrm{d}w}\frac{1}{x},$$

$$\frac{\partial^2\phi}{\partial x^2} = -\frac{\mathrm{d}^2 f}{\mathrm{d}w^2}\frac{\partial w}{\partial x}\frac{y}{x^2} - \frac{\mathrm{d}f}{\mathrm{d}w}\left(-\frac{2y}{x^3}\right)$$

$$= \frac{\mathrm{d}^2 f}{\mathrm{d}w^2}\frac{y^2}{x^4} + 2\frac{\mathrm{d}f}{\mathrm{d}w}\frac{y}{x^3}, \quad \frac{\partial^2\phi}{\partial y^2} = \frac{\mathrm{d}^2 f}{\mathrm{d}w^2}\left(\frac{1}{x}\right)^2.$$

Hence

$$x^2\nabla^2\phi = \frac{\mathrm{d}^2 f}{\mathrm{d}w^2}(1 + w^2) + 2\frac{\mathrm{d}f}{\mathrm{d}w}w = 0.$$

Integrating direct gives

$$\frac{df}{dw}(1 + w^2) = C \text{ (const.)},$$

and

$$f + \text{const.} = \int \frac{C dw}{1 + w^2} = C \tan^{-1} w = C\theta \quad (\text{for } \tan\theta = y/x).$$

This solution of Laplace's equation has considerable practical significance, as we shall see later.

Normally the solution of $\nabla^2\phi = 0$ is $\phi(x, y, z)$, a general function of x, y and z, but it is conceivable that some solutions are of the so-called *separable* form

$$\phi = X(x)Y(y)Z(x)$$

in which each of X, Y, Z is a function of the *one* variable specified. The hope is that X, Y and Z will be determined by ordinary differential equations. How this actually comes about is best seen from a particular example.

Example 8.4 Find separable solutions of the problem of steady heat conduction in the semi-infinite slab shown in section in figure 8.3. The slab is virtually unbounded in the z-direction. The two faces of the slab $y = \pm a/2$ are kept at a fixed temperature ($T = 0$) and heat is supplied to the end-face $x = 0$ in some appropriate fashion. This heat all finds its way out of the top and bottom of the slab and so $T \to 0$ as $x \to \infty$.

This problem can be regarded as a simple model of heat flow in a cooling fin on a motor-cycle engine or a heat exchanger, the fin being so long that its exact length is immaterial.

As T does not depend on z, we take $T = X(x)Y(y)$ for all values of x and y within the slab. Laplace's equation for T then gives

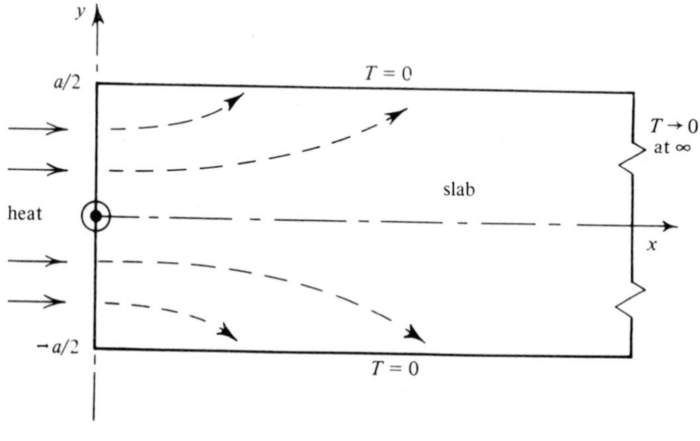

Fig. 8.3. Heat flow in a slab.

$$\frac{d^2X}{dx^2}Y + X\frac{d^2Y}{dy^2} = 0$$

and

$$\frac{d^2X}{dx^2}\bigg/X \text{ (i.e. a function of } x) = -\frac{d^2Y}{dy^2}\bigg/Y \text{ (i.e. a function of } y).$$

As x and y may vary independently within the slab, we may differentiate this equation with respect to x, keeping y constant, with the result that

$$\frac{d}{dx}\left(\frac{d^2X}{dx^2}\bigg/X\right) = 0 \quad \text{and} \quad \frac{d^2X}{dx^2}\bigg/X = \text{const. } (k \text{ say}) = -\frac{d^2Y}{dy^2}\bigg/Y.$$

Hence

$$\frac{d^2X}{dx^2} = kX.$$

This *ordinary* differential equation has the simple solution

$$X = Ae^{(\sqrt{k})x} + Be^{-(\sqrt{k})x}$$

where A and B are constants. The factor X contains the x-dependence of T and T obviously falls monotonically to zero without oscillation as x increases indefinitely. Hence $\sqrt{k}$ is real (for $\sqrt{k}$ complex would imply sinusoidal fluctuation with x) and A must vanish if X is to fall to zero at large x. Replace k by α^2. We can now turn to Y, for which

$$\frac{d^2Y}{dy^2} + \alpha^2 Y = 0 \quad \text{and so} \quad Y = C\sin\alpha y + D\cos\alpha y,$$

in which C and D are constants. If the heat is supplied symmetrically to the end of the slab, Y must be an even function and $C = 0$. Then

$$T = Ee^{-\alpha x}\cos\alpha y$$

if we replace BD by E, for brevity. This kind of exponential/sinusoidal solution of Laplace's equation is one of the simplest forms. It satisfies the boundary conditions $T = 0$ at $y = \pm a/2$ if

$$\cos(\alpha a/2) = 0 \quad \text{or} \quad \alpha = \pi/a,\ 3\pi/a,\ 5\pi/a, \text{ etc.},$$

if we reject the non-event $E = 0$. The problem we have solved is evidently one in which the heat supply at the end is such that (putting $x = 0$) $T = E\cos\pi y/a$ there (or the corresponding results with $3\pi/a$, etc.). The reader should sketch the isothermals and the (orthogonal) heat flow field lines for the above solution. Problem 8.9(*a*) explores the corresponding stream function for this two-dimensional, solenoidal problem. This gives another way of finding the field lines. Problem 8.9(*b*) studies the effect of a surface heat transfer coefficient. It should be evident from this example that a separable solution of a partial differential equation is only possible in those cases where the boundary conditions happen to allow it.

8.6 Fourier's great contribution

There is an obvious need to generalise such separable solutions to suit a wider range of possible boundary conditions. It was this need which led Fourier to invent what later became known as *Fourier Series*, with which we assume the reader is already acquainted. The key points are (*a*) that different solutions of a linear equation such as Laplace's equation can be added or superposed to produce further solutions of the same equation; (*b*) that any distribution of the dependent variable (T in example 8.4) on the boundary can be expressed as the sum of a series of such terms as $E_1 \cos \pi y/a$, $E_3 \cos 3\pi y/a$, etc. (with sine terms as well if the profile is not symmetrical); and (*c*) that each term of this series can be turned into a solution of Laplace's equation by introducing such factors as $e^{-\pi x/a}$, $e^{-3\pi x/a}$, etc. as appropriate.

Example 8.5 Solve the problem posed in example 8.4 in the case where $T = T_0$, a constant, at $x = 0$ for $|y| < a/2$.

The theory of Fourier series enables us to expand T_0 for $|y| < a/2$ in the form

$$T_0 = \frac{4T_0}{\pi}\left\{\cos\frac{\pi y}{a} - \tfrac{1}{3}\cos\frac{3\pi y}{a} + \tfrac{1}{5}\cos\frac{5\pi y}{a}\ldots\right\}$$

and hence deduce the solution for the temperature distribution in the slab

$$T = \frac{4T_0}{\pi}\left\{e^{-\pi x/a}\cos\frac{\pi y}{a} - \tfrac{1}{3}e^{-3\pi x/a}\cos\frac{3\pi y}{a} + \tfrac{1}{5}e^{-5\pi x/a}\cos\frac{5\pi y}{a}\ldots\right\}.$$

It is intriguing to observe how rapidly all but the first term dies out as we move away from the end $x = 0$. When $x = a$ and $y = 0$ the first term is 1600 times bigger than the second term. Whatever the input temperature profile, Nature very swiftly filters it into Her favourite form, the sinusoidal wave.

The reader should note that the solution of Laplace's equation in this example becomes unique once a suitable boundary condition has been specified all round the edge of the region of interest. This is a general theorem in mathematics, but the proof would be out of place here. Problem 8.10 extends example 8.5.

8.7 Laplace's equation in two dimensions

Both the quantities potential ϕ and stream function ψ are relevant in any two-dimensional problem in which div $\mathbf{v} = 0$ and $\mathbf{v} = \text{grad}\,\phi$, where ϕ is the appropriate physical quantity such as $-KT, -\sigma V, -kp^*, -V, -U$, etc. Then Laplace's equation takes its two-dimensional form:

$$\nabla^2\phi = \frac{\partial^2\phi}{\partial x^2} + \frac{\partial^2\phi}{\partial y^2} = 0,$$

there being no z-variation. Alternatively this result also follows from the facts that

$$\frac{\partial^2\phi}{\partial x^2} = \frac{\partial v_x}{\partial x} = \frac{\partial^2\psi}{\partial x\partial y} = \frac{\partial^2\psi}{\partial y\partial x} = -\frac{\partial v_y}{\partial y} = -\frac{\partial^2\phi}{\partial y^2}.$$

The reader should verify by a similar argument that

$$\nabla^2\psi = 0$$

also. Both functions obey the two-dimensional Laplace equation here. They are called *conjugate functions.* Each problem may be expressed and solved in terms of either ϕ or ψ. Once one has been found the other is easily found if required. (See for instance problems 8.6(*b*), 8.8, 8.9, 8.10.) The two families of curves ϕ = const. and ψ = const. form an *orthogonal mesh*, because grad ϕ is perpendicular to lines of constant ϕ and parallel to lines of constant ψ. Figure 8.4(*b*) shows a typical orthogonal laplacian mesh.

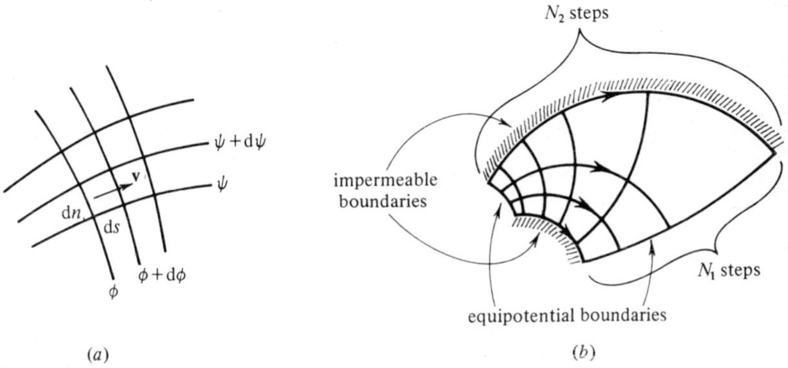

Fig. 8.4. Curvilinear squares.

8.8 Curvilinear squares

Any two-dimensional laplacian mesh has a remarkable property that we can exploit. Figure 8.4(*a*) exhibits a portion of an orthogonal mesh of ϕ and ψ-constant lines, with small intercepts dn and ds as shown. If we choose equal increments of ϕ and ψ all over the mesh, then $d\psi = d\phi$. It follows that $dn = ds$, because $v = \partial\phi/\partial n = \partial\psi/\partial s$. Each quadrilateral is therefore a square in the limit. (All the equations involving small increments are strictly true only in the limit when the increments vanish.) For finite, equal increments in ϕ and ψ, the mesh will be composed of *curvilinear squares*, with right-angled corners and sides as nearly equal as it is possible to make them. Figures 7.5, 8.4(*b*), 8.5 and 8.7 provide examples. The point of this apparently curious idea is that it provides a practical way of getting approximate solutions to Laplace's equation in two dimensions which in skilled hands is remarkably accurate. It is most appropriate for finding the solution in a finite domain with boundaries composed of equipotentials and impermeable walls

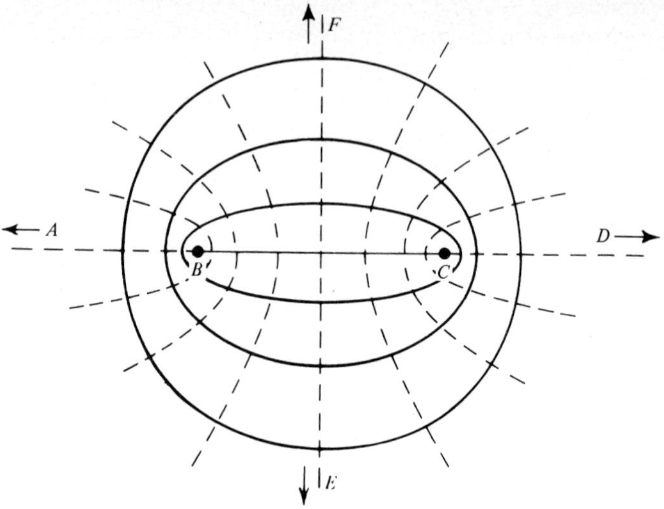

Fig. 8.5. Orthogonal ϕ, ψ mesh. Either the dashed or continuous curves can be the equipotentials. The field lines are the other curves.

along which ψ is constant. The technique merely consists of sketching in an iterative manner with the aid of pencil and rubber (particularly the latter!) until a satisfactory set of curvilinear squares is arrived at. As there is no substitute for the 'do it yourself' approach, the reader should attempt problem 8.11, in doing which he will also begin to acquire an appreciation of the characteristic way in which field lines and equipotentials bunch together and bend in laplacian flow fields. For instance they get closer together on the inside of a bend.

Once the squares are sketched, the practical interpretation is easy. Let the equal increments in ϕ and ψ be δ. If the total flux $\Delta\psi$ through the system is divided into N_1 parts by the field lines and the total potential difference $\Delta\phi$ across the system is divided into N_2 parts by the equipotentials (see figure 8.4(*b*)), then

$$\Delta\psi = \text{total flux (per unit thickness)} = N_1\delta$$

and

$$\Delta\phi = \text{total potential difference} = N_2\delta.$$

Example 8.6 A curvilinear-squares construction has yielded N_1 and N_2 for a problem concerning heat flow through a block of thickness t and conductivity K between two ends at temperatures T_1 and T_2. Find the heat flow rate Q.

Here

$$\phi = -KT \quad \text{and} \quad K(T_1 - T_2) = N_2\delta.$$

Also ψ = heat flow rate per unit thickness, and so

$$Q = tN_1\delta = tK(T_1 - T_2)(N_1/N_2).$$

Notice that it is the ratio N_1/N_2 which is the useful outcome of the curvilinear-squares construction, and that the value of Q is exactly the same as it would have been if the block had been rectangular, with aspect ratio N_1/N_2. In this case the curvilinear squares would be exact squares. Note N_1 or N_2 may be not integral.

One point has been glossed over. If an equipotential boundary has a discontinuity in slope, the field line at this point cannot be simultaneously perpendicular to the boundary on either side of the discontinuity. It can be shown that then the field line merely bisects the angle between the tangents on either side. The nearby curvilinear squares are consequently deformed. A similar result holds for equipotentials at a discontinuity in an impermeable boundary. Finally notice the 'squares' nearest to the singular points B or C in figure 8.5. Whenever an equipotential encounters a field line at an angle (180° at B or C) which is not 90°, as it should be, the nearby squares are severely deformed again.

Now consider any problem such as that in figure 8.4(b) where the equipotential and impermeable boundaries can be interchanged. From the nature of the curvilinear-squares construction, it is obvious that the original squares can be retained and re-interpreted. Then the ϕ and ψ-constant lines of the original problem have been interchanged. The new problem is called the *dual* of the original problem. The crucial ratio N_1/N_2 is replaced by its reciprocal in the dual problem. Problem 8.12 exploits this fact.

These results can be applied to analogue methods where only one of the family of ϕ or ψ-lines can be easily found. If the boundary conditions are interchanged to give the dual problem, the new ϕ or ψ-lines can be interpreted as the ψ or ϕ-lines of the original problem, so generating the complete orthogonal ϕ/ψ mesh for it.

Each pair of solutions can often be given a variety of physical interpretations, as the following example reveals.

Example 8.7 Give physical interpretations to figure 8.5, which shows a typical laplacian mesh of orthogonal lines with ϕ or ψ constant (and which incidentally also illustrates Laplace's equation applied to a domain of infinite extent).

Consider the case where the continuous lines are field lines (ψ const.) and the dashed lines are equipotentials. This could represent the conduction of heat between two constant-temperature fins AB, CD embedded in a conducting medium (or the comparable electric conduction problem); or the electrostatic field between two charged fins AB and CD. If instead only the lower half of the field is used, other possibilities emerge: conduction between two highly conducting sections AB and CD of a wall separated by an insulating section BC, or seepage through flat-topped porous ground under a dam with a flat base BC, surfaces AB and CD each being under a constant head of water. Note that there is no electrostatic case of this kind as no material impermeable to electric flux exists. For magnetism, a superconductor can provide an impermeable wall, however. (The reader should interpret in various ways the case where just the left-hand half of the field is retained and EF is replaced by an equipotential boundary.)

Now consider the dual problem where the dashed lines are field lines and the continuous lines equipotentials. This might represent heat or electric conduction or seepage between two semi-infinite regions through a slot BC in an impermeable wall or a magnetic field passing through a slot in a superconducting sheet. If only the lower half of the field is used, $ABCD$ could be a wall that is impermeable apart from an equipotential section BC at constant temperature, voltage, p^* etc. as appropriate, acting as a distributed source (or sink) of flux. Alternatively EF might be taken as an impermeable boundary for the left-hand half of the field.

Problem 8.13 is a further exercise of this kind.

*8.9 The ϕ, ψ duality

[This section is rather more advanced and may be omitted without detriment to the understanding of later sections. It is a more formal treatment of dual problems.]

Consider a two-dimensional laplacian problem which has boundary conditions of the type illustrated in figure 8.6(a). The domain of interest has a boundary in

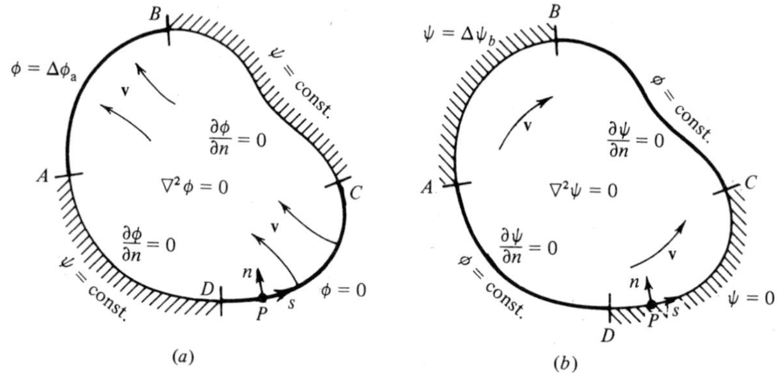

Fig. 8.6. Dual laplacian problems in two dimensions.

four parts, AB and CD being at known, constant potentials (e.g. at constant temperature or voltage) and BC and DA being impermeable or insulated boundaries. The state of the flux might be determined by specifying $\Delta\phi_a$, the difference in potential ϕ between AB and CD. It would then be natural to solve the problem in terms of ϕ, the solution ϕ_a or $\nabla^2\phi = 0$ being determined by the boundary conditions ϕ = const. or $\partial\phi/\partial n = 0$ (n measured normal to the boundary) as indicated in figure 8.6(a). D is taken as datum, where $\phi_a = 0$. DA and BC are of course lines of constant ψ. Note the use of suffices a and b to distinguish quantities associated with the problems posed in figures 8.6(a) and 8.6(b).

Now consider the different problem shown in figure 8.6(b), in which the shape of the domain is the same but the constant-potential and impermeable portions of the boundary have been interchanged, i.e. the *dual problem*. If the state of the flux was determined by specifying $\Delta\psi_b$, the total flux per unit thickness fed through the system from AD to BC, the problem could best be solved in terms of ψ, the solution ψ_b of $\nabla^2\psi = 0$ being determined by the boundary conditions $\psi =$ const. or $\partial\psi/\partial n = 0$ as indicated in figure 8.6(*b*). D is taken as datum, where $\psi_b = 0$. DA and BC are of course lines of constant ϕ.

In the two figures the **v**-arrows indicate the general nature of the vector flux associated with ϕ_a or ψ_b, taking $\Delta\phi_a$ and $\Delta\psi_b > 0$. Note that $\mathbf{v} = \operatorname{grad}\phi$ implies flow from low to high ϕ. There is a comparable sign convention in relation to ψ: advancing from low ψ to high ψ, an observer encounters vector flux from left to right.

The key points to notice are that ϕ_a and ψ_b obey the same differential equation and the same boundary conditions if we make $\Delta\phi_a = \Delta\psi_b$. It then follows from the uniqueness properties of Laplace's equation that ϕ_a and ψ_b are identical.

Consider the point P on boundary DC in both problems. The magnitude of **v** there in problem (*b*) is such that

$$v_b = \frac{\partial\psi_b}{\partial n} = \frac{\partial\phi_b}{\partial s}$$

(if s is distance measured along the field line DC) because of the properties of ϕ and ψ noted earlier. For the same reasons, the magnitude of **v** at P in problem (*a*) is such that

$$v_a = \frac{\partial\phi_a}{\partial n} = -\frac{\partial\psi_a}{\partial s}.$$

(Two points deserve comment here: n, being normal to the boundary, is distance *along* the field lines in problem (*a*), *normal* to the equipotential DC; the minus sign comes from the sign convention for ψ already referred to.) As $\psi_b = \phi_a$ and $\partial\psi_b/\partial n = \partial\phi_a/\partial n$, it follows that

$$\partial\phi_b/\partial s = -\partial\psi_a/\partial s.$$

Integrating along DC reveals that $\Delta\psi_a$, the total flux per unit thickness through system (*a*), equals $\Delta\phi_b$, the difference in potential ϕ across system (*b*). To be consistent with the signs, we must take

$$\Delta\psi_a = \psi_{AD} - \psi_{BC} \quad \text{but} \quad \Delta\phi_b = \phi_{BC} - \phi_{AD}.$$

Finally, by arguments identical to those used to show that $\phi_a = \psi_b$, we infer that $\phi_b = -\psi_a$ if we choose the same datum point, D say, for ϕ_b and ψ_a.

The full duality of these two problems is now apparent. Provided the datums and 'levels' of flux are correctly matched, ϕ for one is ψ for the other, and vice versa, except for the sign change.

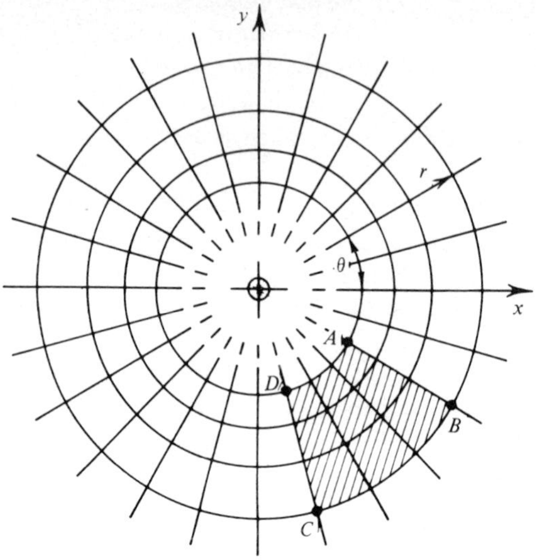

Fig. 8.7

8.10 A particular two-dimensional case

Figure 8.7 shows a particular, circularly symmetric, curvilinear-squares solution of Laplace's equation, comprising radial lines and concentric circles. This case may be simply expressed analytically in terms of polar coordinates. We can choose to take the radial lines (θ = const.) as the field lines, which evidently then emanate from a line source at the origin. Since it involves a source, the problem belongs to the type mentioned in section 8.2, with ψ multi-valued. If the source emits flux S per unit thickness (measured perpendicular to the page) the field strength v is given by the fact that the flux across a circle of radius r is $S = 2\pi rv$. Hence

$$v = S/2\pi r = \mathrm{d}\phi/\mathrm{d}r$$

(for grad ϕ is radial here) and integration gives

$$\phi = (S/2\pi)\log_e r + \text{const.}$$

(Compare problems 3.10(*b*) and 8.6(*b*).) (The equipotentials are circles, r = const., and so ϕ depends only on r.) If we take Ox ($\theta = 0$) as the field line $\psi = 0$, then any other radial line θ = const. is the field line

$$\psi = (S/2\pi)\theta$$

because a fraction $\theta/2\pi$ of the total flux S passes between it and Ox. θ and hence also ψ are multi-valued unless a 'cut' is inserted, e.g. restricting θ between $\pm\,\pi$. Any equipotential line can be replaced by an equipotential boundary; part of the solution could therefore be interpreted as the field between two concentric cylinders (e.g. in the context of heat or electric conduction or electrostatics). Similarly, two radial lines could be replaced by impermeable boundaries, if flux in a sector such as $ABCD$ were of interest. (See problem 8.14.)

Instead we can interpret figure 8.7 by making the concentric circles into the field lines and the radial lines into the equipotentials, the potential now becoming multi-valued. Example 8.3 in fact concerned this case. An obvious physical application is the magnetic field round a long, straight wire carrying a current J, already discussed in example 4.2, where we found that

$$\phi = -U = (J/2\pi)\theta.$$

The corresponding stream function is easily shown to be

$$\psi = -(J/2\pi)\log_e r + \text{const.}$$

ϕ and ψ have become interchanged from the previous case, in the usual way. Notice that this kind of solution cannot apply to any field problem for which the potential is a *physical* quantity, such as $-kp^*$ (in seepage) or $-KT$ (in heat conduction), which is inevitably single-valued.

These two related laplacian solutions will receive further attention in later sections on fluid mechanics, where it transpires that flows other than those in porous media can be described by a potential under certain conditions.

*8.11 Complex variable methods

There is another, even more powerful method of solving Laplace's equation in two dimensions, which deserves a brief mention, even though an exhaustive and rigorous treatment of the method would be out of place here. This section may be omitted by readers who wish to avoid the more mathematical parts of the subject.

If the reader has previously met the one-dimensional wave equation

$$\frac{\partial^2\Omega}{\partial t^2} - c^2\frac{\partial^2\Omega}{\partial x^2} = 0$$

he may already be comparing (and perhaps confusing) it with Laplace's equation. As we shall see later, the wave equation has general solutions of the form $\Omega = \Omega(z)$ where $z = x \pm ct$. (*N.B.* z is *not* now a cartesian coordinate.) Laplace's equation is the case $c = i$, the root of minus one, with y replacing t. Thus the equation

$$\frac{\partial^2\Omega}{\partial x^2} + \frac{\partial^2\Omega}{\partial y^2} = 0$$

must have a general solution $\Omega = \Omega(z)$, where Ω is any well-behaved function of the complex variable $z = x + iy$. To verify that it satisfies Laplace's equation, we note that we may differentiate with respect to x at constant y (or vice versa) with the result that

$$\frac{\partial\Omega}{\partial x} = \frac{d\Omega}{dz}\frac{\partial z}{\partial x} = \frac{d\Omega}{dz} \quad \text{while} \quad \frac{\partial\Omega}{\partial y} = \frac{d\Omega}{dz}\frac{\partial z}{\partial y} = i\frac{d\Omega}{dz},$$

and that

$$\frac{\partial^2\Omega}{\partial x^2} = \frac{\mathrm{d}^2\Omega}{\mathrm{d}z^2}\frac{\partial z}{\partial x} = \frac{\mathrm{d}^2\Omega}{\mathrm{d}z^2} \quad \text{while} \quad \frac{\partial^2\Omega}{\partial y^2} = i\frac{\mathrm{d}^2\Omega}{\mathrm{d}z^2}\frac{\partial z}{\partial y} = -\frac{\mathrm{d}^2\Omega}{\mathrm{d}z^2},$$

and finally that

$$\nabla^2\Omega = 0.$$

In general Ω will be complex, equal to $\phi + i\psi$, say, where ϕ and ψ are real functions of x and y. Then

$$\nabla^2\phi + i\nabla^2\psi = 0$$

which implies that the real quantities $\nabla^2\phi$ and $\nabla^2\psi$ both vanish. Thus any function Ω generates not merely one but two solutions of Laplace's equation! The difficulty lies in finding the Ω appropriate to specified boundary conditions. Also

$$\frac{\partial\phi}{\partial y} + i\frac{\partial\psi}{\partial y} = \frac{\partial\Omega}{\partial y} = i\frac{\mathrm{d}\Omega}{\mathrm{d}z} = i\frac{\partial\Omega}{\partial x} = i\left\{\frac{\partial\phi}{\partial x} + i\frac{\partial\psi}{\partial x}\right\} = -\frac{\partial\psi}{\partial x} + i\frac{\partial\phi}{\partial x}.$$

Since $\partial\phi/\partial y$ etc. are all real we can equate real and imaginary parts and deduce that

$$\frac{\partial\phi}{\partial y} = -\frac{\partial\psi}{\partial x}, \quad \frac{\partial\phi}{\partial x} = \frac{\partial\psi}{\partial y},$$

which are called the *Cauchy–Riemann equations.*

It should now be apparent why the symbols ϕ and ψ were chosen. ϕ and ψ are seen to be related in the correct way to serve as conjugate potential and stream functions for a two-dimensional, conservative and solenoidal vector field. So the complex variable method generates pairs of solutions ϕ, ψ in one fell swoop.

For example, consider

$$\Omega(z) = \mathrm{e}^{-\alpha z} = \mathrm{e}^{-\alpha x}(\cos \alpha y - i \sin \alpha y).$$

Here

$$\phi = \mathrm{e}^{-\alpha x} \cos \alpha y$$

and we have found again the simple solution to the cooling fin problem, discussed in section 8.5, where $\phi = -KT$. Meanwhile

$$\psi = -\mathrm{e}^{-\alpha x} \sin \alpha y.$$

(See problem 8.9(*a*).)

If instead $\Omega = \log z$, it is convenient to turn to polar coordinates r, θ, whereupon

$$z = r\mathrm{e}^{i(\theta+2k\pi)} \quad (k \text{ any integer})$$

and

$$\Omega = \log r + i(\theta + 2k\pi)$$

so that

$$\phi = \log r \quad \text{and} \quad \psi = \theta + 2k\pi.$$

This is evidently the problem discussed in section 8.10, for the case where $S = 2\pi$.

In terms of complex variables, the interchange between the dual solutions discussed in sections 8.8 and 8.9 is very simply achieved by multiplying Ω by $\pm i$. As an example consider the alternative solution

Now
$$\Omega = -i \log z = (\theta + 2k\pi) - i \log r.$$
$$\phi = (\theta + 2k\pi) \quad \text{and} \quad \psi = -\log r.$$

(Compare section 8.10, with $J = 2\pi$.) This is evidently the dual of the previous case, ϕ and ψ having interchanged and one sign having changed. Problems 8.15 and 8.16 give further experience with complex-variable solutions of Laplace's equation.

Problems 8

8.1 Evaluate the following:

(*a*) $\nabla^2(x^2y^2z^2)$,
(*b*) $\nabla^2\{\sin(x+y)\}$,
(*c*) $\nabla^2(\cos x \cosh y)$.

*8.2 ABC is a triangle and P is an internal point which is at perpendicular distances α, β, γ from the sides of the triangle respectively opposite to A, B and C. Show that, summing over three terms,

$$\sum(\alpha \sin A) = \text{const.}$$

Show also that grad α, etc. are unit vectors and $\nabla^2\alpha$, etc. vanish, ∇^2 being two-dimensional. Using example 8.1, show that $\nabla^2(\alpha\beta\gamma) = -2\Sigma(\alpha \cos A)$ and that $\nabla^2(\alpha\beta\gamma)$ is constant only if the triangle is equilateral. Evaluate it for this case and hence solve the (Poisson) equation

$$\nabla^2\phi = -k \text{ (const.)}$$

within an equilateral triangle with sides a at which ϕ vanishes.

8.3 (*a*) For the vector fields

(i) $\mathbf{i}yz + \mathbf{j}xz + \mathbf{k}xy$,
(ii) $\mathbf{i} \sin x \cos y - \mathbf{j} \cos x \sin y$,
(iii) $\mathbf{i}\, e^x \sin y + \mathbf{j}\, e^x \cos y$, and
(iv) $\mathbf{i}x + \mathbf{j}y$,

find the potential and/or stream function, if they exist.
Sketch the field lines and find the flux (per unit z-wise thickness) across the straight line joining the points $A(\pi/2, 0)$ and $B(\pi/2, \pi/2)$ in those cases where ψ exists. Does the answer depend on the straightness or otherwise of AB?
*(*b*) $\mathbf{B}$ and $\mathbf{v}$ are both two-dimensional, solenoidal, vector fields with $B_z = v_z = 0$. The field lines of $\mathbf{v}$ form closed loops. Show that

$$\iint_S \mathbf{v} \times \mathbf{B} \cdot \mathbf{da} = 0,$$

evaluated over the plane domain S enclosed by any field line of $\mathbf{v}$. (*Hint*: Define stream functions ψ_B and ψ_v and take $\mathbf{da}$ as the elementary parallelogram defined by the field lines corresponding to ψ_B, ψ_v, $\psi_B + \mathrm{d}\psi_B$, $\psi_v + \mathrm{d}\psi_v$.)

8.4 (*a*) Show that div $\mathbf{P}$ = net bound charge density; also that div $\mathbf{D} = 0$ in a dielectric devoid of free charge. Deduce that the electrostatic potential V still obeys Laplace's equation inside a linear dielectric for which $\mathbf{D} = k\epsilon_0\mathbf{E}$.
*(*b*) A long cylinder of linear dielectric (in which $\mathbf{D} = k\epsilon_0\mathbf{E}$) of radius a is subjected *in vacuo* to a transverse electric field which is of uniform strength E_0 at large distances from the cylinder. If we take the origin at the centre of the cylinder and the x-axis in the direction of the uniform field E_0, the solution for the electric potential V, in this two-dimensional problem, expressed in polar coordinates, is of the form

$$V = Ar\cos\theta + (B/r)\cos\theta\ (r > a) \quad \text{and} \quad V = Cr\cos\theta\ (r < a).$$

Verify that the following conditions are or can be satisfied:

(i) $\mathbf{E}$ tends to $E_0\mathbf{i}$ for large r.
(ii) V obeys Laplace's equation for $r \gtrless a$.

$$\left(\text{Use either cartesians or the identity } \nabla^2 \equiv \frac{\partial^2}{\partial r^2} + \frac{1}{r}\frac{\partial}{\partial r} + \frac{1}{r^2}\frac{\partial^2}{\partial \theta^2}.\right)$$

(iii) V is continuous across the interface $r = a$. (Is this consistent with the boundary condition that $\mathbf{E}_{\text{tangential}}$ is continuous? See section 6.9(*a*).)
(iv) At the interface $k(\partial V/\partial r)_{\text{inside}} = (\partial V/\partial r)_{\text{outside}}$. (Is this consistent with the boundary condition that $\mathbf{D}_{\text{normal}}$ is continuous? See section 6.9(*a*).) Show that the electric field inside the cylinder is uniform and equals $2E_0\mathbf{i}/(k+1)$.

*(*c*) Repeat example 4.3 using a stream function ψ for magnetic field of form

$$\psi = Ar\sin\theta\ (r < R) \quad \text{and} \quad \psi = (B/r)\sin\theta\ (r > R).$$

Verify that the following conditions are or can be satisfied:

(i) $\mathbf{B}$ tends to zero for large r.
(ii) ψ obeys Laplace's equation for $r \lessgtr R$.
(iii) ψ is continuous across the interface $r = R$. (Is this consistent with the boundary condition that $\mathbf{B}_{\text{normal}}$ is continuous? See section 6.9(*b*).)
(iv) At the interface $(\partial\psi/\partial r)_{\text{inside}} - (\partial\psi/\partial r)_{\text{outside}} = \mu_0 J\sin\theta$. (Is this consistent with the boundary condition for $\mathbf{H}_{\text{tangential}}$? See section 6.9(*b*).)

Also explore the method of solution that uses magnetic potential instead.

A cylindrical former of circular cross-section carries elliptical turns which lie in planes inclined at an angle α to the axis. There are NJ ampere-turns per unit length. Show, by invoking superposition ideas, that the magnetic field inside the tube is uniform and inclined at an angle β to the axis such that $\tan\beta = \frac{1}{2}\cot\alpha$.

*8.5 For problem 5.10 calculate how the surface charge density Q (see section 6.9(*a*)) on the conducting cylinders varies with position. (The electric field *inside* the conductor is zero.) Where is Q greatest?

8.6 (*a*) For what value of m does $\phi = r^m$ satisfy Laplace's equation? (See problem 7.17(*a*).)
(*b*) Consider also the two-dimensional case, where r is the distance from the z-axis, and find a function $\phi(r)$ which satisfies Laplace's equation. What is the corresponding function ψ?

8.7 (*a*) If $\nabla^2\phi = 0$, show that $\oiint \text{grad}\,\phi \cdot \mathbf{da} = 0$, taken over any closed surface.
*(*b*) The mean value $\bar{\phi}$ of ϕ on a spherical surface centred at O is given by $4\pi\bar{\phi} = \oiint \phi\,\mathrm{d}\Omega$, where $\mathrm{d}\Omega$ = solid angle subtended by an area element at O. If the sphere's radius is r, show that $\mathrm{d}\bar{\phi}/\mathrm{d}r = 0$ if $\nabla^2\phi = 0$, and hence that $\bar{\phi} = \phi_0$, the value at O, an interesting general property of laplacian functions.

8.8 Figure 8.8 shows the cross-section of a triangular earth dam, holding back

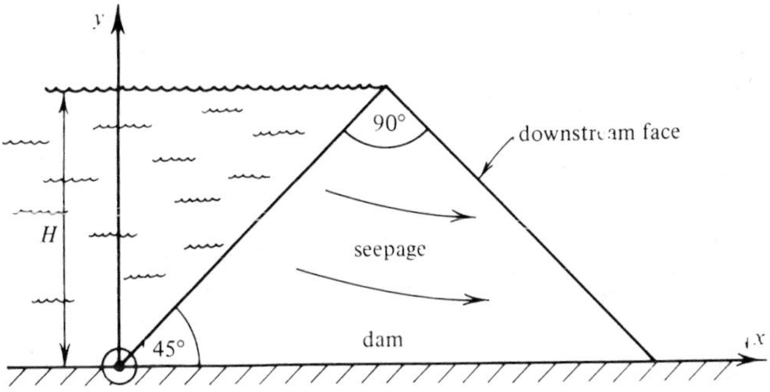

Fig. 8.8. Seepage through a triangular dam.

water up to its crest. Seepage occurs according to the law $\mathbf{v}^* = \text{grad}\,\phi$ where $\phi = -kp^*$ and $\nabla^2\phi = 0$, subject to the following boundary conditions:

(i) On the downstream face ($x + y = 2H$) and at the surface of the upstream water ($y = H$) the water pressure p is zero, if we measure all pressures above atmospheric. The emergent water runs in a thin layer down the downstream face.

(ii) On the upstream face ($x = y$) adjoining static water,

$$p^* = p + \rho g y = \rho g H \text{ (const.)}.$$

(iii) Along the base of the dam $\partial\phi/\partial y = 0$, for the ground below is impermeable.

Verify that Laplace's equation and the boundary conditions are satisfied by

$$-\phi/k = p^* = \rho g\{H-(x^2-y^2)/4H\}.$$

Find the corresponding stream function ψ and the shape of the stream lines and equipotentials. Show that the volumetric flow for unit dam length is $\frac{1}{2}\rho g H k$ (i.e. twice the value in example 7.6 with $h=0$ and $L=2H$).

8.9 (*a*) In the cooling fin problem in example 8.4, the potential $\phi = -KT$. For the input profile $T = E \cos \pi y/a$ at $x = 0$, find the stream function $\psi(x,y)$, and show that the total heat flow per unit z-wise length is $2KE$.
(*b*) Sometimes Laplace's equation has to be solved within boundaries that are neither equipotential nor impermeable, e.g. whenever there is a *surface heat transfer coefficient h* such that at the boundary Q_n, the emergent heat flux per unit area of surface, equals hT, temperature T being measured above ambient.

Repeat the cooling fin problem for this case, with $T = E \cos \alpha y$ at $x = 0$ and $T = 0$ at $x = \infty$. Show that α now satisfies the equation $\alpha \tan(\frac{1}{2}\alpha a) = h/K$.

8.10 (*a*) Repeat problem 8.9(*a*) for the case where the input profile at $x = 0$ is $T = T_0$ (const.). What is now the total heat flow per unit z-wise length?
*(*b*) Repeat the calculation of ϕ and ψ for the case where the heat is supplied with uniform intensity $Q_x = Q_0$ at $x = 0$, $|y| < a$. What is the maximum temperature?

8.11 Copy figure 8.9 on tracing paper and fill it with a mesh of curvilinear squares. It is suggested that three field lines be sketched, joining *AB* to *CD*, dividing the flux into four equal parts. Find the corresponding number of parts into which the difference in potential between *AB* and *CD* is divided

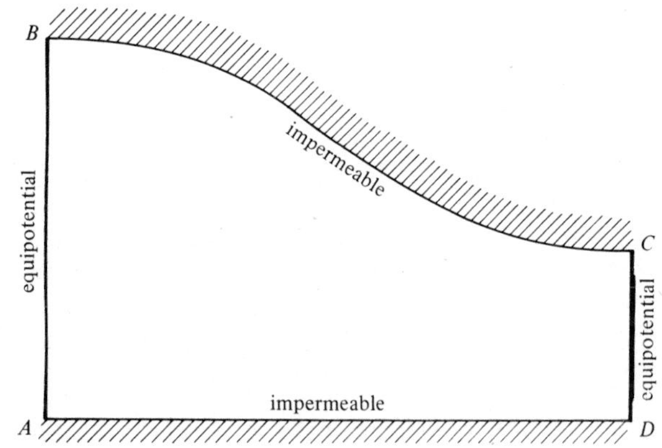

Fig. 8.9

by the equipotentials. Hence calculate the resistance of a metal plate of this shape of thickness t and resistivity τ.

8.12 A uniform metal sheet of any shape, uniform thickness t and electrical conductivity σ has its edge divided into four sections AB, BC, CD, and DA. If AB and CD are connected to highly conducting electrodes and BC and DA are insulated, the resistance between the electrodes is found to be R_1. If the electrode and insulating sections are interchanged, show that the resistance is then equal to $1/(\sigma^2 t^2 R_1)$.

8.13 Give physical interpretations to the two-dimensional laplacian solution appearing in problem 7.14(*b*).

8.14 A hot horizontal pipe of outer diameter d is surrounded by a concentric protective pipe of inner diameter $2d$, the gap being filled with an insulating material of uniform thermal conductivity. For given, uniform temperatures of the two pipes, calculate the percentage rise in heat loss if, owing to a leak, the lower half of the insulator becomes waterlogged and as a result its thermal conductivity is quadrupled. (Check that it is still possible to find a solution with purely radial heat flow, meeting the boundary conditions that there shall be no discontinuity of temperature or *normal* heat flux at the interface between wet and dry insulator.)

*8.15 (*a*) Consider the two-dimensional laplacian solutions generated by taking $\Omega = z^2$ or iz^2, where $z = x + iy$. What are the equipotentials and field lines like? Note the relationship to problems 7.11 and 8.8.
(*b*) Consider $\Omega = z^{1/2}$ and relate it to problems 7.14(*b*) and 8.13.
(*c*) Consider $\Omega = (q/2\pi\epsilon_0)\{\log(z - a) - \log(z + a)\}$ (a real), corresponding to a source and sink at the two points $z = \pm a$ respectively. Find ϕ and ψ and verify that this solution corresponds to the cases already explored in problems 3.10(*c*), 5.10 and 8.5. What shape are the equipotentials and field lines? Draw out these curves for equal increments of ϕ and ψ to see the curvilinear square mesh. A good method is to plot the ϕ (or ψ) lines for the source and sink alone in turn and then to exploit their intersections to find points with the same value of ϕ (or ψ). (See Appendix II, 10(*b*).)
(*d*) How could the above solution be adapted to apply to the steady conduction of heat from a thin, buried cable, releasing heat H per unit length and time, to the plane ground surface, which is at a height h above the cable and at a uniform temperature? What fraction of the heat emerges in a strip of width $2h$, centred above the cable?
(*e*) In terms of Ω and z, investigate the case of the *doublet*, which is the two-dimensional equivalent of a dipole, arrived at by letting a in (*c*) tend to zero with $p = 2qa$ kept constant. What are the equipotentials and field lines like now? (Compare problem 3.10(*c*).)
(*f*) Interpret $\Omega = \sin^{-1} z$ and $\Omega = i \sin^{-1} z$ in relation to figure 8.5. (*Hint*: Exploit $z = \sin \Omega$.)

*8.16 A feature of the problem shown in figure 7.6 is the free surface AB. If the problem is being solved in terms of $\phi\ (= -kp^*)$ and ψ, the boundary conditions on AB are $\psi = \text{const.}$ and $\phi = -k\rho g y$ (for $p = 0$). (Note that *two* independent conditions are both possible and necessary at a boundary whose shape is not known *ab initio*.) If ϕ and ψ are to be found via a complex-variable relation $z = x + iy = f(\Omega)$ where $\Omega = \phi + i\psi$, show that, evaluated along the boundary $\psi = \text{const.}$, $\partial z/\partial \phi = \mathrm{d}z/\mathrm{d}\Omega$ and hence that the imaginary part of $\mathrm{d}z/\mathrm{d}\Omega = -1/\rho g k$. (This result is important in connection with a technique known as the *inverse hodograph method*.)

N.B. Now y is used to denote height so that z can be the complex variable.

Show also that on an Argand diagram $\mathrm{d}z/\mathrm{d}\Omega$ is a vector parallel to $\mathbf{v}^*$ of magnitude $|\mathbf{v}^*|^{-1}$.

9

Four other important partial differential equations

9.1 Fields with divergence

We next turn to those field problems where the vector **v** is still given by the gradient of some potential ϕ but is no longer solenoidal, e.g. because of the presence of sources,

i.e. $\quad \mathbf{v} = \text{grad}\,\phi \quad \text{but} \quad \text{div}\,\mathbf{v} \neq 0.$

To complete the specification of such problems, information concerning the source density div **v** must be available. It may be given explicitly or it may be related in some way to the other variables which appear. There is a great variety of important cases, differing both in physical significance and in mathematical structure.

This chapter refers to four main categories, classified according to their mathematical form:

(i) Poisson's equation, where div **v** is known *ab initio*;
(ii) Helmholtz's equation, where div $\mathbf{v} \propto \phi$;
(iii) the unsteady diffusion equation, where div $\mathbf{v} \propto \partial\phi/\partial t$;
(iv) the wave equation, where again div $\mathbf{v} \propto \partial\phi/\partial t$ but $\partial\mathbf{v}/\partial t$ (rather than **v**) $\propto$ grad ϕ.

At the same time the chapter briefly exhibits some of the simpler mathematical techniques appropriate to such partial differential equations. As with ordinary differential equations, trial and error is often a more productive approach than strict deductive logic.

9.2 Poisson's equation

There are many flux/gradient problems where the source density S is specified in advance as some function of position $S(x, y, z)$. Then

$$\mathbf{v} = \text{grad}\,\phi, \quad \text{div}\,\mathbf{v} = S \quad \text{and} \quad \nabla^2\phi = S,$$

which is *Poisson's equation*. It can arise in electrostatics, in the somewhat rare event that the net charge density $q(x, y, z)$, distributed throughout the region of interest, is known *ab initio*. Then

$$\mathbf{E} = -\text{grad}\,V, \quad \epsilon_0\,\text{div}\,\mathbf{E} = q \quad \text{and} \quad \nabla^2 V = -q/\epsilon_0.$$

This differential equation, together with appropriate boundary conditions, enables the problem to be solved. In any part of the region where q vanishes the equation reverts to Laplace's equation, discussed in the previous chapter. Where a thermally and electrically conducting medium is heated electrically by ohmic dissipation at a known rate R watts per cubic metre, and the heat escapes steadily at intensity $\mathbf{Q}$, we then have

$$\mathbf{Q} = \text{grad}\,(-KT), \quad \text{div}\,\mathbf{Q} = R \quad \text{and} \quad \nabla^2 T = -R/K.$$

Several cases of these kinds, symmetrical or simple enough to be solved by primitive techniques without solving the Poisson equation direct, have already been encountered (in example 5.4 and problems 5.9, 5.15, 7.16). In the examples below and problems 9.1 and 9.2 the Poisson equation has to be solved.

Example 9.1 A tank of nuclear fission products in the form of an ellipsoid (semi-axes a, b and c) releases heat uniformly at a density R, and the heat escapes steadily by thermal conduction (conductivity K) to the tank wall, which is at ambient temperature. Find the maximum temperature, measured above ambient.

Here we must solve the equation $\nabla^2 T = -R/K$ (where T = temperature above ambient) subject to the boundary condition $T = 0$ at the wall, at which

$$\frac{x^2}{a^2} + \frac{y^2}{b^2} + \frac{z^2}{c^2} = 1,$$

if we choose the axes of the ellipsoid as cartesian axes. It so happens that here the solution is available 'by inspection', namely:

$$T = A\left(1 - \frac{x^2}{a^2} - \frac{y^2}{b^2} - \frac{z^2}{c^2}\right) \quad (A \text{ some const.}),$$

which vanishes at the wall and for which

$$-\frac{R}{K} = \nabla^2 T = \sum \frac{\partial^2 T}{\partial x^2} = -2A\left(\frac{1}{a^2} + \frac{1}{b^2} + \frac{1}{c^2}\right).$$

The maximum temperature obviously occurs at the origin and is given by

$$T_{\text{max}} = A = \frac{R}{2K} \bigg/ \left(\frac{1}{a^2} + \frac{1}{b^2} + \frac{1}{c^2}\right).$$

In the case $a = b = c$, this becomes $Ra^2/6K$ as in example 5.4.

The next two examples are more advanced and may be omitted.

***Example 9.2** Solve the two-dimensional Poisson equation $\nabla^2 \phi = -1$ over the square domain bounded by the lines $x = \pm 1, y = \pm 1$ at which $\phi = 0$.

This example illustrates the use of Fourier series. Consider the Poisson equation rewritten as

$$\nabla^2\phi = -\frac{4}{\pi}\sum\frac{(-1)^{(n-1)/2}}{n}\cos\frac{n\pi y}{2} \quad (n \text{ odd}),$$

which is essentially the same series as appeared in example 8.5. This cosine series, equal to -1 in the range $|y| < 1$, is chosen because the problem is symmetrical in the x-axis. An obvious solution of the differential equation is

$$\phi_1 = \frac{16}{\pi^3}\sum\frac{(-1)^{(n-1)/2}}{n^3}\cos\frac{n\pi y}{2} \quad \text{(check by differentiation!)}.$$

This satisfieds the boundary condition $\phi = 0$ at $y = \pm 1$, but not at $x = \pm 1$. Suppose we can find a function ϕ_0 which takes the same boundary values as ϕ_1 but is a solution of Laplace's equation $\nabla^2\phi = 0$ instead. Then $\phi_1 - \phi_0$ is the solution to the problem, for it vanishes at the boundary and

$$\nabla^2(\phi_1 - \phi_0) = \nabla^2\phi_1 = -1.$$

From the previous chapter we know that $e^{\pm(n\pi x/2)}\cos(n\pi y/2)$ is a solution of Laplace's equation. So is the function $\cosh(n\pi x/2)\cos(n\pi y/2)$. which has symmetry of the kind necessary here.

Evidently ϕ_0 is the function

$$\frac{16}{\pi^3}\sum\frac{(-1)^{(n-1)/2}}{n^3}\frac{\cos(n\pi y/2)\cosh(n\pi x/2)}{\cosh(n\pi/2)},$$

where the necessary divisor $\cosh(n\pi/2)$ has been introduced to enable ϕ_0 to match the value of ϕ_1 at $x = \pm 1$. The solution to the original problem is therefore

$$\phi = \frac{16}{\pi^3}\sum\frac{(-1)^{(n-1)/2}}{n^3}\cos\frac{n\pi y}{2}\left\{1 - \frac{\cosh(n\pi x/2)}{\cosh(n\pi/2)}\right\}.$$

The preceding examples concerned Poisson's equation in cases where the source density was uniform. The more general case where it is non-uniform figures in the next example, which concerns an interesting application to flow measurement.

***Example 9.3** Fluid motion along a pipe of radius a under a uniform transverse magnetic field B induces a measurable voltage ΔV between two electrodes XX at the ends of a diameter perpendicular to B as in figure 9.1. Show that

$$\Delta V = 2QB/\pi a$$

where Q = volumetric flow rate, whatever the velocity profile provided it is axisymmetric. The fluid velocity falls to zero at the pipe wall, which is non-conducting.

Take the origin at the centre of the pipe, the x-axis in the **B**-direction and the z-axis in the direction of motion. This is a two-dimensional problem with no z-dependence. The induced e.m.f. $\mathbf{v} \times \mathbf{B}$ is in the y-direction, and induced currents flow in x, y planes. Ohm's law

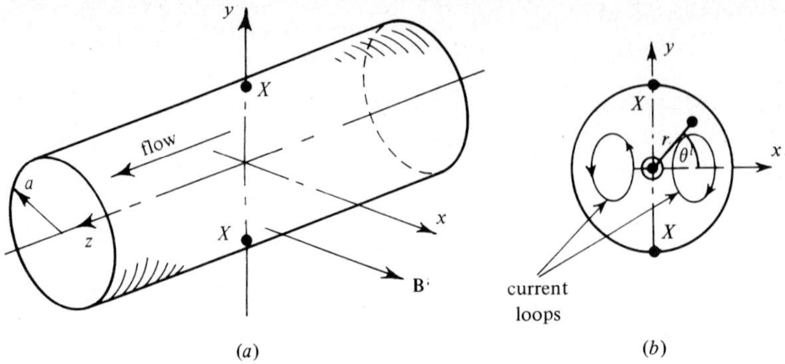

Fig. 9.1. (*a*) Electromagnetic flowmeter. (*b*) Current circulation in a cross-section.

$$\tau \mathbf{j} = -\operatorname{grad} V + \mathbf{v} \times \mathbf{B}$$

has x and y-components:

$$\tau j_x = -\partial V/\partial x \quad \text{and} \quad \tau j_y = -\partial V/\partial y + Bv,$$

V being electric potential. Kirchhoff's law $\operatorname{div} \mathbf{j} = 0$ gives

$$\frac{\partial j_x}{\partial x} + \frac{\partial j_y}{\partial y} = 0 \quad \text{and so} \quad \nabla^2 V = B\frac{\partial v}{\partial y},$$

a Poisson equation, given v. The source density term is essentially $\operatorname{div}(\mathbf{v} \times \mathbf{B})$.

If $v = v(r)$, an axisymmetric velocity profile, we have

$$\nabla^2 V = \left(\frac{\partial^2}{\partial r^2} + \frac{1}{r}\frac{\partial}{\partial r} + \frac{1}{r^2}\frac{\partial^2}{\partial \theta^2}\right) V = B\frac{\mathrm{d}v}{\mathrm{d}r}\sin\theta,$$

(r, θ) being polar coordinates. Try a solution for V of the form $U(r) \sin\theta$, which must satisfy

$$\frac{\mathrm{d}^2 U}{\mathrm{d}r^2} + \frac{1}{r}\frac{\mathrm{d}U}{\mathrm{d}r} - \frac{U}{r^2} = B\frac{\mathrm{d}v}{\mathrm{d}r},$$

after the $\sin\theta$ factor has been removed. This equation may be rewritten as

$$\left(r^2\frac{\mathrm{d}^2 U}{\mathrm{d}r^2} + 2r\frac{\mathrm{d}U}{\mathrm{d}r}\right) - \left(\frac{r\mathrm{d}U}{\mathrm{d}r} + U\right) = B\left\{r^2\frac{\mathrm{d}v}{\mathrm{d}r} + 2rv - 2rv\right\}.$$

Integrating from $r = 0$ to a gives

$$\left[r^2\frac{\mathrm{d}U}{\mathrm{d}r} - rU\right]_0^a = \left[Br^2 v\right]_0^a - 2B\int_0^a rv\,\mathrm{d}r,$$

or

$$aU(a) = 2B\int_0^a rv\,\mathrm{d}r,$$

if we make use of the conditions that v and $\mathrm{d}U/\mathrm{d}r$ vanish when $r = a$. (At the wall, the radial components of **j** and **v** × **B** vanish and therefore so must $\partial V/\partial r$, which equals $(\mathrm{d}U/\mathrm{d}r)\sin\theta$.)

But

$$Q = 2\pi \int_0^a v r \,\mathrm{d}r$$

and the potentials at the electrodes ($r = a, \theta = \pm\pi/2$) are $\pm U(a)$. Hence

$$\Delta V = 2U(a) = 2BQ/\pi a.$$

Thus the output signal from such an electromagnetic flowmeter can be independent of velocity profile, if it is axisymmetric.

The two-dimensional form of Poisson's equation with a *uniform* source density term on the right-hand side plays an important part in various other engineering problems, but for reasons other than the flux/gradient situations discussed in this chapter. For completeness, we merely list the following cases:

(*a*) Rectilinear viscous flow in pipes of arbitrary cross-section, for which the velocity distribution $v_z(x, y)$ is given by

$$\mu\nabla^2 v_z = \partial p/\partial z \text{ (const.)},$$

p being pressure and μ viscosity. (See problem 9.3.) v_z falls to zero at the walls. (An *inviscid* flow case appears in chapter 12.)

(*b*) Torsion of elastic shafts of arbitrary cross-section, for which

$$\nabla^2\psi = -2G\phi \text{ (const.)},$$

ψ being a so-called stress function, G the shear modulus and ϕ the twist per unit length. ψ is zero at the surface.

(*c*) Deflection of a stretched membrane or surface tension film under a pressure difference Δp. If the tension is T per unit length, then the deflection z is given by

$$\nabla^2 z = -\Delta p/T,$$

provided the deflection is small. (This equation is true even if Δp varies; see example 9.5.)

The analogy between these cases can be exploited: each analytical solution can be interpreted physically in several ways; or measurements made on an experiment with one physical realisation can be used to predict the behaviour of another. The *membrane analogy* has been used to solve shaft-torsion problems by experiments on the analogous soap-film problem, for instance.

The electrostatic potential due to a point change Q at a distance r is equal to $-Q/4\pi\epsilon_0 r$ (if we take the datum at infinity). With a continuous distribution of net charge density q, the potentials of individual charged volume elements $\mathrm{d}\tau$ containing net charge $q\,\mathrm{d}\tau$ can be superposed to give solutions of Poisson's equation $\nabla^2 V = -q/\epsilon_0$ in the form

$$V = -\frac{1}{4\pi}\iiint\left(\frac{q}{\epsilon_0}\right)\frac{\mathrm{d}\tau}{r},$$

in which r varies appropriately. The presence of surface charge at boundaries must also be allowed for. There is a branch of applied mathematics called *potential theory* given over to finding solutions to Poisson's equation in this and related forms.

9.3 Helmholtz's equation: criticality

This equation governs steady-state phenomena described by a flux/gradient law whenever the source density for the flux is for some reason proportional to the local value of the potential. In mathematical terms

$$\mathbf{v} = -\operatorname{grad}\phi, \quad \operatorname{div}\mathbf{v} \propto \phi \quad \text{and so} \quad \nabla^2\phi \propto -\phi,$$

which is the Helmholtz form. Note the choice of the negative sign convention for potential here. $\nabla^2\phi$ has the *opposite* sign to ϕ in the physical problems of greatest interest.

One of the most striking applications is to nuclear power. Helmholtz's equation provides the simplest model of nuclear reactor *criticality*. The critical condition is that where the size of the nuclear reactor has been correctly chosen so that the net production rate of neutrons from fission shall be just sufficient to balance leakage and absorption. For simplicity the reactor is treated as homogeneous, with fissile material, moderator, etc. uniformly distributed. Chapter 7 revealed that neutrons leak away or diffuse according to the relation $\mathbf{N} = -D \operatorname{grad} n$, where $\mathbf{N}$, the neutron diffusion intensity, is (number passing per unit area and time), whereas n, the neutron density, is (number present per unit volume). The net rate at which neutrons are steadily released from fission in the face of absorption at each point is proportional to the number present,

i.e. $$\operatorname{div}\mathbf{N} = Cn \quad (C \text{ const.}).$$

The constant coefficients D and C have values that depend on the particular mixture of ingredients in the reactor and could be found from kinetic theory or experiment. Combining the two equations gives a Helmholtz equation

$$D\nabla^2 n + Cn = 0.$$

The specification of the problem must be completed by boundary conditions. An admittedly over-simplified but not wholly misleading condition is to take $n = 0$ at the surface of the reactor, if it is a 'bare' reactor, not surrounded by any neutron-reflecting shield.

The nature of the criticality problem is most readily seen in the following simple example.

Example 9.4 A reactor consists of a bare slab of active material of thickness d and of virtually infinite extent in other directions. Find the critical thickness in terms of the material constants C and D.

The problem is one-dimensional, with all variables dependent only on x measured

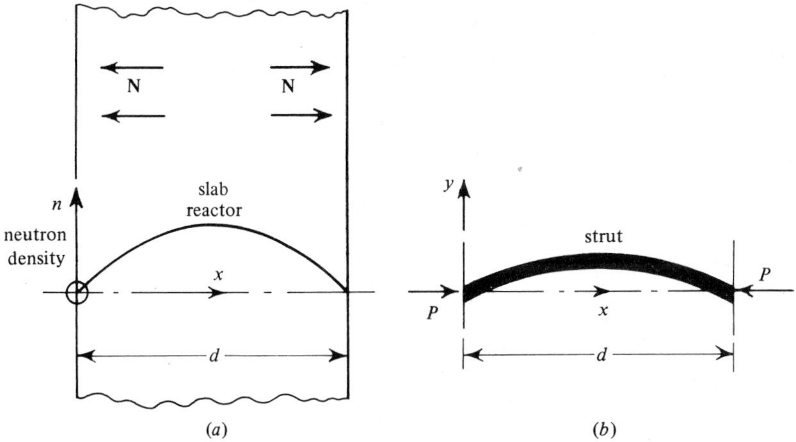

Fig. 9.2. (*a*) Critical slab nuclear reactor. (*b*) Analogous strut-buckling problem.

normal to the slab, and **N** oriented in the $\pm x$-directions (see figure 9.2(*a*)). The Helmholtz equation becomes

$$\frac{\mathrm{d}^2 n}{\mathrm{d}x^2} + \frac{C}{D} n = 0,$$

with $n = 0$ at $x = 0$ or d, if the origin is taken in one face. The simple, ordinary differential equation has the usual sinusoidal solutions. The condition at $x = 0$ limits these to

$$n = A \sin \{(C/D)^{1/2} x\}$$

and the condition at $x = d$ requires that

$$d(C/D)^{1/2} = k\pi$$

(k a positive integer) if we reject the non-event $A = 0$. We must also reject all values of k but unity, because other values allow the sine profile of n to attain negative values within the range $0 < x < d$, which would be physical nonsense. A negative number of neutrons is impossible. It follows that

$$d = \pi(D/C)^{1/2}$$

is the critical thickness which enables the reactor to run steadily with leakage and net production of neutrons just in balance. If

$$d \neq \pi(D/C)^{1/2},$$

the reactor is either inactive ($A = 0$) or in an unsteady state, with its activity falling or rising (the subject of reactor kinetics). Criticality demands that the size be correctly matched to the material characteristics C and D. Problem 9.4 considers some three-dimensional cases.

Any reader familiar with the theory of elastic structures may be struck by the analogy between this slab reactor problem and the buckling of ideal struts. (See figure 9.2(*b*).) The transverse deflection y of a strut of uniform flexural rigidity EI under an axial load P, applied through pin-jointed ends, obeys the bending equation

$$\frac{\mathrm{d}^2y}{\mathrm{d}x^2}+\frac{P}{EI}y = 0$$

for small deflections, x being measured axially from one end. If the length is d, the boundary conditions are $y = 0$ at $x = 0$ or d. The analogy with the reactor problem is now apparent. Non-zero steady deflection is only possible if

$$d(P/EI)^{1/2} = k\pi,$$

and in practice only the case $k = 1$ is of interest. Here criticality demands that the load P be correctly matched to the strut's geometrical and elastic properties d, I and E. The metaphorical term 'buckling' has in fact passed into the jargon of reactor technology on the strength of this analogy.

Mathematically speaking, both problems are *eigenvalue* problems: a non-trivial solution only exists for one or more particular values (eigenvalues) of some governing parameter. Another universal feature of eigenvalue problems is that, in the critical condition, the amplitude of the solution (A for the slab reactor) is not determined by the prevailing conditions. This occurs because the Helmholtz equation and the boundary conditions are homogeneous; any solution may therefore be multiplied by an arbitrary constant, i.e. its amplitude is indeterminate. (In the reactor problem, the amplitude is determined by the reactor's previous history, which makes the control of reactors a particularly interesting problem. There is a loose but instructive analogy with a motor car: the direction in which a car is travelling steadily depends not on the position of the steering wheel but on its previous history of positions since starting out.)

Problem 9.5 explores an example of criticality drawn from another field. Those who remember open coal fires remember this kind of criticality problem – you cannot light a fire with too little fuel! We shall see later that the Helmholtz equation also arises in solving the wave equation.

The example below shows how the idea of a separable solution, as used for Laplace's equation in example 8.4, is also applicable to Helmholtz equations.

Example 9.5 The rigid rim of a light tent roof is a horizontal rectangle with sides of length a and b. When under uniform tension T per unit length and a vertical load W per unit area the fabric deflects a distance z below the plane of the rim according to the equation $\nabla^2 z = -W/T$, *provided z is small compared with a and b*, the x and y axes being adjacent edges of the rim. (Here ∇^2 is two-dimensional.) Show that the roof can be steadily deflected by a pool of rain covering the whole roof provided ρg, the weight per unit volume of the water, equals $\pi^2 T(a^{-2}+b^{-2})$.

In the deflected state, $W = \rho g z$ (for $p + \rho g z$ = const. in the water). The equation

$$\nabla^2 z = -(\rho g/T)z$$

is of Helmholtz form. A separable solution of the form

$$z = X(x)Y(y)$$

would have to satisfy

$$\frac{d^2X}{dx^2}Y + X\frac{d^2Y}{dy^2} = -\frac{\rho g}{T}XY$$

or

$$\frac{1}{X}\frac{d^2X}{dx^2} + \frac{1}{Y}\frac{d^2Y}{dy^2} = -\frac{\rho g}{T}, \quad \text{for all } x, y.$$

By differentiating this equation at constant y or x, it can be seen that

$$\frac{1}{X}\frac{d^2X}{dx^2} \quad \text{and} \quad \frac{1}{Y}\frac{d^2Y}{dy^2} \quad \text{are constant.}$$

That d^2X/dx^2 and d^2Y/dy^2 will be negative is intuitively obvious from the likely curvature of the tent roof. So we take

$$\frac{1}{X}\frac{d^2X}{dx^2} = -\alpha^2, \quad \frac{1}{Y}\frac{d^2Y}{dy^2} = -\beta^2, \quad \text{where } \alpha^2 + \beta^2 = \frac{\rho g}{T}.$$

The boundary conditions require that

$$X = 0 \text{ at } x = 0 \text{ and } a, \quad \text{and} \quad Y = 0 \text{ at } y = 0 \text{ and } b.$$

Thus $\quad X = A \sin \alpha x \quad$ where $\quad \sin \alpha a = 0, \alpha a = \pi$

(for we must reject roots 2π, etc. which would involve negative water depths). Similarly

$$Y = B \sin \beta x \quad \text{where} \quad \beta b = \pi.$$

We conclude that

$$\frac{\rho g}{T} = \frac{\pi^2}{a^2} + \frac{\pi^2}{b^2}.$$

This is the 'criticality' condition which correctly matches ρg, T and the size of the roof, for small deflections.

9.4 The unsteady diffusion equation

We now move on to a third class of field problems in which **v** still equals ± grad ϕ and div **v** is not zero but instead is proportional to the rate of change of ϕ in time. Such relations arise naturally in connection with many unsteady conduction or diffusion problems. Consider for example unsteady heat conduction in the absence of any heat release other than that associated with the material's own heat capacity.

Then $\quad \mathbf{Q} = -K \text{ grad } T \quad$ and $\quad \text{div } \mathbf{Q} = -C\partial T/\partial t,$

C being the volumetric heat capacity. Eliminating $\mathbf{Q}$ gives the scalar equation

$$\frac{\partial T}{\partial t} = \frac{K}{C}\nabla^2 T.$$

This form of equation is called the *unsteady diffusion equation* or alternatively the *heat conduction equation*, if it concerns heat flow.

In ground water flows, with $\mathbf{v}^* = -k \operatorname{grad} p^*$, the compressibility of the fluid and the ground can sometimes be modelled by the equation

$$\operatorname{div} \mathbf{v}^* = -S\partial p/\partial t = -S\partial p^*/\partial t$$

(z being a constant during time-differentiation) and then

$$\frac{\partial p^*}{\partial t} = \frac{k}{S}\nabla^2 p^*,$$

another equation of the same type. In the case of diffusion of a foreign species through a host medium (e.g. dopant in a semiconductor) we have similarly

$$\operatorname{div} \mathbf{N} = -\partial n/\partial t \quad \text{and} \quad \mathbf{N} = -D \operatorname{grad} n,$$

so that $\partial n/\partial t = D\nabla^2 n$, often called Fick's second law. In all these cases of the unsteady diffusion equation, the coefficient K/C, k/S, D, etc. has dimensions $(\text{length})^2(\text{time})^{-1}$ and is described as a *diffusivity*. Later on we shall find that even electromagnetic fields obey the same equation under certain conditions, while problem 9.6 investigates a *viscous* diffusion phenomenon. Problem 9.7 studies some other thermal diffusion problems.

For the present we will limit consideration to an example which quickly reveals the essential nature of all the phenomena governed by the unsteady diffusion equation. It also enables us to demonstrate another widely applicable technique for rendering partial differential equations more tractable by reducing them to ordinary differential equations. For simplicity we consider a one-dimensional problem, i.e. one in which spatial dependence is confined to the x-coordinate only, say, so that

$$\frac{\partial T}{\partial t} = \alpha \frac{\partial^2 T}{\partial x^2}.$$

If we take a heat conduction problem then α is the *thermal diffusivity* K/C.

Example 9.6 A uniformly conducting medium of thermal diffusivity α occupies the semi-infinite region $x > 0$. Initially the medium is at a uniform temperature $T = 0$. For $t \geqslant 0$, the surface $x = 0$ is suddenly raised to and held at a constant temperature T_0, by a suitable source of heat, uniformly distributed over the surface. Show that each level of temperature has penetrated a distance proportional to $t^{1/2}$ after time t.

We would expect the subsequent temperature profiles to be of the form sketched

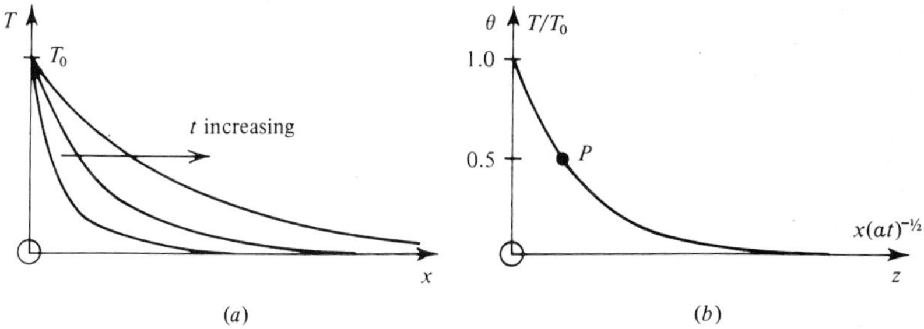

Fig. 9.3. (*a*) Temperature penetrations at various times. (*b*) Universal, dimensionless plot of (*a*).

in figure 9.3(*a*), each curve corresponding to an instant in time. T asymptotes to zero deep in the medium.

Dimensional analysis provides the way ahead. (Dimensional arguments are not the sole prerogative of the experimentalist, as is sometimes supposed!) This problem has only two characteristic parameters, namely T_0 and α. There is, for instance, no length scale. Therefore, at the plane defined by x at the instant defined by t, the temperature T is given by some function

$$T = f(x, t, \alpha, T_0).$$

As the *five* variables mentioned are expressible in terms of *three* basic units, length, time and temperature, only *two* independent dimensionless groups, T/T_0 and $x(\alpha t)^{-1/2}$, say, can be formed from them. Buckingham's 'pi' theorem therefore implies that

$$T/T_0 = \theta \text{ (say)} = F(z), \quad \text{where} \quad z = x(\alpha t)^{-1/2}.$$

This equation tells us that if we replot figure 9.3(*a*) as a graph of T/T_0 against $x(\alpha t)^{-1/2}$, as in figure 9.3(*b*), then all the profiles will fall on to one curve as shown; the one curve describes the profile at all instants. For this reason, such solutions of partial differential equations are described as *self-similar* or *similarity* solutions. θ, being a function of one variable only, must be governed by an *ordinary* differential equation.

Consider a given level of temperature, $\frac{1}{2}T_0$ say, indicated by P in figure 9.3(*b*). Evidently, after a time t, this level has penetrated a distance x into the medium that is proportional to (and probably of the same order as) $(\alpha t)^{1/2}$, i.e. penetration is proportional to $t^{1/2}$.

The further solution of this case is relegated to problem 9.8(*a*) since it concerns mathematical technique rather than new ideas in vector field theory. (The *ordinary* differential equation relating θ to z is to be formed and solved.)

This example has exposed the essential characteristic of the unsteady diffusion equation: the effect of a disturbance propagates distances proportional to the square root of time elapsed since the event. (This makes a dramatic contrast with the property of the wave equation, discussed later in the chapter, namely, that disturbances propagate over distances proportional to the time itself.) Notice also that the penetration distance is proportional to the square root of the diffusivity. Proper exploitation of the properties of this differential equation is half the art of good cooking, whether of Sunday joints or doped semiconductors!

Another general property of the diffusion equation which is not so easily demonstrated is that any disturbance causes small perturbations *immediately* at great distances (whereas with the wave equation, no perturbation *whatever* spreads faster than the prevailing wave velocity).

Another thermal case is that where the boundary condition at the surface states that T varies sinusoidally in time. This is a problem of interest to mechanical engineers concerned with fluctuating temperatures in engines, and to foundation engineers, agriculturalists, and others concerned with diurnal or annual fluctuations in ground temperature. If the frequency of fluctuation is ω, the characteristic time is $1/\omega$ and the ideas above strongly suggest that fluctuations will penetrate a distance of the order of $(\alpha/\omega)^{1/2}$. We postpone further discussion of this case until chapter 11, where under the heading 'electromagnetic skin effect' we will solve the equation for the case of a sinusoidal boundary condition, and confirm this estimate of penetration distance.

Problem 9.9 explores a variant of the unsteady diffusion equation which is non-linear. There are many other flux/gradient field problems where the source density is related to the potential ϕ in even more subtle ways. Problem 9.10 concerns an example that arises in connection with thermionic devices, where again the differential equation is non-linear. In contrast the Laplace, Poisson, Helmholtz, diffusion and wave equations are all linear, i.e. the potential ϕ and its derivatives appear only as first-degree terms.

Before we can study the wave equation, the chapter's fourth and last kind of linear partial differential equation, we must pause briefly to extend our treatment of fluid mechanics.

9.5 Inertial fluid mechanics: the D/D*t* and (v · grad) operators

So far we have mounted only a limited attack on fluid mechanics, confining attention to hydrostatics and flow in a porous medium, both of them cases where the inertia of the fluid plays no part. To cope with more general problems involving inertia, we need a mathematical formulation of fluid acceleration, the rate at which the velocity changes in time as experienced by a small travelling fluid element. Such a time-differentiation is called *Lagrangian*; we denote this differentiation with the notation $\mathrm{D}/\mathrm{D}t$. For example $\mathrm{D}p/\mathrm{D}t$ is the rate at which the pressure p of a travelling fluid element is changing. That it is not the same as $\partial p/\partial t$, the rate of change in time at a fixed point in space, is evident from the case of steady compressible flow, in which $\partial p/\partial t = 0$ but a fluid element expands or contracts as it travels. In contrast, problem 9.6 provides an example where $\mathrm{D}v/\mathrm{D}t$ and $\partial v/\partial t$ are equal, v being the speed of the fluid. The example below clarifies the situation.

Example 9.7 If the density ρ is a given function of x, y, z and t, express $\mathrm{D}\rho/\mathrm{D}t$ in terms of the derivatives of this function.

For a travelling fluid element, ρ is changing because all of x, y, z and t are changing. $\mathrm{D}\rho/\mathrm{D}t$ is therefore given by the total differential theorem:

$$\frac{\mathrm{D}\rho}{\mathrm{D}t} = \frac{\partial \rho}{\partial t}\frac{\mathrm{D}t}{\mathrm{D}t} + \frac{\partial \rho}{\partial x}\frac{\mathrm{D}x}{\mathrm{D}t} + \frac{\partial \rho}{\partial y}\frac{\mathrm{D}y}{\mathrm{D}t} + \frac{\partial \rho}{\partial z}\frac{\mathrm{D}z}{\mathrm{D}t}.$$

$\mathrm{D}t/\mathrm{D}t$ is unity and $\mathrm{D}x/\mathrm{D}t$ for a travelling fluid element is v_x, the x-component of velocity, and so on. If Σ denotes a sum over three terms,

$$\frac{\mathrm{D}\rho}{\mathrm{D}t} = \frac{\partial \rho}{\partial t} + \sum v_x \frac{\partial \rho}{\partial x}.$$

Problem 9.11(*a*) applies the corresponding result for a fluid's temperature variation.

An equivalent form is

$$\mathrm{D}\rho/\mathrm{D}t = \partial\rho/\partial t + \mathbf{v} \cdot \text{grad}\ \rho.$$

The last term measures the contribution due to spatial non-uniformity. The bigger the component of **v** in the grad ρ direction, the faster the element moves into regions where ρ is significantly different. The group (**v** · grad) can be regarded as the operator $v\partial/\partial s$, which equals the magnitude of **v** times the derivative with respect to distance s measured along a field line (or streamline, in the present case), for $\partial\rho/\partial s$ is the component of grad ρ in the **v**-direction. Evidently we can write the operator $\mathrm{D}/\mathrm{D}t$ in the compound form

$$\mathrm{D}/\mathrm{D}t = (\partial/\partial t + \mathbf{v} \cdot \text{grad}).$$

The mass conservation equation in section 7.10 can now take the slightly terser form

$$\mathrm{D}\rho/\mathrm{D}t = -\rho\ \text{div}\ \mathbf{v}.$$

This reveals that div $\mathbf{v} = 0$ and **v** is solenoidal if each travelling fluid element is incompressible, i.e. incapable of changing its density so that $\mathrm{D}\rho/\mathrm{D}t = 0$. Notice that the fluid density may nevertheless be non-uniform, perhaps because of variations in the upstream supply, as in problem 9.11(*b*). Problem 9.12 uses the $\mathrm{D}/\mathrm{D}t$ operator in the formulation of energy statements.

Turning again to the acceleration **a** and taking each component in turn, we see that

$$\mathbf{a} = \frac{\mathrm{D}\mathbf{v}}{\mathrm{D}t} \equiv \frac{\partial \mathbf{v}}{\partial t} + \sum v_x \frac{\partial \mathbf{v}}{\partial x} \equiv \frac{\partial \mathbf{v}}{\partial t} + (\mathbf{v} \cdot \text{grad})\mathbf{v}$$

if we generalise the idea of the (**v** · grad) operator (see section 7.10) to apply to a vector field. It is essential to retain the bracket, because **v** · grad **v**, like grad **v** itself, is meaningless. As $\mathbf{v} = \Sigma \mathbf{i} v_x$ and **i**, etc. are constant vectors, the x-component of $\mathrm{D}\mathbf{v}/\mathrm{D}t$ is

$$\partial v_x/\partial t + \mathbf{v} \cdot \text{grad } v_x;$$

which needs no brackets, as v_x is a scalar. An expression such as $(\mathbf{u} \cdot \text{grad})\mathbf{v}$ means the rate of change of $\mathbf{v}$ with distance s taken in the $\mathbf{u}$-direction times the magnitude of $\mathbf{u}$, i.e. $u\partial\mathbf{v}/\partial s$, evaluated at a given instant. Note that $\mathbf{v}$ may be changing in direction as well as magnitude (see problem 9.13).

We are now equipped to state Newton's law for fluid-mechanical situations in which viscous stresses and forces other than pressure gradients are negligible. For a small travelling fluid element of any shape, volume ϵ and mass $\rho\epsilon$, the net force due to non-uniform pressures is $-(\text{grad } p)\epsilon$, as usual, and so

$$\rho\epsilon\, \mathrm{D}\mathbf{v}/\mathrm{D}t = -(\text{grad } p)\epsilon \quad \text{and} \quad \rho\, \mathrm{D}\mathbf{v}/\mathrm{D}t + \text{grad } p = \mathbf{0},$$

which is *Euler's equation.* The extension of this equation to include gravity appears as problem 9.14. It is a very intractable differential equation, chiefly because the $(\mathbf{v} \cdot \text{grad})\mathbf{v}$ part of the $\mathrm{D}\mathbf{v}/\mathrm{D}t$ is non-linear, i.e. quadratic, in $\mathbf{v}$ and its derivatives.

In general, flow is unsteady; the velocity field changes from one instant to the next. Then the streamline pattern, composed of the field lines of the velocity field, describes the directions of flow *at an instant*; it does not describe the loci of particular fluid particles (which will come under the influence of a whole series of different streamline patterns at successive instants); nor does any streamline coincide with the instantaneous state of a line of dye, say, injected into the flow at an upstream point. Such a line, called a *streak-line*, connects all fluid elements which passed through that point at earlier instants, when the streamline pattern was different. Fluid particle loci and streak-lines are also different in unsteady flow. Problem 9.15 gives an example.

Steady flow is more straightforward, however. There the three families of lines (streamlines, fluid loci, streak-lines) all become the same and it becomes considerably simpler to identify the fluid acceleration $\mathbf{a}$. Since $\partial\mathbf{v}/\partial t = \mathbf{0}$ now,

$$\mathbf{a} = (\mathbf{v} \cdot \text{grad})\mathbf{v} \quad \text{or} \quad v\partial\mathbf{v}/\partial s,$$

where s is distance measured along a streamline. Thinking of the streamline as the locus of a fluid element now, we may apply the results that the acceleration has a component $v\partial v/\partial s$ along the streamline and a component v^2/R perpendicular to it, along the principal normal, R being the radius of curvature. (See problem 1.12(*c*).) The reader should note that these two components of $v\partial\mathbf{v}/\partial s$ are respectively due to the changing *magnitude* of $\mathbf{v}$ and to its changing *direction* associated with the curvature.

The streamwise component of Euler's equation in steady flow is

$$\rho v\partial v/\partial s + \partial p/\partial s = 0,$$

which may be integrated along the streamline to give *Bernoulli's equation*

$$\tfrac{1}{2}\rho v^2 + p = \text{const.} \quad \text{along a streamline, if } \rho \text{ is constant.}$$

In steady flow, the transverse component of Euler's equation is

$$\rho v^2/R + \partial p/\partial n = 0,$$

if n is measured along the principal normal, on the concave side. This equation does not have a general integral. Problems 9.16 and 9.17 respectively explore some consequences of these streamwise and transverse equations. They are most useful in elementary problems where the streamline pattern is known *ab initio* because of symmetry or other constraints.

9.6 Perturbations in a compressible fluid; the scalar wave equation

Progress can often be made with non-linear problems such as the general fluid-mechanical problem by turning them into linear ones by a process of approximation called *linearisation*. This section exhibits this very important general mathematical technique in the context of the propagation of small disturbances (sound waves) in a compressible fluid which is otherwise uniform (gravity being neglected). The linearisation process then constitutes the *acoustic approximation*.

The non-linear equations that govern the problem are:

Mass conservation

$$\partial\rho/\partial t + \operatorname{div} \rho\mathbf{v} = 0 \quad \text{and}$$

Euler's equation

$$\rho\partial\mathbf{v}/\partial t + \rho(\mathbf{v}\cdot\operatorname{grad})\mathbf{v} + \operatorname{grad} p = \mathbf{0}.$$

We also need a *p/ρ relation* derived from thermodynamics to describe how change of pressure causes change of density. As sound waves are reversible, adiabatic and therefore isentropic, we shall use the positive thermodynamic property $(\partial p/\partial\rho)_s$ to relate changes of p and ρ.

Linearisation depends on the key idea that if each variable such as density ρ is treated as being equal to a *uniform* value ρ_0 (corresponding to the unperturbed state) plus a small *perturbation* ρ' then we can ignore terms in the equations which are of more than first degree in the dashed, perturbation quantities.

Example 9.8 Find the linearised forms of the terms $\rho\mathbf{v}$ and $(\mathbf{v}\cdot\operatorname{grad})\mathbf{v}$.

$$\rho\mathbf{v} = (\rho_0 + p')(\mathbf{v}_0 + \mathbf{v}') = \rho_0\mathbf{v}_0 + \rho_0\mathbf{v}' + \rho'\mathbf{v}_0 + \rho'\mathbf{v}'$$

which approximates to

$$\rho_0\mathbf{v}_0 + \rho_0\mathbf{v}' + \rho'\mathbf{v}_0,$$

linear in $\mathbf{v}'$ and ρ', if we neglect the quadratic term $\rho'\mathbf{v}'$. The intervention of differentiation does not alter the degree of a term, but differentiation applied to a constant quantity like ρ_0 removes it. Therefore the term

$$(\mathbf{v}\cdot\operatorname{grad})\,\mathbf{v} = (\mathbf{v}\cdot\operatorname{grad})\mathbf{v}' = (\mathbf{v}_0\cdot\operatorname{grad})\mathbf{v}' + (\mathbf{v}'\cdot\operatorname{grad})\mathbf{v}',$$

which linearises to $(\mathbf{v}_0\cdot\operatorname{grad})\mathbf{v}'$, because the second term is of second degree in $\mathbf{v}'$.

Problem 9.18 pursues the case of steady flow with $\mathbf{v}_0 \neq \mathbf{0}$. If instead the fluid is at rest in the unperturbed state, with $\mathbf{v}_0 = \mathbf{0}$, then after linearisation $\rho\mathbf{v}$ becomes $\rho_0\mathbf{v}'$ and $(\mathbf{v} \cdot \text{grad})\mathbf{v}$ becomes zero in Euler's equation. The mass conservation equation degenerates to

$$\partial\rho'/\partial t + \rho_0 \operatorname{div} \mathbf{v}' = 0, \qquad (\alpha)$$

since ρ_0 is independent of t, x, y or z. The equation (α) expresses the source density in the velocity field and is analogous to the conservation equation relating $\partial T/\partial t$ and div $\mathbf{Q}$ in the case of heat conduction. Euler's equation enables us to relate the scalar ρ' and the vector $\mathbf{v}'$ in another way so as to complete the set of differential equations. In Euler's equation $\rho\partial\mathbf{v}/\partial t$ linearises to $\rho_0\partial\mathbf{v}'/\partial t$ while

$$\text{grad}\, p = \text{grad}\, p' = a_0^2 \,\text{grad}\, \rho'$$

because, for small perturbations from the unperturbed state,

$$\frac{p'}{\rho'} = \frac{\delta p}{\delta \rho} = a_0^2, \quad \text{the value of } \left(\frac{\partial p}{\partial \rho}\right)_s \text{ in the unperturbed state.}$$

(Note that pressure perturbations p' are readily expressed as $a_0^2\rho'$.) So Euler's equation degenerates to

$$\rho_0\partial\mathbf{v}'/\partial t + a_0^2 \,\text{grad}\, \rho' = \mathbf{0} \qquad (\beta)$$

i.e. it is $\partial\mathbf{v}'/\partial t$ rather than $\mathbf{v}$ itself which is proportional to the gradient of the scalar. This is what makes this case so different from heat conduction, where $\mathbf{Q}$ is proportional to grad T.

Because the order of time and space differentiation can be exchanged,

$$\operatorname{div} (\partial\mathbf{v}'/\partial t) = (\partial/\partial t) \operatorname{div} \mathbf{v}'.$$

This allows elimination of $\mathbf{v}'$ from (α) and (β) to yield

$$\partial^2\rho'/\partial t^2 = a_0^2\nabla^2\rho',$$

the partial differential equation that governs the variations ρ' in the scalar variable ρ. It is an example of the *three-dimensional scalar wave equation*.

Yet again the operator ∇^2 finds an important application. By involving a *second* derivative with respect to time, the wave equation completes an interesting progression alongside the two others which we met earlier in the chapter:

$$\phi = \pm \text{ const.} \times \nabla^2\phi \quad \text{(Helmholtz's equation)},$$

$$\frac{\partial\phi}{\partial t} = \text{const.} \times \nabla^2\phi \quad \text{(unsteady diffusion equation)},$$

and

$$\frac{\partial^2\phi}{\partial t^2} = \text{const.} \times \nabla^2\phi \quad \text{(wave equation)}.$$

The behaviour of solutions of the wave equation is very different from that of solutions of the diffusion equation, as was remarked in section 9.4.

9.7 The one-dimensional wave equation

This is not the place for a full discussion of the wave equation in three-dimensions. A short review of the properties of the one-dimensional version of the wave equation is worthwhile, however, for readers who have not met the subject before. It will enable us to show why the equation is called the *wave* equation.

In a one-dimensional problem, all variables depend on time and on one space coordinate only, x say, being uniform over each plane $x = \text{const.}$ at each instant, for all values of y and z. Then the equation for acoustic disturbances becomes

$$\frac{\partial^2 \rho'}{\partial t^2} = a_0^2 \frac{\partial^2 \rho'}{\partial x^2},$$

the *one-dimensional wave equation*, which in section 8.11 we contrasted with Laplace's equation in two dimensions. Of all the commoner partial differential equations, it is the easiest to solve, for the general solution is

$$\rho' = f(x - a_0 t) + g(x + a_0 t),$$

in which f and g are *any* functions. Note that $f(x - a_0 t)$ is a particular kind of function of x and t, namely a function of w, the combination $x - a_0 t$. The f and g parts are different solutions of the equation and their sum is one also, because the equation is linear. Function f satisfies the equation because $\rho' = f(w)$ implies that

$$\frac{\partial \rho'}{\partial x} = \frac{\mathrm{d}f}{\mathrm{d}w}\frac{\partial w}{\partial x} = \frac{\mathrm{d}f}{\mathrm{d}w} \quad \text{and} \quad \frac{\partial^2 \rho'}{\partial x^2} = \frac{\mathrm{d}^2 f}{\mathrm{d}w^2}\frac{\partial w}{\partial x} = \frac{\mathrm{d}^2 f}{\mathrm{d}w^2}$$

whereas

$$\frac{\partial \rho'}{\partial t} = \frac{\mathrm{d}f}{\mathrm{d}w}\frac{\partial w}{\partial t} = -a_0\frac{\mathrm{d}f}{\mathrm{d}w}$$

and

$$\frac{\partial^2 \rho'}{\partial t^2} = -a_0\frac{\mathrm{d}^2 f}{\mathrm{d}w^2}\frac{\partial w}{\partial t} = a_0^2\frac{\mathrm{d}^2 f}{\mathrm{d}w^2} = a_0^2\frac{\partial^2 \rho'}{\partial x^2}.$$

A solution of the form

$$\rho' = f(w), \quad w = x - a_0 t,$$

can easily be seen to represent a wave. Figure 9.4 shows part of a graph of the function f, relating ρ' to w, Q being the origin. It is also a graph of ρ' against position x for the instant $t = 0$. Now take a point O on the negative w-axis such that $OQ = a_0 t$, which increases linearly with time at a speed a_0. For any point P, $OP = w + a_0 t = x$. Figure 9.4 with O as origin can therefore be regarded, at each instant t, as a graph of ρ' against position x. We can take O as the fixed origin and allow Q together with the profile $\rho' = f(w)$, fixed relative to Q, to move at constant speed a_0 to the right. This confirms that $\rho' = f(x - a_0 t)$ represents a *wave*, i.e. a distribution travelling with an unchanging profile at a constant speed equal to the thermodynamic property

$$a_0 = \{(\partial p/\partial \rho)_s\}^{1/2}.$$

Problem 9.19 explores these ideas from other aspects.

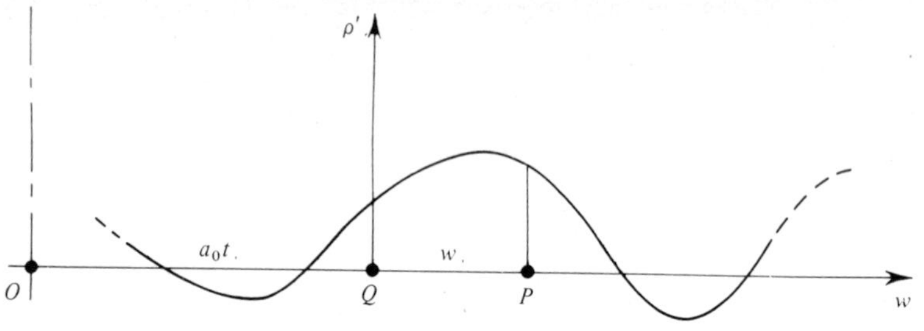

Fig. 9.4

In the one-dimensional case, equation (α) in section 9.6 becomes

$$\partial\rho'/\partial t + \rho_0 \partial v_x'/\partial x \;=\; 0.$$

Corresponding to the right-travelling wave $\rho' = f(w)$ there will be a travelling distribution $v_x' = h(w)$, say. Then

$$\frac{\partial v_x'}{\partial x} = \frac{\mathrm{d}h}{\mathrm{d}w}\frac{\partial w}{\partial x} = \frac{\mathrm{d}h}{\mathrm{d}w}, \quad \text{so} \quad a_0\frac{\mathrm{d}f}{\mathrm{d}w} = \rho_0\frac{\mathrm{d}h}{\mathrm{d}w}.$$

Integrating gives $a_0 f = \rho_0 h$, for ρ' and v_x' vanish together in the unperturbed state. In other words $a_0\rho' = \rho_0 v_x'$, i.e. the wave profiles for density and velocity are proportional to each other. Problem 9.20 concerns the same calculations for a left-travelling wave.

The full general solution includes both $f(x - a_0t)$ and $g(x + a_0t)$. Combining these terms means that the fluctuating ρ' and v_x' distributions can be regarded as the superposition of two waves, travelling in opposite directions through each other.

Finally notice that equation (β) in section 9.6 indicates that there are no acceleration components in the y or z-directions (for grad ρ' is in the x-direction here). This means that v_y and v_z never change from their unperturbed values of zero. The vector variable $\mathbf{v}'$ in sound waves is therefore aligned in the direction of propagation. For this reason they are described as *longitudinal* waves, to be contrasted with *transverse* waves such as occur on a plucked guitar string, the velocity vector then being perpendicular to the direction of propagation.

The linear wave equation describes waves whose speed or profile does not change. The reader should note that 'non-linear' waves also occur. Then the speed or profile *does* change – we have all seen ocean waves breaking on a beach. Problem 9.21 provides an intriguing example.

9.8 Wave reflection and standing waves

Two-directional waves often occur because of the reflection of unidirectional waves off a plane boundary such as a wall. In mathematical terms, the reflected wave is created to help meet some boundary condition with which the incident wave is incompatible on its own. Consider for instance a wave travelling leftwards along a

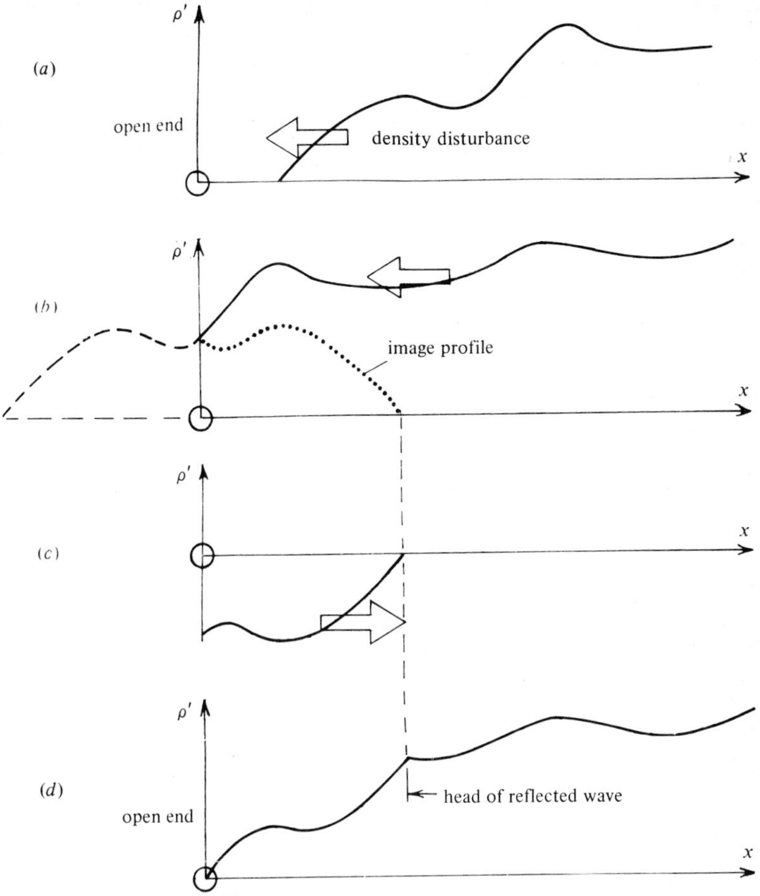

Fig. 9.5. (*a*) Incident wave before reflection. (*b*) Incident wave during reflection. (*c*) Reflected wave at the same instant as (*b*). (*d*) Total profile at the same instant as (*b*).

pipe and encountering its open end, as occurs in an open organ pipe. (See figure 9.5.) At the end, the external atmosphere constrains the pressure and therefore also the density ρ to be constant, i.e. the boundary condition there is $\rho' = 0$. Figure 9.5 shows how a reflected wave can combine with the incident wave so as to meet this condition. Profile (*d*) is the sum of profiles (*b*) and (*c*) at the same instant with due attention to sign. To keep $\rho' = 0$ at the end at all times, the reflected wave (*c*) is the negative reverse of the defunct part of the incident wave. In (*b*) this defunct part and its image in the ρ' axis are shown dashed to aid the construction of (*d*), which is the difference in height of the two curves in the right-hand half of (*b*). Observe how a *compression* wave, which raises the density as in (*a*), reflects off an *open* end as the *expansion* wave (*c*). Problem 9.22 explores reflection off a closed end where v must vanish and problem 9.23 extends the ideas to waves in solids.

Sound waves are often sinusoidal in structure, with a profile of form

$$\rho' = A \sin\{2\pi(x \pm a_0 t)/\lambda\},$$

in which A and λ are constants. A is the amplitude and λ is the wavelength, for an increase of λ at a given instant implies an increase of 2π in the argument of the sine function. The frequency is equal to a_0/λ, the number of wavelengths passing a point in unit time.

Repeated reflection can allow sinusoidal waves of equal amplitude travelling in opposite directions to build up. The typical solution is then

$$\begin{aligned}\rho' &= A \sin\left\{\frac{2\pi(x+a_0 t)}{\lambda}\right\} + A \sin\left\{\frac{2\pi(x-a_0 t)}{\lambda}\right\}\\ &= 2A \sin\frac{2\pi x}{\lambda}\cos\frac{2\pi a_0 t}{\lambda}.\end{aligned}$$

The reader should check directly that this expression satisfies the wave equation. The interference pattern produced by superposing these two waves does not give any impression of propagation in either direction. The amplitude of the profile $2A\sin(2\pi x/\lambda)$ merely fluctuates in time according to the extra cosine factor. Such a combined wave is called a *standing wave*. It forms the basis of most musical instruments: the various natural notes of a wind instrument or organ pipe come from standing acoustic waves.

Example 9.9 Longitudinal acoustic standing waves occur in an open-ended pipe of length L. Investigate what frequencies can occur.

The open-end boundary condition $\rho' = 0$ applies at the ends $x = 0$ and L, if x is measured from one end. Then the standing-wave solution given above can be applied provided that

$$\sin 2\pi L/\lambda = 0 \quad \text{and} \quad 2\pi L/\lambda = k\pi \quad (k \text{ integral}).$$

Thus $\lambda = 2L/k$ and the corresponding frequencies are $a_0/2L$, $2a_0/2L$, $3a_0/2L$, $4a_0/2L$, etc. The musical note of lowest frequency (the 'fundamental') has a wavelength equal to twice the pipe length. The next possible notes have twice the frequency (an 'eighth', i.e. 'octave' higher) and three times the frequency (a 'twelfth' higher), etc. (These numbers 8 and 12 come from the musician's habit of counting the notes up the diatonic scale.)

Pleasant harmony is made by blending frequencies that are related by simple arithmetic ratios – another outlet for Fourier series! Different instrumental tone qualities are largely explained by the different relative magnitudes of the various 'harmonics' (waves of different frequencies). The clarinet has only the *odd* harmonics, with frequencies in the ratios 1:3:5, etc.

Problem 9.24 explores various waves and oscillations of a musical kind. They serve as a useful introduction to the more complex vibration phenomena that concern engineers.

A wind instrument is an example of an important device known as a *resonant cavity*, much exploited not only in music and acoustics but also in laser and microwave engineering for promoting selected frequencies. Example 9.9 reveals that only certain frequencies are capable of reflecting within a given cavity in such a way as to generate a standing wave. Problem 9.25 adopts a slightly different approach.

Similar ideas apply to the three-dimensional wave equation,

$$\partial^2 \rho'/\partial t^2 = a_0^2 \nabla^2 \rho'.$$

If we postulate a solution

$$\rho' = \phi(x, y, z) \cos \omega t,$$

i.e. a standing wave in three dimensions, where the amplitudes everywhere oscillate in phase or antiphase at frequency ω, then

$$\nabla^2 \phi + (\omega^2/a_0^2)\phi = 0,$$

which is Helmholtz's equation again. Given adequate boundary conditions, the usual eigenvalue or criticality problem is generated; a non-zero amplitude distribution ϕ can only exist provided the coefficient ω^2/a_0^2 is correctly matched to one of a series of values that depend on the shape and nature of the cavity. In other words, since a_0 is known, standing waves are only possible for a definite sequence of values for ω, the *resonant frequencies*, just as in example 9.9 for the one-dimensional case. A two-dimensional example follows and problem 9.26 concerns a three-dimensional case.

***Example 9.10** A long circular pipe of radius R contains air in which the sound speed is a_0. Investigate the resonant frequencies of sound waves that travel *radially*.

In this problem ϕ in the differential equation above depends only on r, the distance from the axis. Now $\nabla^2\phi$ becomes $\mathrm{d}^2\phi/\mathrm{d}r^2 + (1/r)\mathrm{d}\phi/\mathrm{d}r$ and ϕ obeys the equation

$$\frac{\mathrm{d}^2\phi}{\mathrm{d}r^2} + \frac{1}{r}\frac{\mathrm{d}\phi}{\mathrm{d}r} + \frac{\omega^2}{a_0^2}\phi = 0,$$

the solutions of which are a separate class of functions called Bessel functions (of zero order). Tabulated values are available, just as for the more common functions. The standard solution which is finite at $r = 0$ is

$$\phi = AJ_0(\omega r/a_0),$$

in which A is an unknown amplitude. ($J_0 = 1$ when $r = 0$.) We must also satisfy the boundary condition that the air velocity is always zero at the wall. The equation

$$\rho_0 \partial \mathbf{v}/\partial t + a_0^2 \,\mathrm{grad}\, \rho' = \mathbf{0}$$

indicates that, when $r = R$ at the wall, grad $\rho' = \mathbf{0}$, i.e. that $\mathrm{d}\phi/\mathrm{d}r = 0$. It is a fact that

$$(\mathrm{d}/\mathrm{d}x)J_0(x) = -J_1(x),$$

where J_1 is a Bessel function of the first order. Hence we must have

$$J_1(\omega R/a_0) = 0.$$

This is the eigenvalue condition which correctly matches ω, R and a_0 so that non-zero solutions can exist.

Figure 9.6 shows the nature of the functions J_0 and J_1. They are rather like damped sinusoids but they do *not* have regular periodicity. $J_1(x)$ vanishes when $x = 3.83, 7.02, 10.17$, etc. Thus the resonant frequencies are

$$3.83\ a_0/2\pi R, \quad 7.02\ a_0/2\pi R, \quad 10.17\ a_0/2\pi R, \text{ etc.}$$

which are not related in any simple arithmetic way.

This kind of general behaviour is typical of circular or cylindrical systems. Another example is the vibration of a circular drumhead. (See problem 9.27.) It is the lack of a simple arithmetic relationship between the prevailing frequencies that makes a drum sound dissonant as compared with any 'one-dimensional' system like a clarinet or a violin string.

We shall take up some of these ideas again in chapter 11 in connection with electromagnetic waves, which are an intrinsically linear phenomenon and can be analysed without any linearisation procedure, however large their amplitude.

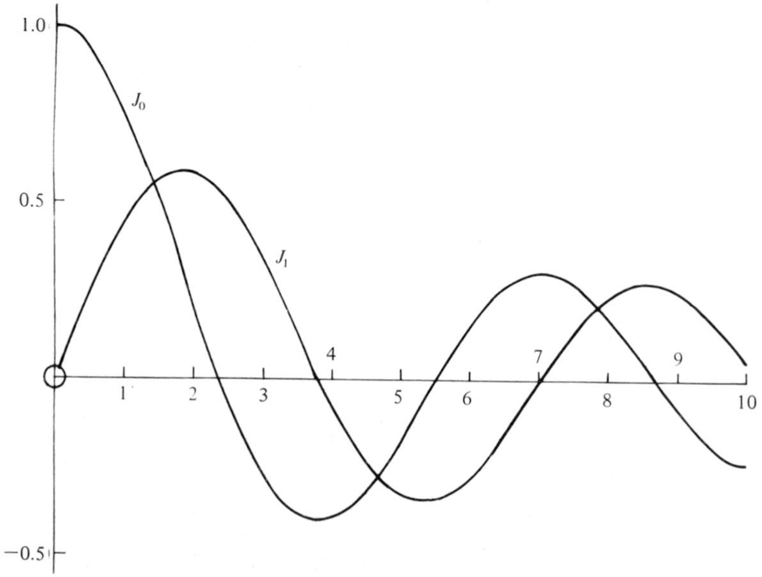

Fig. 9.6. Bessel functions of the first kind of orders zero and one.

9.9 Summary of the main ideas of chapters 7, 8 and 9

These chapters have ranged widely over phenomena of physical and engineering interest, but the material has all been united by one pervading theme, namely, that all these phenomena feed off and are describable in terms of the non-uniformities of vector and scalar fields. To quantify these non-uniformities we were led to invent the mathematical concepts of div and grad. Though the physical range was dauntingly broad, the reader should be comforted by the observation that relatively few new mathematical concepts were needed, because of the frequent and instructive mathematical analogies between apparently diverse phenomena. The economy and other advantages of an approach which turns readily from fluid mechanics via electromagnetism to nuclear technology should by now be apparent.

In these chapters we have stayed mostly in Zone 2 of figure 2.1, drawing inspiration from such topics as heat or fluid flow, but short forays into Zone 3 have occurred whenever we applied the div and grad concepts to electric or magnetic fields.

Problems 9

9.1 A large slab of material of thickness d contains net charge q distributed according to the equation $q = A \sin \pi x/d$, where A is a constant and x is measured normally into the slab from one face. Find the distribution of potential V in the slab if $V = 0$ at both faces and verify that the electric fields at the surfaces are consistent with Gauss's law applied to the slab.

9.2 Repeat problem 5.15 and example 5.4 by expressing them in terms of Poisson's equation and noting that $\nabla^2\phi = \mathrm{d}^2\phi/\mathrm{d}r^2 + (1/r)\mathrm{d}\phi/\mathrm{d}r$ when ϕ depends only on the plane, polar coordinate r, and $\nabla^2\phi = \mathrm{d}^2\phi/\mathrm{d}r^2 + (2/r)\mathrm{d}\phi/\mathrm{d}r$ when ϕ depends on radial distance r from the origin in three dimensions.

9.3 A fluid of viscosity μ undergoes rectilinear laminar flow along a pipe of elliptic cross-section with semi-axes a and $2a$ under a pressure gradient $\partial p/\partial z$, z being the streamwise coordinate. Show that the volumetric flow rate is $(2\pi a^4/5\mu)(-\partial p/\partial z)$. (Compare example 9.1.)

9.4 (*a*) Find the edge length d of a bare, cubic homogeneous reactor if it is critical when made of material with the characteristic properties C and D defined in section 9.3. Assume a neutron density n of the form $X(x)Y(y)Z(z)$, referred to three adjacent edges as cartesian axes.

Verify that in the steady (critical) state the rate at which neutrons leak out of the faces equals the total rate of net neutron production. Relate this result to Gauss's theorem.

*(*b*) Given that $\nabla^2\phi$ can be written as $(1/r)(\mathrm{d}^2/\mathrm{d}r^2)(r\phi)$ whenever there is spherical symmetry about the origin, find the critical radius of a bare, spherical nuclear reactor in terms of the coefficients C and D. The neutron density n is finite at the origin. (r = distance from origin.)

If the reactor is cooled by its own uniform thermal conductivity and the heat release rate is proportional to n, show that the excess temperature above ambient is also proportional to n.

9.5 A horizontal layer of smouldering peat of depth d and thermal conductivity K lies on flat, non-conducting ground. Its upper surface is at ambient temperature. Combustion releases heat at a rate equal to CT per unit volume and time, where C is a constant which models the chemical kinetics and T is excess temperature above ambient. If conditions are steady, find d. (In a more realistic model of the process, what other diffusion phenomena should be considered?)

9.6 A uniform fluid of density ρ and viscosity μ is at rest adjoining a plane wall. Suddenly (at $t = 0$) the wall is set in motion at a steady velocity u in its own plane. Viscosity causes the fluid to move in the direction of u, the fluid velocity v being a function of y (distance normal to the wall) and t. The pressure is uniform.

By considering the dynamics of a rectangular slice of fluid of thickness $\mathrm{d}y$ show that $\partial\tau/\partial y = \rho\partial v/\partial t$, where τ is the viscous shear stress $\mu\partial v/\partial y$. Deduce that v obeys the one-dimensional unsteady diffusion equation, with a *diffusivity* equal to $\nu = \mu/\rho$, the *kinematic viscosity*. Develop a result analogous to that of example 9.6.

(This problem is called the *Rayleigh problem*. Rayleigh was interested in the question why viscous effects are often confined near the wall in unsteady flows along pipes, e.g. in acoustic waves or pulsatile blood flow.)

9.7 (*a*) The temperature T in a uniform thermally conducting medium depends on the coordinate x according to a relation $T = F \sin mx$ in which F is a function of time. If the thermal conductivity is K and the volumetric heat capacity is C, find how long it takes for F to decay in the ratio e:1.
(*b*) A large plane slab of thickness a of the same material is initially at a uniform temperature T_0. If its surfaces are both held at $T = 0$, find how the slab's temperature evolves as a Fourier series in terms of time t and x, distance measured from the centre-plane.

*9.8 (*a*) Complete the solution of the problem posed in example 9.6 by deducing the ordinary differential equation relating θ to z, integrating it twice to find firstly $\mathrm{d}\theta/\mathrm{d}z$ and then θ. (The function $\mathrm{erf_c} x = 1 - (2/\sqrt{\pi}) \int_0^x \mathrm{e}^{-x^2} \mathrm{d}x$ takes the values 1 and 0 at $x = 0$ and ∞ respectively.)
(*b*) Starting again from dimensional considerations, solve the following one-dimensional unsteady heat flow problem. Initially an amount of heat H per unit area is deposited in a very thin, plane layer inside an infinite uniformly conducting material, which elsewhere has $T = 0$ initially. How does the temperature at the centre of the layer vary when $t > 0$? (When $t = 0$ it is unbounded.)

Note that CT has the dimensions of heat per unit volume. The most convenient dimensionless dependent variable is $\theta = CT(\alpha t)^{1/2}/H$. In solving the differential equation for θ, use the facts that $(d/dz)(z\theta) = z\,d\theta/dz + \theta$ and $\partial T/\partial x = 0$ when $x = 0$, from symmetry, for $t > 0$. (*N.B.* $\int_{-\infty}^{\infty} e^{-x^2}dx = \sqrt{\pi}$.)

*9.9 A compressible gas flows in a uniform porous medium according to the law $\mathbf{v}^* = -k$ grad p, gravity being negligible. If the uniform void fraction in the medium is m, the mass conservation equation becomes $m\partial\rho/\partial t +$ div $\rho\mathbf{v}^* = 0$, and $\rho = Cp$ (C const.) if isothermal conditions prevail. Show that $\rho\mathbf{v}^* = -\frac{1}{2}Ck$ grad (p^2) and hence that $m\partial p/\partial t = \frac{1}{2}k\nabla^2(p^2)$, a non-linear differential equation. (Note that in the steady state p^2 obeys Laplace's equation, for which the usual techniques apply.) Find one-dimensional, unsteady solutions of the forms:

$$\left.\begin{aligned} &\text{(i)} \quad p = Ax + Bt \\ \text{or} \quad &\text{(ii)} \quad p = A/t^{1/3} - Bx^2/t. \end{aligned}\right\} \quad A \text{ and } B \text{ const.}$$

Sketch how these distributions evolve. What do they represent? Note that the behaviour is quite different from the *linear* unsteady diffusion equation.

*9.10 In a thermionic device, electrons are uniformly emitted from a plane cathode surface into vacuum and accelerate away under a normal electric field, which is itself affected by the electron charge present. If the uniform current density is j, relate the electron velocity v at a distance x from the cathode to q, the charge density, and V the electric potential, taking v and V both zero at the cathode. Show that

$$\frac{d^2V}{dx^2} = -\frac{j}{\epsilon_0}\left(\frac{m}{2eV}\right)^{1/2}$$

(where m = electron mass, $-e$ = electron charge) and verify that $V = kx^{4/3}$ is a solution. Find k and the relation between j and the potential V_0 at the anode, a parallel surface at a distance d from the cathode. (This result is known as *Child's law*.)

9.11 (*a*) A fluid's velocity and temperature are given by the equations

$$\mathbf{v} = (4x^2y^2/t)\mathbf{i} - (2y/t)\mathbf{j} + 0\mathbf{k}$$

and

$$T = y^2(1 + t^4) - 1/x.$$

Investigate how the temperature of any individual travelling fluid particle varies.

*(*b*) Fluid is emitted uniformly from the plane $x = 0$ with an inlet density ρ_0 which varies in time so that $\rho_0 = ae^{-bt}$. The fluid then moves x-wards with a velocity given by

$$v_x = ce^{-fx+gt} + b/f, \quad v_y = 0, \quad v_z = 0,$$

a, b, c, f and g being constants. Verify that the inlet condition and the mass-conservation equation are satisfied by the solution

$$\rho = a\mathrm{e}^{-bt+fx}. \tag{α}$$

Find the rate of change of the density of a travelling fluid particle and deduce, as a function of time $\rho_P(t)$, the density of a travelling fluid particle P which is emitted from the plane $x = 0$ at time $t = 0$.

By solving the equation $\mathrm{d}x/\mathrm{d}t = v_x$, show that the position of P is given by

$$x = \frac{1}{f}\log\left\{\frac{cf\mathrm{e}^{gt}}{g-b} + \left(1 - \frac{cf}{g-b}\right)\mathrm{e}^{bt}\right\}.$$

By inserting this in the general relation (α), confirm your result for $\rho_P(t)$.

9.12 Throughout this question, viscous, gravity, and other body forces should be neglected.

(*a*) For a finite volume V of fluid enclosed within a surface S *moving with* a stream of the same fluid, express $\dot{W}$ (the rate at which the fluid in V does work on the surrounding fluid) as a surface integral over S involving pressure p and velocity $\mathbf{v}$, and convert this to a volume integral by Gauss's theorem.

(*b*) Show that $\operatorname{div} \mathbf{v} = \rho(\mathrm{D}/\mathrm{D}t)(1/\rho)$ and $\mathbf{v} \cdot \operatorname{grad} p = -\rho(\mathrm{D}/\mathrm{D}t)(\frac{1}{2}\mathbf{v}^2)$.

(*c*) Show that $\dot{W}$ equals the integral over V of

$$\dot{w} = \rho\left\{p\frac{\mathrm{D}}{\mathrm{D}t}\left(\frac{1}{\rho}\right) - \frac{\mathrm{D}}{\mathrm{D}t}(\tfrac{1}{2}\mathbf{v}^2)\right\},$$

the work rate per unit volume. Interpret the two terms.

(*d*) If $\mathbf{Q}$ is the heat flow field, deduce from the first law of thermodynamics (expressed per unit volume for a fluid element $\mathrm{d}\tau$) that

$$\operatorname{div} \mathbf{Q} + \dot{w} + \rho(\mathrm{D}/\mathrm{D}t)(u + \tfrac{1}{2}\mathbf{v}^2) = 0,$$

in which u = internal energy per unit mass. Invoking problem 7.4(*b*), deduce that

$$\operatorname{div} \mathbf{Q} + \rho T(\mathrm{D}s/\mathrm{D}t) = 0,$$

(s = entropy per unit mass) which expresses the second law of thermodynamics, viscous dissipation being absent.

(*e*) If $\mathbf{Q} = -K \operatorname{grad} T$ and C = volumetric heat capacity $\rho T(\mathrm{d}s/\mathrm{d}T)$ (the effect of other variables on s being neglected) deduce the *heat convection equation*

$$C\mathrm{D}T/\mathrm{D}t = K\nabla^2 T.$$

How does this differ from the heat *conduction* equation?

(*f*) If the flow is steady, deduce from (*d*) the *steady-flow energy equation* in the form

$$\text{div}\,(\rho \mathbf{v} h_0 + \mathbf{Q}) = 0,$$

where $h_0 = u + p/\rho + \frac{1}{2}\mathbf{v}^2$, the *stagnation enthalpy*.
(*g*) If the entropy of the fluid is constant, show that Euler's equation may be written

$$\mathrm{D}\mathbf{v}/\mathrm{D}t = -\,\text{grad}\,h.$$

9.13 Given that $\mathbf{u} = y\mathbf{i} - x\mathbf{j} + 0\mathbf{k}$ and $v = x\mathbf{i} + y\mathbf{j} + 0\mathbf{k}$, find the components of $(\mathbf{u} \cdot \text{grad})\mathbf{v}$ and find its direction relative to $\mathbf{v}$. Interpret this result by considering how $\mathbf{v}$ changes between adjacent points on a $\mathbf{u}$-field line.

Repeat the process for $(\mathbf{u} \cdot \text{grad})\mathbf{u}$, $(\mathbf{v} \cdot \text{grad})\mathbf{u}$ and $(\mathbf{v} \cdot \text{grad})\mathbf{v}$.

9.14 Modify Euler's equation to allow for uniform gravity, assuming ρ is constant. Show that the effect is merely to replace p by $p^* = p + \rho g z$, if z = height. (The implication is that gravity has no effect on such fluid motion whenever p does not enter into any other condition, such as the boundary condition at a free surface where p is constant.)

9.15 An unsteady two-dimensional fluid velocity field is given by

$$\mathbf{v} = x\mathbf{i} + (y - t)\mathbf{j}.$$

By solving the equations $\mathrm{d}x/\mathrm{d}t = v_x$ and $\mathrm{d}y/\mathrm{d}t = v_y$, find the location at time $t = t_1$ of a particle which passes through the point (1, 1) at time $t = t_0$. Hence show that

(*a*) the locus of a particle which passes through the point (1, 1) at time $t = 0$ is given by $y = 1 + \log x$, and
(*b*) the streak-line at $t = 0$ through the point (1, 1) is given by $y = 1 + x \log x$. (t_0 is the parameter for this curve.)
(*c*) Also show that the streamline at $t = 0$ through the point (1, 1) is given by $y = x$.
(*d*) Sketch and contrast these three curves.

9.16 (*a*) Show that Bernoulli's equation for a compressible fluid becomes

$$\int \frac{\mathrm{d}p}{\rho} + \tfrac{1}{2}\mathbf{v}^2 = \text{const. along a streamline.}$$

If the entropy is constant, show that this is equivalent to

$$h_0 \text{ (stagnation enthalpy)} = h + \tfrac{1}{2}v^2 = \text{const.},$$

h being enthalpy per unit mass.

In isentropic flow along a streamtube, show that $v = a = \{(\partial p/\partial \rho)_s\}^{1/2}$ where the cross-sectional area of a streamtube is a minimum, provided the fluid velocity is still changing there. (In section 9.7, a is identified as the speed of sound.)
(*b*) Referring to problem 9.14 show that Bernoulli's equation (with ρ constant) becomes $\frac{1}{2}v^2 + p/\rho + gz$ = const. in the presence of gravity, z being measured vertically upwards.

A floating boat has a hole in the bottom, through which the sea-water enters, forming a vertical free jet. Show that the jet just reaches the level of the surrounding sea, if Bernoulli's equation is applicable for the flow into the jet.

9.17 (*a*) Liquid of uniform density ρ is rotating uniformly with angular velocity Ω about a vertical axis. The streamlines are horizontal circles centred on the axis. Such a motion is called a *forced vortex*. Find how the pressure p varies with distance r from the axis and with height z, allowing for gravity **g**. At the free surface, $p = \text{const}$. Show that it (and indeed all constant pressure surfaces) are paraboloids of revolution as asserted in problem 5.3.

Can a small body (lighter than the fluid) float at rest relative to the fluid at the free surface?

(*b*) Repeat the calculation for a *free vortex*, which is a circular flow in which $v = A/r$ ($A = \text{const.}$). Sketch the shape of the free surface, noting its behaviour as $r \to 0$.

(*c*) Fluid of uniform density ρ rotates at uniform angular velocity Ω in a centrifuge. If gravity is negligible, use the corollary to Gauss's theorem to show that the total pressure force on an immersed body, held fixed relative to the fluid, is equal to $-\rho\Omega^2\iiint \mathbf{r}\,\mathrm{d}\tau$, taken over the body's volume. (**r** is here the *two*-dimensional radius vector, perpendicular to the axis, but the body can be three-dimensional.)

If the body were replaced by fluid, can the above result be reconciled with the 'centrifugal force' on this fluid, acting at its centroid?

9.18 Perform the acoustic linearisation process for the case of steady flow ($\partial/\partial t = 0$) where the velocity in the unperturbed state (upstream) is uniform and equal to v_0 in the x-direction. Show that

$$a_0^2\nabla^2\rho' = v_0^2(\partial^2\rho'/\partial x^2).$$

If $v_0 > a_0$ (supersonic flow) show that two-dimensional wave-type solutions of the form $\rho' = f(x \pm \beta y)$ are possible, where $\beta^2 = (v_0/a_0)^2 - 1$. Can you interpret these? (The nature of the differential equation and its solutions changes completely as v_0 passes a_0. The subsonic and supersonic versions are known respectively as *elliptic* and *hyperbolic* equations.) What (elliptic) equation results when $v_0 \ll a_0$?

9.19 (*a*) In relation to the solution $\rho' = f(w)$ of the wave equation, with $w = x - a_0t$, a point P at which ρ' is constant must also have w constant. Deduce the velocity of P (which is *not* a fluid particle).

(*b*) Show that the operator $(\partial/\partial t + \mathbf{u}\cdot\text{grad})$ applied to a scalar field $s(x, y, z, t)$ yields its rate of change $\mathrm{d}s/\mathrm{d}t$ as seen by an observer moving with velocity **u**.

In (*a*) what is the rate of change of ρ' as seen by an observer travelling with velocity a_0 in the x-direction?

9.20 In relation to the expression $\rho' = g(u)$, where $u = x + a_0t$, show that (*a*) it satisfies the equation

$$\frac{\partial^2 \rho'}{\partial t^2} = a_0^2 \frac{\partial^2 \rho'}{\partial x^2},$$

(*b*) it represents waves travelling towards negative x, and
(*c*) the associated velocity perturbation v_x' in a sound wave is given by $a_0 \rho' = -\rho_0 v_x'$.

9.21 Motorway traffic flow provides a one-dimensional, unsteady field problem if the individual vehicles are 'blurred out' into a mean velocity v and a density n (vehicles per unit road length). A reasonable model of driver behaviour is given by a law of the form $v = f(n)$. Show that conservation of cars implies that

$$\partial n/\partial t + (\partial/\partial x)(nv) = 0,$$

if x is distance. Show further that an observer travelling at a velocity

$$u = v + dv/dn$$

(see problem 9.19(*b*)) sees traffic of constant density (and so also speed) and is therefore himself travelling at constant speed. How can this result be exploited to allow prediction of the evolution of traffic whose state is given at some initial instant? (The loci on an x, t diagram of the constant speed 'observers' are called *characteristics.*) When are pile-ups probable? (This is a non-linear wave problem. The traffic waves (of constant speed and density) can only propagate in one direction. Do they travel forwards or backwards relative to the traffic? Because inertia plays no part they are called *kinematic waves*. A difference from *linear* wave theory is that the profile changes in time.)

9.22 A particular one-dimensional sound wave of speed a_0 is a step wave that propagates into still air and rapidly raises the density by $\delta\rho$ to a new steady value. Find the velocity behind it. If the wave encounters a plane, normal wall, find the total increase in density behind the reflected wave and show that the wave reflection increases the pressure at the wall by $2a_0^2\,\delta\rho$.

9.23 Longitudinal waves in elastic solids are similar to those in fluids. Consider the bouncing of a uniform vertical elastic rod of length l, density ρ_0 and cross-sectional area A on a horizontal rigid floor. Let the wave speed be a_0 and the uniform vertical velocity of the rod be $u(\ll a_0)$ before impact. Show that after a step wave has travelled once up and down the rod, reflecting off the free end, the whole rod is travelling upwards again at velocity u. Calculate the pressure at the bottom end while contact is maintained and verify that its impulse over the contact period accounts for the reversal of the rod's momentum. What is the maximum shortening of the rod?

9.24 (*a*) The transverse deflection y of a string under tension T and mass per unit length ρ, stretched between two fixed points A and B a distance L apart, obeys the wave equation

$$\frac{\partial^2 y}{\partial t^2} = a^2 \frac{\partial^2 y}{\partial x^2} \quad \left(a^2 = \frac{T}{\rho}\right),$$

where x is measured from A towards B, provided $y \ll L$. Find standing wave solutions of the form $y = X(x)T(t)$, noting that $y = 0$ at $x = 0$ and L. What oscillation frequencies are possible?
*(*b*) The string is initially plucked into a form consisting of two straight lines with a maximum deflection $y = \epsilon$ at $x = L/2$, and then released from rest. Express the subsequent behaviour in terms of a Fourier series, and find the ratio of the amplitudes of successive 'harmonics'. Are all possible frequencies present?
(*c*) A clarinet behaves like a pipe with one end open and the other closed. Show that the possible frequencies for acoustic, longitudinal, standing waves form a sequence in the ratios 1:3:5, etc.

9.25 A pipe of uniform cross-section and of length L has one fixed plane end and one movable plane end, both normal to the axis. The movable end is given an axial velocity $A \sin \omega t$, where the amplitude A is small. Find a standing wave solution for the acoustic wave equation in terms of axial velocity inside the pipe, and show that its amplitude increases without limit as $\omega \to n\pi a_0/L$ (n integral), a_0 being the speed of sound. Take the origin at the fixed end. (This demonstrates how a cavity *resonates* at certain frequencies. In practice the resonant amplitude is limited by dissipative or non-linear effects.)

9.26 Find the first four possible frequencies for (resonant) standing waves inside a cubical cavity of edge length d, taking the solution of the three-dimensional wave equation to be of the separable form

$$\rho' = X(x)Y(y)Z(z) \cos \omega t.$$

Take three adjacent edges as cartesian axes and the sound speed as a_0 and note that the gradient of ρ' normal to a fixed boundary must be zero. (Can you prove this?)

Contrast this problem with problem 9.4(*a*).

9.27 (*a*) A drumhead is a plane, circular membrane of radius R, tension T per unit length and mass ρ per unit area. During vibration its deflection z is a function of time t and of x, y (cartesian coordinates in the equilibrium plane, $z = 0$). The restoring force due to curvature and tension is $-T\nabla^2 z$ per unit area (compare example 9.5). Deduce that z obeys the two-dimensional wave equation

$$\frac{\partial^2 z}{\partial t^2} = \frac{T}{\rho} \nabla^2 z,$$

in which the laplacian is two-dimensional.
*(*b*) Find circularly symmetric solutions of this equation of the form $z = \phi(r) \sin \omega t$, satisfying the edge condition $z = 0$ at $r = R$, and deduce

the first three corresponding natural frequencies. (Compare example 9.10.) (The first roots of $J_0(x) = 0$ are 2.405, 5.52, 8.65.)

10

Rotationality and the curl operator in fluid mechanics

10.1 General fluid mechanics

It is now time to reconsider general inertial fluid mechanics, lifting any restriction that perturbations are small. When the changes of velocity, pressure, etc. in a fluid are large, linearisation is unacceptable and we must face the true non-linearity of fluid mechanical behaviour, implied by Euler's equation (see section 9.5),

$$\rho\left(\frac{\partial \mathbf{v}}{\partial t} + (\mathbf{v}\cdot\text{grad})\mathbf{v}\right) + \text{grad}\, p = \mathbf{0}.$$

Another difficulty presented by this equation is that it expresses the variations of **v**, the main variable of interest, in terms of yet another unknown, p, a situation which seems to be rather futile. Compressible flow is an even worse prospect, it would seem, because then ρ is not known either. Euler's equation appears to be relatively useless as a starting point.

We shall find that there is a way to break out of this impasse by making dynamical statements in a form that does not involve the unhelpful variable p. (It is perhaps paradoxical that, although fluid pressure is the cause of fluid motion, we make mathematical progress best by eliminating it from the analysis!) A whole new array of mathematical ideas turns out to be involved. These then find very fruitful applications outside fluid mechanics as well.

10.2 Vorticity, the key to fluid dynamics

The problem, then, is how to make statements about the velocity field which do not also involve the extra unknown, pressure p. The solution is to concentrate on the way that fluids *rotate*. If the reader is in the habit of using his powers of observation upon natural phenomena, he may already have been struck by the interesting properties of fluid rotation. Fluid that is not spinning is reluctant to start, while fluid that is spinning is reluctant to stop. Demonstration 5(*b*) in Appendix II makes the point dramatically.

It is not difficult to see qualitatively why such behaviour occurs, if viscous shear stresses are unimportant. Consider a sphere of uniform fluid immersed in a stream of the same fluid. It is acted on by normal pressures which if non-uniform will cause the sphere to accelerate. However, all these pressure forces act through the centre (and centre of mass) of the sphere and therefore apply no turning moment such as would cause it to change its state of rotation. In elementary mechanics, if we want to avoid bringing an unknown force into consideration, we take moments about a point on its line of action; here, taking moments about the centre of the sphere of fluid should enable us to leave all the pressure forces out of consideration.

Deeper reflection reveals that the situation must be more subtle; a fluid element is not a rigid body and is in general deforming as it is swept along. What exactly is meant by the phrase 'state of rotation'? To be more precise about this is clearly our first task. We may be optimistic that accomplishing this will then render possible kinetic statements that do not involve the pressure.

For a start, consider two-dimensional flow, invariant in the z-direction and with no velocity components in that direction. Then if any fluid element rotates, it can do so only about an axis parallel to the z-direction. Consider two adjacent points P and Q in an x, y plane, separated by a small displacement $\mathbf{dr}$ of length ds, with $\mathbf{dr}$ inclined at an angle θ to the x-axis as in figure 10.1. If the velocity components at

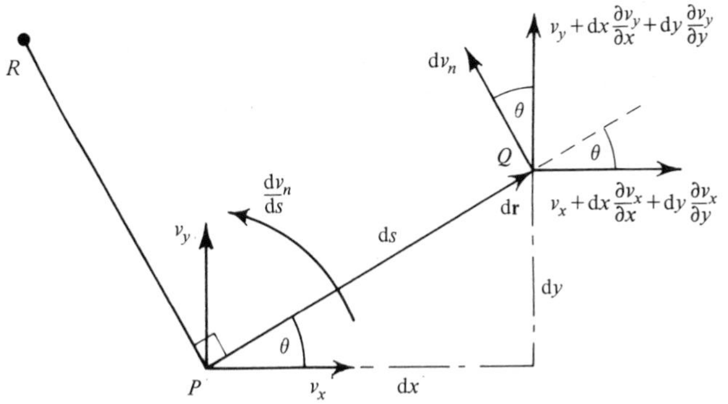

Fig. 10.1

P are v_x and v_y, we can express the slightly different velocities at Q in terms of the increments dx and dy of x and y by means of Taylor's series, retaining only first order terms, as recorded in the figure. The quantities v_x, $\partial v_x/\partial x$ etc. are all to be evaluated at P. The same result can be put into vector terms by expressing the difference in velocities between Q and P as

$$\mathbf{dv} = \left(dx\frac{\partial}{\partial x} + dy\frac{\partial}{\partial y}\right)\mathbf{v} = (\mathbf{dr}\cdot\text{grad})\mathbf{v},$$

for $\mathbf{dr}$ has components dx and dy (and $dz = 0$). This is a generalisation for a vector $\mathbf{v}$ of the result $d\phi = \mathbf{dr}\cdot\text{grad}\,\phi$ for a scalar ϕ, and is valid also in three dimensions, where a further term $dz(\partial\mathbf{v}/\partial z)$ would appear.

If dv_n is the component of $\mathbf{dv}$ normal to PQ, the anticlockwise angular velocity of PQ is dv_n/ds, and dv_n can be found from resolving the changes in v_x and v_y in a direction normal to PQ. At the same time inserting $dx = ds\cos\theta$ and $dy = ds\sin\theta$, we find that

$$dv_n = \left(\frac{\partial v_y}{\partial x}ds\cos\theta + \frac{\partial v_y}{\partial y}ds\sin\theta\right)\cos\theta - \left(\frac{\partial v_x}{\partial x}ds\cos\theta + \frac{\partial v_x}{\partial y}ds\sin\theta\right)\sin\theta$$

and so the angular velocity of PQ is given by

$$\frac{dv_n}{ds} = \frac{\partial v_y}{\partial x}\cos^2\theta - \frac{\partial v_x}{\partial y}\sin^2\theta + \frac{1}{2}\left(\frac{\partial v_y}{\partial y} - \frac{\partial v_x}{\partial x}\right)\sin 2\theta.$$

This result shows that the fluid's angular velocity at P is not unique but depends on θ, the orientation of the particular line PQ chosen for evaluating it. This happens because the fluid near P is deforming in general and the angle between any two lines through P is changing. The obvious next move is to evaluate the *average* angular velocity, taken over all values of θ from 0 to 2π. The mean value of both $\cos^2\theta$ and $\sin^2\theta$ is $\frac{1}{2}$, and that of $\sin 2\theta$ is zero. Therefore

$$\text{mean. ang. vel. at } P = \frac{1}{2}\left(\frac{\partial v_y}{\partial x} - \frac{\partial v_x}{\partial y}\right), \quad \text{or} \quad \tfrac{1}{2}\mathbf{k}\left(\frac{\partial v_y}{\partial x} - \frac{\partial v_x}{\partial y}\right),$$

if we represent it by a vector in the z-direction in the obvious way. (The righthand screw convention has been observed correctly, for anticlockwise rotation in figure 10.1 is clockwise when viewed in the z-direction, with right-handed axes.)

It is standard practice in fluid mechanics to avoid constant insertion of the factor $\frac{1}{2}$ by introducing a new vector $\boldsymbol{\omega}$, called the *vorticity*, defined as twice the mean angular velocity or

$$\boldsymbol{\omega} = \mathbf{k}\left(\frac{\partial v_y}{\partial x} - \frac{\partial v_x}{\partial y}\right),$$

in the two-dimensional case. It can also be arrived at in another way: if θ is replaced by $(\theta + \pi/2)$ in the formula for the angular velocity of PQ, we find that

$$\text{ang. vel. of } PR = \frac{\partial v_y}{\partial x}\sin^2\theta - \frac{\partial v_x}{\partial y}\cos^2\theta - \frac{1}{2}\left(\frac{\partial v_y}{\partial y} - \frac{\partial v_x}{\partial x}\right)\sin 2\theta,$$

where PR is perpendicular to PQ. If we evaluate the *sum* of the angular velocities of the two mutually perpendicular lines PR, PQ, this also proves to be equal to $\mathbf{k}\{(\partial v_y/\partial x) - (\partial v_x/\partial y)\}$ if expressed as a vector, irrespective of θ, i.e. again the vorticity. We shall make use of this approach in sections 10.4 and 10.5 since it is easier than taking the mean angular velocity for all orientations. Note that it is immaterial which pair of perpendicular lines we choose. If the sum of the angular velocities of two perpendicular lines is independent of their orientation, then this sum is also the mean value of all such sums, which is obviously the sum of the mean angular velocities of the two lines individually, i.e. twice the angular velocity of a single line, averaged over all orientations.

10.3 The three-dimensional case; the curl operator

The generalisation of vorticity for fully three-dimensional flows is

$$\boldsymbol{\omega} = \mathbf{i}\left(\frac{\partial v_z}{\partial y} - \frac{\partial v_y}{\partial z}\right) + \mathbf{j}\left(\frac{\partial v_x}{\partial z} - \frac{\partial v_z}{\partial x}\right) + \mathbf{k}\left(\frac{\partial v_y}{\partial x} - \frac{\partial v_x}{\partial y}\right),$$

in which the last term is already familiar, and the others are obtainable by cyclic permutation of x, y and z. The reader may be happy to accept this on trust and can then omit the derivation given in section 10.5, which generalises the two-dimensional version.

The vector $\boldsymbol{\omega}$ is a new kind of measure of the non-uniformity of the vector field **v**. For a velocity field **v** it measures its 'rotational' quality, but obviously the same differential operation can be applied to any vector field such as the electric field **E** so as to generate a new vector field. Whether such a derivation has any physical interpretation or importance remains to be seen. The shorthand symbol that is used for the operation is curl (or rot, short for rotation, in some texts), both evocative names. The general mathematical definition then is

$$\text{curl } \mathbf{v} = \mathbf{i}\left(\frac{\partial v_z}{\partial y} - \frac{\partial v_y}{\partial z}\right) + \mathbf{j}\left(\frac{\partial v_x}{\partial z} - \frac{\partial v_z}{\partial x}\right) + \mathbf{k}\left(\frac{\partial v_y}{\partial x} - \frac{\partial v_x}{\partial y}\right)$$

for a vector field **v**. Like any other vector field, curl **v** has its own field lines which in the case of fluid motion are called *vorticity lines*. These can be at any inclination to the original field lines or streamlines. But in a two-dimensional field $\mathbf{v}(x, y)$ with v_z zero, curl **v** has only a z-component and is thus perpendicular to **v**. Example 10.1 below explores another case, while problem 10.1(b) verifies the relationship to angular velocity in the case where this is uniform.

Example 10.1 For the vector field **v** with components ($\tanh z$, $\operatorname{sech} z$, 0) find the angle between the vector and its curl. What is the field like?

curl **v** has components ($\tanh z \operatorname{sech} z$, $\operatorname{sech}^2 z$, 0). We observe that curl $\mathbf{v} = \mathbf{v} \operatorname{sech} z$, i.e. **v** times a scalar. It follows that this vector field is parallel to its own curl.

For large positive z, **v** is essentially the uniform field **i**, and for large negative z, **v** is essentially $-\mathbf{i}$. In the vicinity of the plane $z = 0$, however, **v** veers round rapidly as z varies, still retaining unit magnitude. When $z = 0$, **v** is in fact the vector **j**. The field lines are straight lines, lying in planes z = const. in progressively different orientations.

(This veering or shearing kind of configuration is very characteristic of fields that are parallel to their own curl.)

In terms of the del operator, $\nabla = \mathbf{i}\partial/\partial x + \mathbf{j}\partial/\partial y + \mathbf{k}\partial/\partial z$, it is clear that curl $\mathbf{v} = \nabla \times \mathbf{v}$. This expression is not an ordinary vector product, however, as is evidenced by the fact that it is not perpendicular to **v** in general. On the analogy with the determinant form for vector products we can also write

$$\text{curl } \mathbf{v} = \begin{vmatrix} \mathbf{i} & \mathbf{j} & \mathbf{k} \\ \partial/\partial x & \partial/\partial y & \partial/\partial z \\ v_x & v_y & v_z \end{vmatrix},$$

which is extremely useful as a mnemonic. It is easy to distinguish curl from grad and div because it neither operates on nor generates a *scalar* field; it turns a *vector* field into another *vector* field. It also looks very different, with its six cartesian terms, instead of the three of div or grad.

A vector field whose curl vanishes everywhere is called *irrotational*, even when the field is not a velocity field and so rotation is not literally relevant.

10.4 Some simple flows

Examining a few simple two-dimensional examples will help to impart an initial grasp of the nature of fluid angular velocity and vorticity. Problem 10.2 concerns another case.

Example 10.2 Figure 10.2 shows the velocity profile of a *rectilinear shear flow* (a motion in which the streamlines are straight and parallel, and one layer of fluid slides over the next to give a shearing action). The velocity v_x is given as a function of y. Find the vorticity and relate it to the angular velocities of line elements parallel and perpendicular to the flow.

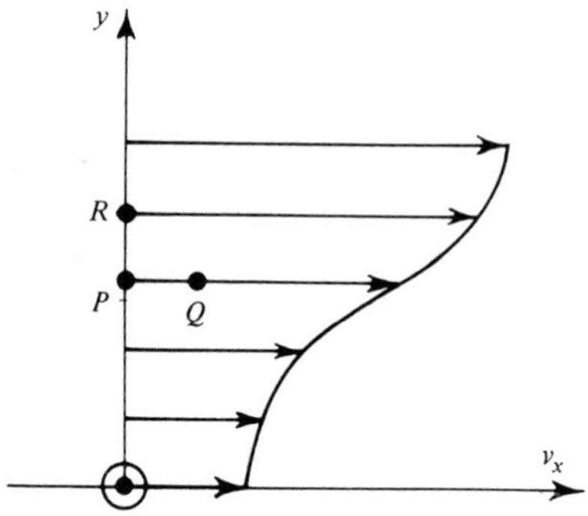

Fig. 10.2. Velocity profile of rectilinear flow, $v_x = f(y)$.

The vorticity $\boldsymbol{\omega}$ is given by $\boldsymbol{\omega} = \text{curl}\, \mathbf{v} = -\mathbf{k}\, \mathrm{d}v_x/\mathrm{d}y$, in which the minus sign indicates clockwise rotation in the figure, where $\mathrm{d}v_x/\mathrm{d}y$ is positive. But a fluid line element *PQ* is swept along in its own direction without rotation, whereas the angular velocity of a line element *PR* is $\mathrm{d}v_x/\mathrm{d}y$. The angle *QPR* is in the process of changing as each bit of fluid is sheared out of shape. The need to work in terms of a *mean* angular velocity should be obvious. For the two perpendicular lines *PQ*, *PR*, this mean is $\frac{1}{2}(\mathrm{d}v_x/\mathrm{d}y)$ which is half the magnitude of the vorticity, as it should be.

The above example reveals another point, namely, that vorticity is not necessarily associated with streamline curvature; here we have rectilinear flow with non-zero vorticity. As a contrast next consider some examples in which the streamlines *are*

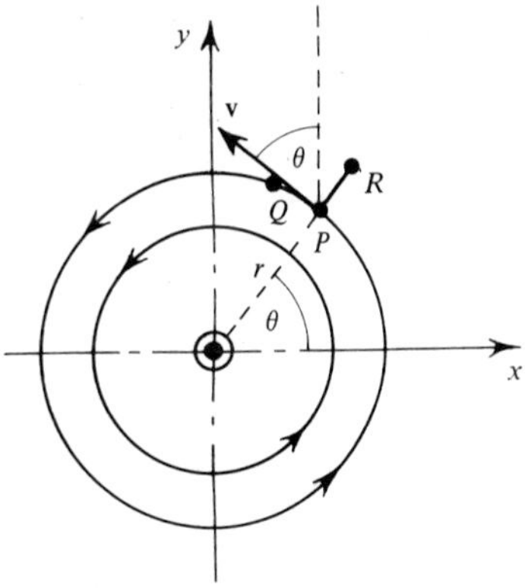

Fig. 10.3

curved, where in fact they are concentric circles as shown in figure 10.3 and the velocity depends only on distance r from the origin, although it could vary with r in any fashion. The one set of field lines can be inhabited by an infinity of different vector fields.

Example 10.3 Because of viscosity, fluid left in a steadily rotating tank ultimately rotates like a solid body, at uniform angular velocity Ω. Find the cartesian components of the velocity and the vorticity $\boldsymbol{\omega}$ and relate $\boldsymbol{\omega}$ to the angular velocities of line elements in the fluid.

Take axes as shown in figure 10.3. The fluid velocity $r\Omega$ has components v_x and v_y equal to $-r\Omega \sin\theta$ and $r\Omega\cos\theta$, or $-\Omega y$ and Ωx, respectively. The vorticity $\boldsymbol{\omega}$, found by differentiating these components, is $2\mathbf{k}\Omega$ i.e. twice the angular velocity, as it should be. The vorticity equals the sum of the angular velocities of any two perpendicular line elements, for the angular velocity of *any* line element is Ω. There is no fluid deformation occurring, since the fluid moves as if rigid. This motion is called a *forced vortex*.

Example 10.4 Repeat the calculation with a velocity field which has the same field lines but for which v varies like r^{-1} instead of r, i.e. $v = A/r$.

Now

$$v_x = -v\sin\theta = -Ay/(x^2+y^2) \quad \text{and} \quad v_y = v\cos\theta = Ax/(x^2+y^2),$$

since $r^2 = x^2 + y^2$. Hence

$$\frac{\partial v_y}{\partial x} = \frac{A}{(x^2+y^2)} - \frac{2Ax^2}{(x^2+y^2)^2} = \frac{A(y^2-x^2)}{(x^2+y^2)^2} = \frac{\partial v_x}{\partial y}, \quad \text{similarly.}$$

So the vorticity is zero, i.e. the flow is irrotational. Although a fluid particle P

circulates around O, it does this without itself rotating! How this can occur is clearer if we consider line elements PQ and PR in figure 10.3. PQ is swept round on the circular streamline through P and clearly has *anticlockwise* angular velocity, but PR's inner end P is moving faster transversely than its outer end R (because $v \propto r^{-1}$) and so PR has *clockwise* angular velocity. Thus zero mean angular velocity becomes possible. This is the kind of motion that occurs naturally when fluid devoid of rotation is caused to follow circular streamlines. It is called a *free vortex.*

This example provides further evidence that vorticity and streamline curvature are independent ideas; here there is curvature but no vorticity. Notice however that the vorticity is not zero but increases without limit on the z-axis (where $r = 0$), which is a singularity of the velocity field. Such a singularity is called a *line vortex.* In practice viscosity or cavitation always inhibits such extreme behaviour but nevertheless it is often valid and convenient to model the line vortex mathematically as a singularity. A better model is to replace the fluid in a concentric cylindrical core of small radius a by a section of a forced vortex of uniform angular velocity A/a^2, so that the peripheral speed A/a is correctly matched to the free vortex flow with $v = A/r$ outside the core. In the core $v = Ar/a^2$.

Line vortices need not be straight. A familiar example of a curved line vortex is provided by a smoke-ring, where the core or line vortex forms a closed, circular loop.

It is important to distinguish between a vorticity line and a line vortex. The latter is an extreme case of the former, which is normally merely one of the field lines of a non-singular vorticity field. The line vortex is often surrounded by vorticity-free fluid.

Finally let us return to figure 10.2. It is quite common for the velocity gradient dv_x/dy to become locally very large, so allowing v_x to change by a large amount over a short interval in the y-direction. This band of values of y then defines what is called a *shear layer* or, if it adjoins a solid wall, a *boundary layer.* In the mathematical modelling of a physical problem it is often valid and convenient to treat these layers as having vanishing thickness, so that both dv_x/dy and the vorticity increase without limit there. The layer is then called a *sheet vortex* or *vortex sheet*. Shear layers are very unstable and prone to turbulence.

A familiar example of the vortex sheet idea occurs when fluid is treated as if it were in motion virtually up to a wall, as for instance when fluid in a pipe is assumed to have uniform velocity. There is then a vortex sheet (a boundary layer of negligible thickness) adjoining the wall because the fluid right at the wall is stationary. Vortex sheets need not be plane, just as line vortices need not be straight.

*10.5 A derivation of vorticity in three dimensions

This section may be omitted by readers less concerned with the mathematical points. Alternative treatments are to be found in books which use tensors to analyse the deformation and rotation of continuous media. Here we merely generalise the two-dimensional treatment given in section 10.2. We first record a purely mathematical identity.

*10.5 *Derivation of vorticity in three dimensions*

If $\mathbf{v}$, $\mathbf{I}_1$ and $\mathbf{I}_2$ are any three vectors, then

$$\mathbf{I}_1 \times \mathbf{I}_2 \cdot \text{curl}\,\mathbf{v} \equiv \mathbf{I}_2 \cdot \{(\mathbf{I}_1 \cdot \text{grad})\mathbf{v}\} - \mathbf{I}_1 \cdot \{(\mathbf{I}_2 \cdot \text{grad})\mathbf{v}\}$$

where, as usual, an operator $(\mathbf{u} \cdot \text{grad})$ means $\Sigma u_x \partial/\partial x$. This result is easily verified in terms of the cartesian components of the three vectors. It is the analogue of the algebraic result

$$(\mathbf{a} \times \mathbf{b}) \cdot (\mathbf{c} \times \mathbf{d}) \equiv \mathbf{b} \cdot (\mathbf{a} \cdot \mathbf{c})\mathbf{d} - \mathbf{a} \cdot (\mathbf{b} \cdot \mathbf{c})\mathbf{d}$$

(see problem 1.10) with $\mathbf{c}$ replaced by ∇.

We shall apply it in the case where $\mathbf{I}_1$ and $\mathbf{I}_2$ are any two perpendicular unit vectors localised at a point P in the fluid. Together with $\mathbf{I}_3 = \mathbf{I}_1 \times \mathbf{I}_2$ they form a triad of perpendicular unit vectors as in figure 10.4.

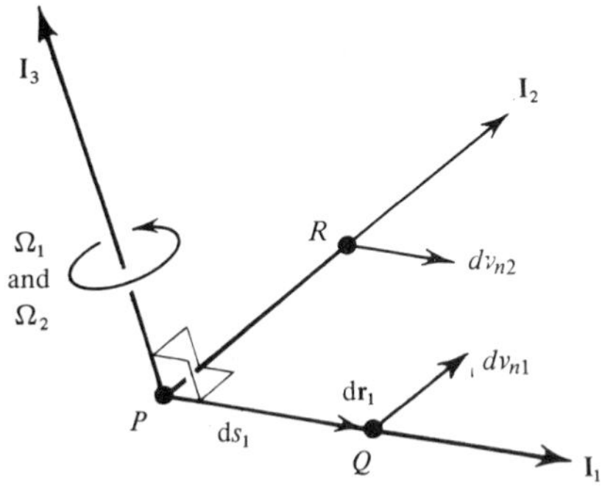

Fig. 10.4. Right-handed triad of unit vectors at P. $\mathbf{I}_1$ is inclined out of the paper, $\mathbf{I}_2$ into it.

Consider a short line element $\overrightarrow{PQ}$ or $d\mathbf{r}_1$ of length ds_1 lying in the direction of $\mathbf{I}_1$, with $d\mathbf{r}_1 = ds_1\mathbf{I}_1$. The velocity of Q relative to P is the difference in velocity between the ends of the vector $d\mathbf{r}_1$. This equals $(d\mathbf{r}_1 \cdot \text{grad})\mathbf{v}$ or $ds_1(\mathbf{I}_1 \cdot \text{grad})\mathbf{v}$, to the first order of small quantities. The transverse component of this relative velocity (dv_{n1} in the figure) in the direction of $\mathbf{I}_2$ can be found by scalar-multiplying by $\mathbf{I}_2$. So

$$dv_{n1} = ds_1\, \mathbf{I}_2 \cdot \{(\mathbf{I}_1 \cdot \text{grad})\mathbf{v}\},$$

which is a generalisation of the dv_n term in section 10.2. This can be attributed to an angular velocity Ω_1 about the $\mathbf{I}_3$ axis equal to dv_{n1}/ds_1 or $\mathbf{I}_2 \cdot \{(\mathbf{I}_1 \cdot \text{grad})\mathbf{v}\}$. Similarly, for $\overrightarrow{PR}$ lying in the $\mathbf{I}_2$ direction, the transverse relative velocity dv_{n2} of R is attributable to an angular velocity $\Omega_2 = -\mathbf{I}_1 \cdot \{(\mathbf{I}_2 \cdot \text{grad})\mathbf{v}\}$ about the $\mathbf{I}_3$ axis. (The minus sign is consistent with the righthand screw convention.) As in the earlier two-dimensional cases we take the sum of the angular velocities of these two perpendicular lines PQ and PR about $\mathbf{I}_3$ as a measure of the fluid's rotation about the $\mathbf{I}_3$ axis. From the vector identity this is seen to be $\mathbf{I}_3 \cdot \text{curl}\,\mathbf{v}$, the component of curl $\mathbf{v}$ in the $\mathbf{I}_3$ direction. This result is independent of our particular choice of the lines PQ and PR in the plane perpendicular to $\mathbf{I}_3$. As the average angular velocity of PQ

or of *PR* for all orientations in the plane perpendicular to $\mathbf{I}_3$ will obviously be the same, $\mathbf{I}_3 \cdot$ curl **v** will be twice this mean angular velocity. We thus see that in three dimensions curl **v** does indeed measure fluid rotation in the sense that its component in any direction equals twice the mean angular velocity in the plane perpendicular to that direction. The meaning of curl **v** is quite independent of the choice of cartesian axes. Like grad and div, curl is invariant to change of axes.

Problem 10.3 takes an alternative view of the way that curl **v** represents a fluid's state of rotation.

10.6 Some important vector identities

We have already encountered a second-order vector operator in the shape of div grad $\phi \equiv \nabla^2\phi$. Two other ones of great importance are curl grad ϕ and div curl **v**. The combinations grad grad, grad curl, div div, curl div are all nonsensical because grad operates only on scalars whereas div and curl operate only on vectors, while div is a scalar but curl and grad are vectors. Chapter 11 refers to grad div and curl curl.

The *x*-component of curl grad ϕ is

$$\frac{\partial}{\partial y}\frac{\partial \phi}{\partial z} - \frac{\partial}{\partial z}\frac{\partial \phi}{\partial y},$$

which vanishes, as do the other components. Thus we have the identity

$$\text{curl grad } \phi \equiv \mathbf{0}.$$

It follows that a *conservative* vector field **v** (with a potential ϕ such that $\mathbf{v} = \text{grad } \phi$) is always *irrotational,* e.g. an electrostatic field **E** obeys curl $\mathbf{E} = \mathbf{0}$. Already our new curl concept is showing intriguing relationships to earlier notions. Problems 10.4(*b*) and (*c*) put this result to work. Now consider

$$\text{div curl } \mathbf{v} = \frac{\partial}{\partial x}\left(\frac{\partial v_z}{\partial y} - \frac{\partial v_y}{\partial z}\right) + \frac{\partial}{\partial y}\left(\frac{\partial v_x}{\partial z} - \frac{\partial v_z}{\partial x}\right) + \frac{\partial}{\partial z}\left(\frac{\partial v_y}{\partial x} - \frac{\partial v_x}{\partial y}\right).$$

The terms cancel in pairs, because

$$\frac{\partial^2 v_z}{\partial x \partial y} = \frac{\partial^2 v_z}{\partial y \partial x}, \quad \text{etc.}$$

Hence $\quad \text{div curl } \mathbf{v} \equiv 0,$

i.e. *a curl field is always solenoidal* and has no sources. This result means that a curl field has the usual properties of a solenoidal field that its flux linked by a loop is a valid, unambiguous concept and that its field-line pattern can indicate intensity as well as direction; a curl field gets stronger where its field lines gather closer together.

The identity is often interpreted as implying that vorticity lines can never end and must form closed loops, but this is not true. A solenoidal field only forms finite closed loops in general in two-dimensional problems or ones with high symmetry; moreover vorticity lines *can* end at a free surface exposed to vacuum, for consider such a surface on liquid that is rotating uniformly about a vertical axis, in which the vorticity lines are vertical. But vorticity lines do not end where moving fluid adjoins another solid or fluid medium, nor can they end in the middle of continuous fluid. At a stationary wall they simply bend as necessary and may accumulate densely in a boundary layer (or vortex sheet if the layer is modelled as having negligible thickness). Then the close packing of the lines indicates the high intensity of vorticity prevailing in the layer. If the wall or adjoining medium is moving in a rotational manner, the vorticity lines carry on into it. Example 10.5 illustrates some of these ideas and problem 10.5 provides further cases for investigation.

Example 10.5 A fluid velocity field is given by the equation

$$\mathbf{v} = -\Omega y\mathbf{i} + \Omega x\mathbf{j} + 0\mathbf{k}$$

where Ω is a function of z. Examine the form of the streamlines and vorticity lines, considering the cases $\Omega = z^2$ and $\Omega = \tanh z$.

Since $v_z = 0$, the flow is in horizontal planes, if we take z as vertical. A streamline is given by

$$\frac{\mathrm{d}x}{-\Omega y} = \frac{\mathrm{d}y}{\Omega x},$$

which leads to $x^2 + y^2 = \text{const}$. The streamlines are horizontal circles, centred on the z-axis as in figure 10.5. The flow pattern is a generalisation of the forced vortex discussed in example 10.3.

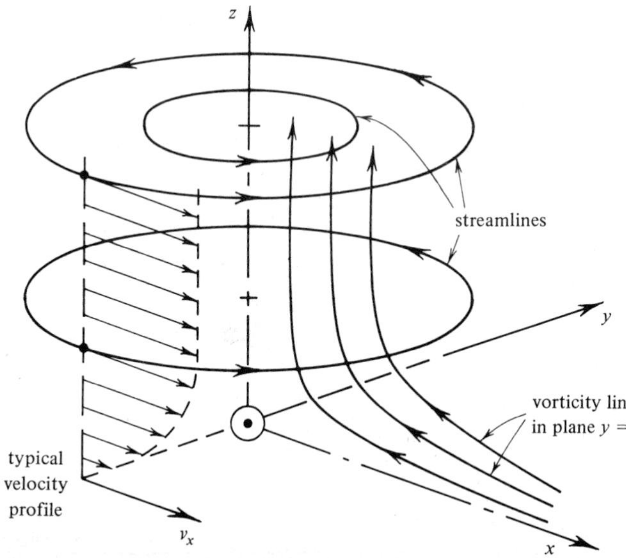

Fig. 10.5

The vorticity $\boldsymbol{\omega}$ is given by

$$\boldsymbol{\omega} = \text{curl } \mathbf{v} = -x\frac{\mathrm{d}\Omega}{\mathrm{d}z}\mathbf{i} - y\frac{\mathrm{d}\Omega}{\mathrm{d}z}\mathbf{j} + 2\Omega\mathbf{k},$$

which satisfies div $\boldsymbol{\omega} = 0$, of course. Along a vorticity line

$$\frac{\mathrm{d}x}{-x} = \frac{\mathrm{d}y}{-y} = \frac{\mathrm{d}z}{2\Omega}\frac{\mathrm{d}\Omega}{\mathrm{d}z} = \frac{\mathrm{d}\Omega}{2\Omega}.$$

Integrating gives

$$x/y = \text{const.} \quad \text{and} \quad x\Omega^{1/2} = \text{const.}$$

The first equation tells us that the vorticity lines lie in vertical planes through the z-axis and the second equation expresses their shape in such a plane, e.g. the plane $y = 0$, shown in the figure. The shape depends on the function Ω. If $\Omega = z^2$, the vorticity lines are rectangular hyperbolas, with xz = const.

The figure shows the case $\Omega = \tanh z$. For large z, Ω is uniform and we have a forced vortex, in which the vorticity lines are vertical. Near the base plane, $\mathbf{v}$ and Ω fall rapidly to zero and there is a boundary layer in which the vorticity lines become nearly horizontal and more closely packed.

The configuration is axisymmetric about the z-axis. Consider a 'tube of vorticity', defined by rotating a vorticity line about the z-axis. Its surface is composed of vorticity lines and any horizontal cross-section of it is a circle of radius r such that $r\Omega^{1/2}$ is constant. This implies that $2\Omega\pi r^2$ = constant, which confirms that vorticity flux is being conserved within the tube, for 2Ω is the vertical component of vorticity.

We shall need to use in later sections another important vector identity, which provides a further link between the vector operators curl and grad. This result is

$$(\mathbf{v} \cdot \text{grad})\mathbf{v} \equiv (\text{curl } \mathbf{v}) \times \mathbf{v} + \text{grad } (\tfrac{1}{2}\mathbf{v}^2).$$

The reader should satisfy himself of its truth by the use of cartesian components.

Example 10.6 Show that for any irrotational field of magnitude v, $v/R = \partial v/\partial n$, where R = radius of curvature of a field line and n = distance measured along its principal normal on the concave side.

Since curl $\mathbf{v} = \mathbf{0}$,

$$(\mathbf{v} \cdot \text{grad})\mathbf{v} = \text{grad } (\tfrac{1}{2}v^2) = v \text{ grad } v.$$

The left-hand side has a component v^2/R along the principal normal to the field line and grad v has a component $\partial v/\partial n$ in that direction. So

$$v^2/R = v\partial v/\partial n \quad \text{and} \quad v/R = \partial v/\partial n.$$

This general result for irrotational fields reveals how curvature and non-uniformity are related. It is particularly obvious in connection with the curvilinear squares discussed in section 8.8. (See for instance figures 8.4, 8.7.) The result is also true for any field parallel to its own curl.

10.7 An application to fluid mechanics; irrotational flow with uniform density

It was speculation about fluid mechanics that led us to develop all this mathematical apparatus associated with the curl operator. It is high time to investigate whether the hoped-for dividend in the shape of improved understanding of fluid dynamics is forthcoming.

For a start, consider a restricted case, namely, any steady flow with uniform density in which all streamlines originate upstream in a region of flow that is devoid of vorticity $\boldsymbol{\omega}$. Notice that we are here excluding flows where there are closed eddies such as occur behind bluff bodies involving streamlines that do not originate upstream. Notice also that in using the word 'originate' we are implicitly invoking the fact that, in steady flow, fluid particles do travel in time along the streamlines.

This case is very important in aeronautics for it describes flight through a uniform stationary atmosphere, the motion being turned into a steady flow by referring it to a frame moving with the aeroplane.

If we ignore viscosity and gravity, Euler's equation in steady flow is

$$\rho(\mathbf{v}\cdot\text{grad})\mathbf{v} + \text{grad}\, p = \mathbf{0},$$

as the part $\partial\mathbf{v}/\partial t$ of the acceleration $\mathrm{D}\mathbf{v}/\mathrm{D}t$ vanishes. Exploiting the vector identity for $(\mathbf{v}\cdot\text{grad})\mathbf{v}$ and the fact that ρ is constant, we have

$$(\text{curl}\,\mathbf{v})\times\mathbf{v} + \text{grad}\,(\tfrac{1}{2}\mathbf{v}^2) + \text{grad}\,(p/\rho) = \mathbf{0}$$

or

$$\boldsymbol{\omega}\times\mathbf{v} + \text{grad}\,(\tfrac{1}{2}\mathbf{v}^2 + p/\rho) = \mathbf{0}.$$

Hence grad $(\frac{1}{2}\mathbf{v}^2 + p/\rho)$ is perpendicular to $\mathbf{v}$, i.e. it has no component in the $\mathbf{v}$-direction, and so $(\frac{1}{2}\mathbf{v}^2 + p/\rho)$ is constant along streamlines. This is the familiar Bernoulli theorem, proved in a new way.

Far upstream $\boldsymbol{\omega}$ and therefore also grad $(\frac{1}{2}\mathbf{v}^2 + p/\rho)$ vanish and $(\frac{1}{2}\mathbf{v}^2 + p/\rho)$ is uniform, *taking the same value on all streamlines.* This implies that it is the same throughout the field of flow, all parts of which can be reached along streamlines that originate in the upstream region, i.e. grad $(\frac{1}{2}\mathbf{v}^2 + p/\rho)$ and hence also $\boldsymbol{\omega}\times\mathbf{v}$ vanish everywhere. Since $\mathbf{v}\neq\mathbf{0}$ in general, $\boldsymbol{\omega}$ is either zero or parallel to $\mathbf{v}$. We can express this as $\boldsymbol{\omega} = \alpha\mathbf{v}$, in which α is some scalar field, which may be zero. But div $\boldsymbol{\omega} = 0$ (for div curl $\equiv 0$) and so

$$0 = \text{div}\,\alpha\mathbf{v} = \alpha\,\text{div}\,\mathbf{v} + \mathbf{v}\cdot\text{grad}\,\alpha = \mathbf{v}\cdot\text{grad}\,\alpha,$$

since div $\mathbf{v} = 0$, the density being constant. The implication is that grad α is perpendicular to $\mathbf{v}$, i.e. that α does not vary along streamlines. But α is zero on all streamlines upstream, where $\boldsymbol{\omega} = \mathbf{0}$ but $\mathbf{v}\neq\mathbf{0}$, and so α vanishes everywhere. This implies that curl $\mathbf{v}$ is zero everywhere. It has been formally proved that, in this case at least, fluid which is initially without rotation continues in that state.

There is considerable interest, especially in aeronautics, in irrotational motions of this kind with curl $\mathbf{v} = \mathbf{0}$. Observe how successfully the invention of vorticity to describe fluid rotation has enabled us, for such flows, to replace the formidable Euler's differential equation for $\mathbf{v}$, which involves the further unknown p, by the simpler one, curl $\mathbf{v} = \mathbf{0}$, which is the dynamic statement that we have been seeking in a form which is independent of p.

There is a further unexpected benefit: the original non-linear differential equation for $\mathbf{v}$ has been replaced by the *linear* one, curl $\mathbf{v} = \mathbf{0}$, which should therefore be easier to solve. One may wonder where the essential non-linearity of fluid dynamics has disappeared to. The answer is Bernoulli's equation; after the velocity field has been found from curl $\mathbf{v} = \mathbf{0}$, the *non-linear* equation $\frac{1}{2}\mathbf{v}^2 + p/\rho = \text{const.}$ enables the pressure to be calculated. In aeronautics, for instance, the pressure is the variable of greatest interest for it determines the lift on the aeroplane.

We next turn briefly to the case where the vorticity is not zero.

Example 10.7 Show that in an inviscid, steady, constant-density flow $\frac{1}{2}\mathbf{v}^2 + p/\rho =$ constant along a vorticity line. Investigate whether this is ever true for the motion given in example 10.5.

The equation

$$\boldsymbol{\omega} \times \mathbf{v} + \text{grad}\,(\tfrac{1}{2}\mathbf{v}^2 + p/\rho) = \mathbf{0}$$

indicates that grad $(\frac{1}{2}\mathbf{v}^2 + p/\rho)$ is perpendicular to $\boldsymbol{\omega}$ and hence that $(\frac{1}{2}\mathbf{v}^2 + p/\rho)$ is constant along vorticity lines (as well as streamlines).

If the flow in example 10.5 is steady, with fluid particles travelling in horizontal circles of radius r with angular velocity Ω, the acceleration of a fluid particle is merely the usual centripetal term $r\Omega^2$. Evidently the pressure gradient must be horizontal and radial, i.e. the pressure depends only on r. Newton's law or Euler's equation requires that

$$\mathrm{d}p/\mathrm{d}r = \rho r\Omega^2 \quad \text{and} \quad p = p_0 + \tfrac{1}{2}\rho r^2\Omega^2,$$

where p_0 is the pressure at the z-axis, which must be uniform. The velocity is $r\Omega$ and so $(\frac{1}{2}\mathbf{v}^2 + p/\rho) = p_0/\rho + r^2\Omega^2$. Along a vorticity line $r^2\Omega = \text{const.}$, and $(\frac{1}{2}\mathbf{v}^2 + p/\rho)$ can be constant along a vorticity line only if $\Omega = \text{const.}$, i.e. if the flow is a pure forced vortex.

If instead Ω does vary with z, the flow cannot be steady. The radial pressure gradients could not balance the different 'centrifugal forces' prevailing at different levels.

Problem 10.6 concerns some generalisations to compressible flow with and without vorticity.

It is important to stress the fact that attention has been restricted so far to flows in which the only force acting on the fluid, namely grad p, was irrotational (for curl grad $p \equiv \mathbf{0}$). Under such conditions the fluid tends to preserve its state of rotation (or non-rotation). The addition of gravity, which is an irrotational force field, if the fluid has uniform density, does not alter this situation, because p may then be replaced by p^*. (See problem 9.14.) However, as soon as there is a rotational force field present due to viscous stress or $\mathbf{j} \times \mathbf{B}$ forces, etc., each fluid element's rotation is constantly changing. Notice that it is the curl of the force field that is the primary influence on the motion, for it measures the extent to which the force field can elude the efforts of the irrotational pressure gradient field to nullify it. This perhaps begins to reveal why the curl concept is so fundamental to any serious study of fluid mechanics.

The example below and problem 10.8 concern two simple cases which reveal how rotational forces can alter a fluid's state of rotation. Both concern rectilinear motion in the x-direction, with all quantities varying only with y and time t, and with the vorticity $\boldsymbol{\omega}$ merely in the z-direction.

Example 10.8 Fluid of uniform electrical conductivity σ moves with a velocity v (in the x-direction) which depends only on y and t. A uniform magnetic field of magnitude B is applied in the y-direction but there is no electric field. Show that the induced $\mathbf{j} \times \mathbf{B}$ forces cause the vorticity of any fluid element to decay in the ratio e:1 in a time $\rho/\sigma B^2$.

As

$$\mathbf{E} = \mathbf{0}, \mathbf{j} = \sigma(\mathbf{v} \times \mathbf{B}) \quad \text{and} \quad \mathbf{j} \times \mathbf{B} = \sigma(\mathbf{v} \times \mathbf{B}) \times \mathbf{B} = \sigma((\mathbf{B} \cdot \mathbf{v})\mathbf{B} - B^2\mathbf{v})$$

(from the standard vector identity) $= -\sigma B^2 \mathbf{v}$ (for $\mathbf{B}$ and $\mathbf{v}$ are perpendicular). The acceleration is merely $\partial \mathbf{v}/\partial t$ (for $(\mathbf{v} \cdot \text{grad})\mathbf{v} = \mathbf{0}$ here) and so Euler's equation, with magnetic force added, is

$$\rho \frac{\partial \mathbf{v}}{\partial t} + \text{grad}\, p = -\sigma B^2 \mathbf{v}.$$

That the irrotational pressure gradient force cannot balance $\mathbf{j} \times \mathbf{B}$ is revealed if we 'take the curl' of both sides, noting that curl grad $\equiv \mathbf{0}$ and that curl $(\partial \mathbf{v}/\partial t) = (\partial/\partial t)$ curl $\mathbf{v}$. Then

$$\rho \partial \boldsymbol{\omega}/\partial t = -\sigma B^2 \boldsymbol{\omega},$$

if we insert $\boldsymbol{\omega}$ for curl $\mathbf{v}$. The vorticity obviously decays exponentially with an 'e:1 time' equal to $\rho/\sigma B^2$.

Problem 10.8 reveals viscosity having a similar effect. However, it is possible for $\mathbf{j} \times \mathbf{B}$ or viscous forces to create rather than destroy vorticity under other circumstances.

Our satisfaction at the effectiveness of the curl operator in helping to solve fluid-dynamical problems needs tempering with the observation that, for flows that are *not* irrotational, the elimination of pressure from Euler's equation by the use of the identity curl grad $p \equiv \mathbf{0}$, giving

$$\text{curl}\, \rho\{\partial \mathbf{v}/\partial t + (\mathbf{v} \cdot \text{grad})\mathbf{v}\} = \mathbf{0}$$

if no extra rotational forces are present, does not in general eliminate the formidable non-linearities. Irrotational flow is an untypically tractable, freak case from a mathematical point of view, even though it is quite common in practice. Many books give it excessive attention, to the exclusion of rotational flow. We shall postpone further discussion of flows with rotation until chapter 12, where it emerges that conservation of 'state of rotation' does not simply mean conservation of vorticity, but is considerably more subtle. (See also problem 10.7.)

10.8 Irrotationality and potential

If we exploit the identity curl grad $\equiv \mathbf{0}$, a remarkable comparison between irrotational flow and flow in a porous medium emerges, despite the fact that the two cases, the one inviscid and inertia-dominated, the other viscous-dominated and inertia-free, are so different physically. In a uniform porous medium, $\mathbf{v}^* = -\text{grad}\, kp^*$, and the identity curl grad $\equiv \mathbf{0}$ implies that curl $\mathbf{v}^* = \mathbf{0}$, i.e. this flow is also irrotational. By this particular instance we are merely reaffirming the conclusion in section 10.6 that any conservative vector field, with a potential ϕ such that $\mathbf{v} = \text{grad}\, \phi$, is irrotational. The converse is also true: if curl $\mathbf{v} = \mathbf{0}$, then $\mathbf{v}$ is conservative and has a potential ϕ. The formal proof of this must await Stokes's theorem in section 10.10.

In short, the terms *conservative* and *irrotational* are synonymous. One might ask why both epithets are still in use if they mean the same thing. Conventional usage often has it that 'irrotational' refers normally to velocity fields and 'conservative' refers to force fields such as $\mathbf{E}$ or gravity.

One immediate consequence is that a ready test for conservative fields becomes available; it is merely necessary to check that curl vanishes everywhere. (See problem 10.9. Problem 10.10 considers the implications for perfect differentials.)

Example 10.9 Test whether the force field $\mathbf{F} = xy^2\mathbf{i} + x^2y\mathbf{j} + z\mathbf{k}$ (the subject of problem 3.9(*b*)) is conservative, and if so find its potential ϕ.

$$\text{curl}\, \mathbf{F} = \left\{\frac{\partial}{\partial y}(z) - \frac{\partial}{\partial z}(x^2y)\right\}\mathbf{i} + \left\{\frac{\partial}{\partial z}(xy^2) - \frac{\partial}{\partial x}(z)\right\}\mathbf{j}$$

$$+ \left\{\frac{\partial}{\partial x}(x^2y) - \frac{\partial}{\partial y}(xy^2)\right\}\mathbf{k} = \mathbf{0},$$

and so $\mathbf{F}$ is indeed conservative, and equal to grad ϕ. The equations

$$\frac{\partial \phi}{\partial x} = xy^2, \quad \frac{\partial \phi}{\partial y} = x^2y, \quad \frac{\partial \phi}{\partial z} = z$$

can be integrated to give

$$\phi = \tfrac{1}{2}x^2y^2 + F(y,z), \quad \phi = \tfrac{1}{2}x^2y^2 + G(x,z), \quad \phi = \tfrac{1}{2}z^2 + H(x,y),$$

where F, G, H are unknown functions. The only compatible result is

$$\phi = \tfrac{1}{2}x^2y^2 + \tfrac{1}{2}z^2 + \text{const.}$$

In irrotational flow curl $\mathbf{v} = \mathbf{0}$ and therefore some scalar field $\phi(x, y, z)$ exists such that $\mathbf{v} = \text{grad}\,\phi$ and $\int_{\text{datum}} \mathbf{v} \cdot \mathbf{dr} = \phi$. (Note that no minus sign has been inserted.) The important quantity ϕ is called the *velocity potential*. It provides a convenient way of expressing the *vector* velocity field in terms of a *scalar*. Irrotational flow is often called *potential flow* because of the existence of ϕ, which is a mathematical abstraction, not a measureable physical quantity such as $-kp^*$, the potential for flow in a porous medium. Being abstract, ϕ can be multi-valued in the manner of magnetic potential U in section 4.3 in multiply-connected domains, penetrated by other regions R in which curl $\mathbf{v}$ is not zero. The important consequence of ϕ being multi-valued is that $\int \mathbf{v} \cdot \mathbf{dr}$, the change in ϕ, may fail to vanish when the line integral returns to its starting point, closing the loop; the circulation $\oint \mathbf{v} \cdot \mathbf{dr}$ round loops which link the regions R can be non-zero. An example will show how this happens.

Example 10.10 Consider a free vortex with $v = A/r$ except in its forced vortex core of radius a in which $v = Ar/a^2$, as described in section 10.4. Find the velocity potential ϕ in the irrotational free vortex part of the flow. Is it multivalued? What is the magnetic analogue of this case?

In the free vortex,

$$\partial\phi/\partial x = v_x = -Ay/(x^2 + y^2), \quad \partial\phi/\partial y = v_y = Ax/(x^2 + y^2).$$

Integrating gives

$$\phi = A \tan^{-1}(y/x) = A\theta,$$

if we choose ϕ zero where the polar coordinate θ is zero. θ is of course multi-valued and ϕ is seen to increase by $2\pi A$ for every circuit round the core, i.e. $\oint \mathbf{v} \cdot \mathbf{dr}$ on any loop that links the core equals $\pm 2\pi A$ (the sign depending on the direction taken round the loop). A free vortex is seen to have a unique, characteristic circulation associated with it. Its value is obvious if a concentric circular loop of length $2\pi r$ is chosen. $\oint \mathbf{v} \cdot \mathbf{dr} = 0$ for any loop that does not link the core.

The magnetic analogue is the circular field due to uniform currents in a long, straight wire of radius a, where $B \propto 1/r$ outside but $\propto r$ inside the wire. (See examples 4.1 and 4.2 and section 8.10.) Problem 10.12(*c*) treats this case by complex variable methods and chapter 12 takes the matter further.

Although we have not yet proved it, potential flow that is unsteady (see problem 10.11) or in which the density varies because of compressibility is possible. But if the density is uniform, the mass conservation equation div $\mathbf{v} = 0$ combines with the equation $\mathbf{v} = \text{grad}\,\phi$ to give a new outlet for Laplace's equation

$$\text{div grad } \phi = \nabla^2\phi = 0.$$

All the mathematical apparatus available for laplacian problems can therefore be brought to bear upon constant-density potential flows. If they are two-dimensional, the conjugate functions ϕ and ψ are applicable as usual and curvilinear squares or complex variable methods may be used. Many of the cases in chapter 8 can be interpreted as potential flows. Problem 10.12 gives examples involving complex variables. Another instructive example follows.

***Example 10.11** What motion can be represented by the laplacian velocity potential

$$\phi = Ae^{ky} \sin k(x - ct)? \quad (A, k, c \text{ const.})$$

This is evidently an *unsteady* potential flow, for time t appears. The flow is in vertical planes, if we take y as vertical. The velocities are

$$v_x = \partial\phi/\partial x = kAe^{ky} \cos k(x - ct),$$

$$v_y = \partial\phi/\partial y = kAe^{ky} \sin k(x - ct).$$

From the relations $v_x = \partial\psi/\partial y$ and $v_y = -\partial\psi/\partial x$, the stream function is seen to be

$$\psi = Ae^{ky} \cos k(x - ct).$$

It too obeys Laplace's equation. Figure 10.6(a) shows typical streamlines $\psi = \text{const.}$ at the instant $t = 0$. Note that they are *not* the loci of fluid particles, for this is an unsteady flow. (Particle loci are in fact closed curves, approximately circular in shape. Some are sketched in the figure.)

The motion may instead be referred to a frame of reference that is moving x-wise at velocity c and coincides with the first frame at $t = 0$. Relative to the new frame, $(x - ct)$ is replaced by x and v_x is reduced by c, so that

$$v_x = kAe^{ky} \cos kx - c, \quad v_y = kAe^{ky} \sin kx, \quad \psi = Ae^{ky} \cos kx - cy.$$

This is now a steady flow, with the streamlines sketched in figure 10.6(b) for the instant $t = 0$ when the two frames coincide. The curves are not exact sinusoids. If the fluid above a certain streamline is removed, figure 10.6(b) could be regarded as representing free-surface waves travelling rightwards relative to a stream which is moving leftwards at sufficient speed to keep the waves stationary. Then figure 10.6(a) represents the same waves advancing through otherwise stationary water, a subject of great interest in connection with ocean wave power.

The condition on pressure at the surface has still to be considered. In the steady flow, Bernoulli's equation with gravity (see problem 9.16(b)) must be satisfied. At the free surface $p = \text{const.}$ and so $\frac{1}{2}v^2 + gy = \text{const.}$ there. But $v^2 = v_x^2 + v_y^2$ and so

$$\tfrac{1}{2}(k^2A^2e^{2ky} + c^2 - 2kcAe^{ky} \cos kx) + gy$$

must be constant along the surface streamline $Ae^{ky} \cos kx - cy = \text{const.}$ This is possible if A is small enough for the A^2 term to be neglected and if $c^2 = g/k$. Thus

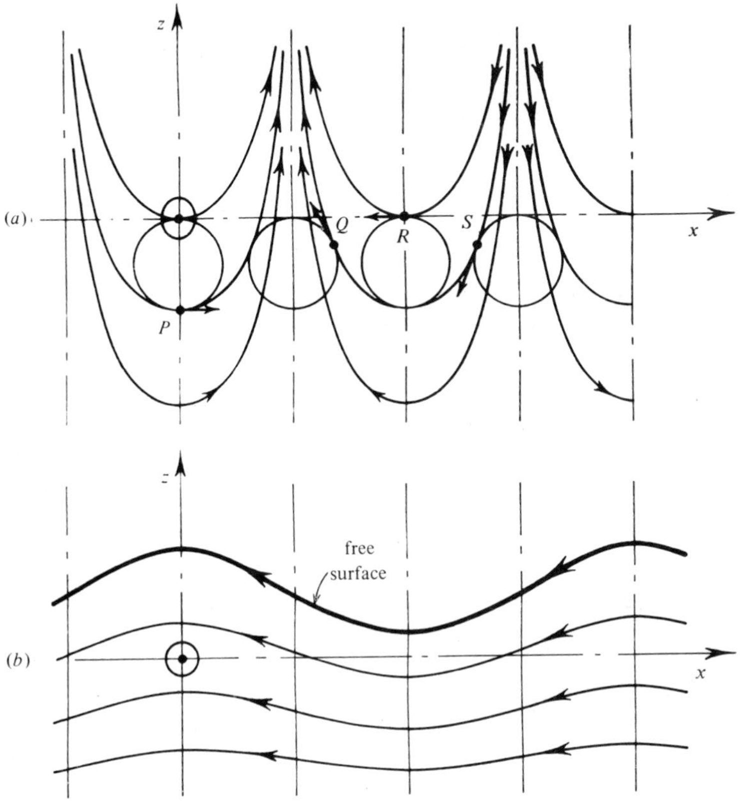

Fig. 10.6. (*a*) Streamlines at the instant $t = 0$. The small closed loops are the paths of the particles P, Q, R and S, shown at the instant $t = 0$. (*b*) Streamlines and particle paths relative to the moving frame at the instant $t = 0$.

the wave speed has been determined. Note that the wave speed is $(\lambda g/2\pi)^{1/2}$, where λ is the wave length, equal to $2\pi/k$.

10.9 Curl and circulation

The fact that curl $\mathbf{v} = \mathbf{0}$ if $\oint \mathbf{v} \cdot d\mathbf{r} = 0$ round all loops (for then $\mathbf{v} = \text{grad } \phi$) strongly suggests that it would be worth looking for some general relationship between circulation $\oint \mathbf{v} \cdot d\mathbf{r}$ and curl $\mathbf{v}$.

As a first stage we shall investigate the circulation $d\Gamma$ round the rim of the small surface element $d\mathbf{a}$, shown in figure 10.7, by taking an origin O in the vicinity and expanding the vector $\mathbf{v}$ and its components as Taylor series, retaining only first-order small quantities. This presupposes differentiability and an absence of singularities in the vector field. Then

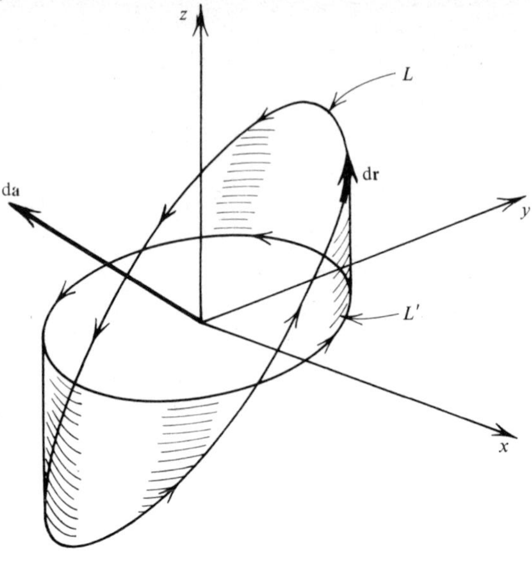

Fig. 10.7

$$v_x = v_{x0} + \left(\frac{\partial v_x}{\partial x}\right)_0 x + \left(\frac{\partial v_x}{\partial y}\right)_0 y + \left(\frac{\partial v_x}{\partial z}\right)_0 z, \quad \text{etc.}$$

In this expression the suffix 0 denotes values at the origin, which may therefore be taken outside the integrals when the circulation is evaluated:

$$d\Gamma = \oint \mathbf{v} \cdot d\mathbf{r} = \oint \sum v_x \, dx = v_{x0} \oint dx + \left(\frac{\partial v_x}{\partial x}\right)_0 \oint x \, dx$$
$$+ \left(\frac{\partial v_x}{\partial y}\right)_0 \oint y \, dx \ldots + \left(\frac{\partial v_y}{\partial x}\right)_0 \oint x \, dy \ldots$$

For brevity only representative terms have been recorded. Integrals such as $\oint dx$ and $\oint x \, dx = \frac{1}{2} \oint d(x^2)$ all vanish when taken round a closed loop.

Before commenting on the other integrals we must establish a clear sign convention, relating **dr** and **da**. We adopt the usual righthand screw rule, namely, that viewed in the positive **da** direction positive **dr** gives *clockwise* circulation. Figure 10.7 reveals how circulation round the loop L according to this convention corresponds to a clockwise circulation round its projection L' on to the x, y plane, when viewed in the z-direction, if da_z, the z-component of **da**, is positive. But da_z, $\oint x \, dy$ and $-\oint y \, dx$ all represent the area of L', the projected area of L in the z-direction. (An unavoidable source of confusion is the fact that conventionally the x, y plane is viewed in the *negative* z-direction, i.e. with Ox to the right and Oy upwards. In such a view the area of L' is $\oint x \, dy$ for *anticlockwise* circulation round L'.)

In similar fashion we have

$$da_y = \oint z \, dx = -\oint x \, dz$$

and

$$da_x = \oint y \, dz = -\oint z \, dy.$$

It follows that

$$d\Gamma = \left(\frac{\partial v_z}{\partial y} - \frac{\partial v_y}{\partial z}\right) da_x + \left(\frac{\partial v_x}{\partial z} - \frac{\partial v_z}{\partial x}\right) da_y + \left(\frac{\partial v_y}{\partial x} - \frac{\partial v_x}{\partial y}\right) da_z$$

which is recognisable as the scalar product curl $\mathbf{v} \cdot \mathbf{da}$, if we drop the suffix 0, curl $\mathbf{v}$ being evaluated in the vicinity of L. Problem 10.13 is a more primitive version of this argument.

In words the result $d\Gamma =$ curl $\mathbf{v} \cdot \mathbf{da}$ states that $\mathbf{v}$'s circulation round a small loop equals the flux of curl $\mathbf{v}$ through it. This important conclusion gives us a wholly new interpretation of curl, much more generalised and widely applicable than its meaning in relation to the rotational quality of velocity fields in fluids, explored in problem 10.14. It is evident that curl is some kind of *circulation density per unit area*, to be contrasted with div which is source density per unit volume. It measures the vigour with which the field departs from the state of being conservative; in the case of a force field such as $\mathbf{E}$ it can best be described as measuring the extent to which useful work could be extracted from the field in a given locality by repeated excursions round a closed path. A search coil could identify this in the case of an electric field. The reader should not rest content with curl as a lifeless mathematical concept, a mere permutation of a del, a cross and a vector, but should come to terms with this new physical idea of circulation density.

A curious feature of this situation is that curl is a vector but circulation is a scalar. How then can curl be a circulation density? The explanation lies in the directionality of the area element $\mathbf{da}$. The point is that the circulation $\oint \mathbf{v} \cdot \mathbf{dr}$ round $\mathbf{da}$ is proportional to the component of curl $\mathbf{v}$ in the direction of $\mathbf{da}$. We can twirl $\mathbf{da}$, as an electrical engineer might turn a search coil at a given point in an electric field in search of the maximum induced e.m.f. (see Demonstration 6 in Appendix II), until the circulation is a maximum. The vector $\mathbf{da}$ is then pointing in the direction of curl $\mathbf{v}$. For other orientations at the same point, if $\mathbf{da}$ is inclined at an angle θ to curl $\mathbf{v}$, the circulation is reduced by a factor cos θ, and the familiar cosine-quality of a vector is being exhibited. For all orientations for which $\theta = \pi/2$, no curl flux is intercepted by $\mathbf{da}$ and the circulation is zero.

Another noteworthy point is that, as curl is circulation *per unit area*, doubling the linear dimensions of the small loop quadruples the circulation round it rather than merely doubling it, as might have been expected. The reason is that $\oint \mathbf{v} \cdot \mathbf{dr}$ is non-zero only because $\mathbf{v}$ is non-uniform and so contributions from opposite sides of the loop do not cancel. Doubling the linear dimensions doubles not only the elements $\mathbf{dr}$ but also the differences between the values of $\mathbf{v}$ on opposite sides of the loop and a quadrupled circulation results. This explains why a limiting value for (circulation) ÷ (loop area) exists as the loop becomes vanishingly small. Notice that circulation, like curl itself, is another way of measuring the non-uniformity of a vector field.

Problem 10.15 exemplifies these various ideas.

10.10 Stokes's theorem

The element $\mathbf{da}$ could be part of any finite surface S spanning a given loop L. If the surface is divided completely into small elements such as $\mathbf{da}$, as suggested in figure 10.8, and the circulation $d\Gamma$ round the elements are all added together, the contributions from boundary lines between pairs of adjoining elements such as P and Q

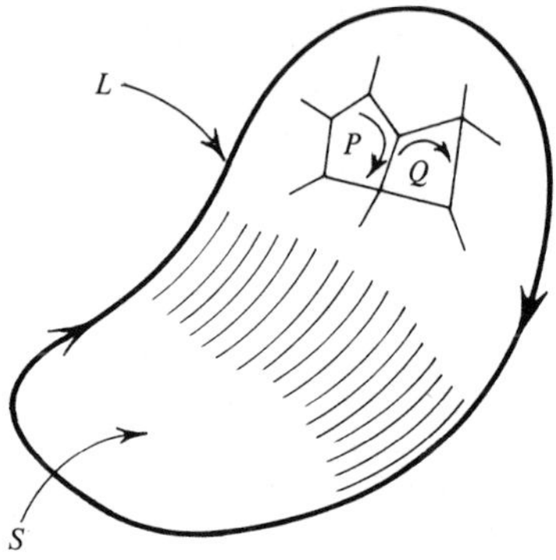

Fig. 10.8

cancel out because each such boundary is traversed once in each direction and so only the contributions from those parts of element boundaries which form part of L survive. Thus the sum of all circulations $d\Gamma$ equals $_L\oint \mathbf{v}\cdot d\mathbf{r}$, the circulation Γ round the edge of the whole surface. But the circulation $d\Gamma$ round $\mathbf{da}$ is curl $\mathbf{v}\cdot\mathbf{da}$ and therefore the sum of all circulations, in the limit when all elements are made vanishingly small, becomes the surface integral $_S\iint$ curl $\mathbf{v}\cdot\mathbf{da}$. We conclude that

$$\Gamma = \oint_L \mathbf{v}\cdot d\mathbf{r} = \iint_S \text{curl}\,\mathbf{v}\cdot\mathbf{da},$$

which is *Stokes's theorem* for any surface S spanning a loop L. It states that the circulation of a vector field round a loop equals the flux of its curl through the loop. The use of the phrase 'flux through a loop' is justified because curl flux linkage is a valid concept as a curl field is solenoidal. (See also problem 10.16.)

Example 10.12

(*a*) Find the circulation of the vector field

$$\mathbf{v} = (x+y)\mathbf{i} + (x-y)\mathbf{j} + xy\mathbf{k}$$

round the circle $x^2 + y^2 = 1$ in the plane $z = 1$.

(*b*) Find the flux of the same vector field through the same circle by expressing it as curl $\mathbf{u}$.

(*a*) curl $\mathbf{v} = x\mathbf{i} - y\mathbf{j} + 0\mathbf{k}$. This has no flux through any plane $z =$ const. and so the circulation required is zero, by Stokes's theorem.

(*b*) The equation $\mathbf{v} = \text{curl } \mathbf{u}$ does not determine $\mathbf{u}$ uniquely. For $\mathbf{u}$ to exist, div $\mathbf{v}$ must vanish, which is the case. A simple choice of $\mathbf{u}$ is one independent of z, e.g.

$$\mathbf{u} = 0\mathbf{i} + \tfrac{1}{2}x^2y\mathbf{j} + \{xy + \tfrac{1}{2}(y^2 - x^2)\}\mathbf{k}.$$

Then the required flux is

$$\iint \text{curl } \mathbf{u} \cdot \mathbf{da} = \oint \mathbf{u} \cdot \mathbf{dr} = \oint \tfrac{1}{2}x^2y \, dy$$

evaluated round the circle. If we put $x = \cos\theta, y = \sin\theta$, the integral becomes

$$\tfrac{1}{2}\int_0^{2\pi} \cos^3\theta \sin\theta \, d\theta,$$

which is zero. The reader should verify these results by direct calculation, which is not difficult in these particular cases.

Stokes's theorem immediately confirms the truth of the earlier assertion that if curl $\mathbf{v} = \mathbf{0}$ everywhere, then $\oint \mathbf{v} \cdot \mathbf{dr} = 0$ round any loop, i.e. *irrotational* implies *conservative.* It also reveals how, in a multiply-connected domain in which curl $\mathbf{v}$ vanishes, penetrated by regions R in which curl $\mathbf{v} \neq \mathbf{0}$, $\oint \mathbf{v} \cdot \mathbf{dr}$ may be non-zero on loops that link R, as remarked in section 10.8 and illustrated below.

Example 10.13 Relate Stokes's theorem to a line vortex and a sheet vortex.

Example 10.10 considered an irrotational free vortex with a forced-vortex core of radius a, in which the vorticity (normal to the planes of flow) is $2A/a^2$ (twice the angular velocity). The total vorticity flux through any loop linking the core all comes from the core region and equals its area πa^2 times $2A/a^2$, i.e. $2\pi A$, the circulation, consistently with Stokes's theorem. Letting $a \to 0$, so as to generate a line vortex, does not disrupt the result.

Figure 10.9 shows a cross-section of a sheet vortex, taken normal to the vorticity. The sheet separates two irrotational streams with velocities v_1 and v_2. *ABCD* is a small loop L of length l straddling the vortex sheet. Stokes's theorem gives:

$$\text{vorticity flux through } ABCD \text{ (all in the sheet)} = \oint_L \mathbf{v} \cdot \mathbf{dr}$$

$$= v_1 l - v_2 l = (v_1 - v_2)l.$$

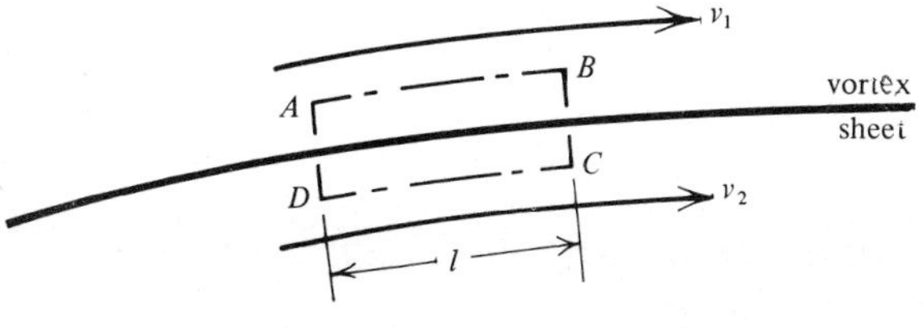

Fig. 10.9. Cross-section of vortex sheet, taken perpendicular to the vorticity.

We conclude that the vorticity flux per unit length in the sheet vortex equals the jump in velocity across it. (There is an obvious parallel with the comparable result for **H** in section 6.9.)

Problem 10.5 exploits these results. These cases typify the use of Stokes's theorem in the presence of regions of rapid spatial change, which may often be modelled as mathematical singularities. Problems 10.17 and 10.18 deploy Stokes's theorem in various other ways.

The righthand screw convention continues to apply for Stokes's theorem: positive $\mathrm{d}\mathbf{r}$ must give a clockwise circuit of L when viewed in the positive $\mathrm{d}\mathbf{a}$ direction.

It is appropriate here to contrast the grad, curl and div concepts and their differential/integral properties. Listed together in order of increasing dimensionality, the generalised interpretations of grad, div and curl are:

grad = change per unit distance (for $\mathrm{d}\phi = \mathrm{d}\mathbf{r}\cdot\operatorname{grad}\phi$),
curl = circulation per unit area (for $\mathrm{d}\Gamma = \mathrm{d}\mathbf{a}\cdot\operatorname{curl}\mathbf{v}$),
div = net efflux (source strength) per unit volume (for $\mathrm{d}\Phi = \mathrm{d}\tau\operatorname{div}\mathbf{v}$),

if $\mathrm{d}\Phi$ is used to denote the net flux from the volume element $\mathrm{d}\tau$. The equivalent progression in integral form for finite domains is:

$$\Delta\phi = \int \operatorname{grad}\phi\cdot\mathrm{d}\mathbf{r},$$

$$\Gamma = \oint \mathbf{v}\cdot\mathrm{d}\mathbf{r} = \iint \operatorname{curl}\mathbf{v}\cdot\mathrm{d}\mathbf{a} \quad \text{(Stokes's theorem)}$$

and

$$\Phi = \oiint \mathbf{v}\cdot\mathrm{d}\mathbf{a} = \iiint \operatorname{div}\mathbf{v}\,\mathrm{d}\tau \quad \text{(Gauss's theorem).}$$

Stokes's theorem indicates that, in fluid mechanics, the rather unnatural, abstract quantity, the velocity circulation $\oint \mathbf{v}\cdot\mathrm{d}\mathbf{r}$, can be regarded as an alternative measure of the rotational quality of the motion. The consequences of this for fluid mechanics are so profound that we postpone them to a separate, final chapter. The next chapter is devoted to exploring the remarkable consequences of applying the curl idea to electromagnetism. The reader interested primarily in fluid mechanics can go to chapter 12 instead at this point if he so prefers.

*10.11 Some more mathematical results

The final section of this chapter concerns various further mathematical results associated with the curl operator. These have many interesting physical applications, but the less mathematical reader may choose to omit this section and the many related problems.

In section 10.6 we met three important vector identities, relating curl to div or grad. Two further important identities are

$$\operatorname{div}(\mathbf{u}\times\mathbf{v}) \equiv \mathbf{v}\cdot\operatorname{curl}\mathbf{u} - \mathbf{u}\cdot\operatorname{curl}\mathbf{v}$$

and

$$\text{curl}\, s\mathbf{v} \equiv s\, \text{curl}\, \mathbf{v} + \text{grad}\, s \times \mathbf{v}.$$

The latter may be compared with the result

$$\text{div}\, s\mathbf{v} \equiv s\, \text{div}\, \mathbf{v} + (\text{grad}\, s) \cdot \mathbf{v},$$

discussed in section 7.10. The reader should verify the two new results by using cartesian components. They look more plausible in the 'del' notation. Problems 10.19 and 10.20 provide exercises on them.

If we set **v** equal to **I**, a uniform unit vector, then

$$\text{curl}\, s\mathbf{I} = \text{grad}\, s \times \mathbf{I},$$

and Stokes's theorem gives

$$\mathbf{I} \cdot \oint_L s\, \text{d}\mathbf{r} = \oint_L s\mathbf{I} \cdot \text{d}\mathbf{r} = \iint_S (\text{curl}\, s\mathbf{I}) \cdot \text{d}\mathbf{a} = \iint_S \text{grad}\, s \times \mathbf{I} \cdot \text{d}\mathbf{a}$$

$$= \mathbf{I} \cdot \iint_S \text{d}\mathbf{a} \times \text{grad}\, s$$

for a surface S spanning a loop L, if we transform the triple scalar product. But **I** can be taken in turn in the x, y and z directions, and so

$$\oint_L s\text{d}\mathbf{r} \equiv \iint_S \text{d}\mathbf{a} \times \text{grad}\, s.$$

This identity finds outlets in problems 10.21 and 10.22.

If we set **u** equal to **I**, a uniform unit vector, then

$$\text{div}\, (\mathbf{I} \times \mathbf{v}) = -\mathbf{I} \cdot \text{curl}\, \mathbf{v},$$

and Gauss's theorem gives

$$-\mathbf{I} \cdot \iiint_V \text{curl}\, \mathbf{v}\text{d}\tau = -\iiint_V \mathbf{I} \cdot \text{curl}\, \mathbf{v}\, \text{d}\tau = \oiint_S \mathbf{I} \times \mathbf{v} \cdot \text{d}\mathbf{a}$$

$$= \oiint_S \mathbf{I} \cdot \mathbf{v} \times \text{d}\mathbf{a} = -\mathbf{I} \cdot \oiint_S \text{d}\mathbf{a} \times \mathbf{v}$$

for a volume V enclosed by a surface S. Hence

$$\iiint_V \text{curl}\, \mathbf{v}\, \text{d}\tau \equiv \oiint_S \text{d}\mathbf{a} \times \mathbf{v}.$$

Compare this with:

$$\iiint_V \text{div}\, \mathbf{v}\, \text{d}\tau \equiv \oiint_S \mathbf{v} \cdot \text{d}\mathbf{a} \quad \text{(Gauss's theorem)}$$

and

$$\iiint_V \text{grad}\, \phi\, \text{d}\tau \equiv \oiint_S \phi\, \text{d}\mathbf{a} \quad \text{(Gauss's corollary).}$$

This sequence of three identities, each relating a volume to a surface integral, should be contrasted with the other sequence of integrals involving curl, div and grad in the previous section. Problem 10.23 makes use of the new result for $\oiint \mathrm{d}\mathbf{a} \times \mathbf{v}$, as does the next example.

Example 10.14 For a finite sphere of volume V and radius R, drawn in a moving fluid of uniform density ρ, a mean vorticity $\boldsymbol{\omega}_1$ is defined by taking

$$\boldsymbol{\omega}_1 = \iiint \operatorname{curl} \mathbf{v}\, \mathrm{d}\tau / V$$

integrated over the sphere. Show that the vector moment of momentum of a thin concentric spherical shell of the fluid of radius R and uniform thickness $\mathrm{d}R$ is equal to $\frac{1}{2}\boldsymbol{\omega}_1 I_1$, if I_1 is the moment of inertia of the shell about any diameter.

Let $\mathbf{R}$ be a radius vector of the sphere. Then $\mathbf{R}\mathrm{d}a = R\mathrm{d}\mathbf{a}$, if $\mathrm{d}\mathbf{a}$ is a surface element at the extremity of $\mathbf{R}$. The corresponding element of the shell has a mass $\rho \mathrm{d}R\mathrm{d}a$ and so

$$\text{mom. of mom.} = \oiint \rho \mathrm{d}R \mathrm{d}a\, \mathbf{R} \times \mathbf{v} = \rho R \mathrm{d}R \oiint \mathrm{d}\mathbf{a} \times \mathbf{v}$$

$$= \rho R \mathrm{d}R \iiint \operatorname{curl} \mathbf{v}\, \mathrm{d}\tau.$$

But $I_1 = \frac{8}{3}\pi\rho R^4 \mathrm{d}R$ and $V = \frac{4}{3}\pi R^3$ and so moment of momentum $= \frac{1}{2}\boldsymbol{\omega}_1 I_1$.

This result throws further light on the fact that (vorticity) ÷ 2 equals mean angular velocity at a point. Notice that in this example it was not necessary to take a *small* fluid element and use a truncated Taylor series, as in sections 10.2 and 10.5. Problem 10.24 is a generalisation of this example.

Problems 10

10.1 (*a*) Find the curl of the vector fields:

(i) $yz\mathbf{i} + xz\mathbf{j} + xy\mathbf{k}$,

(ii) $f(x,y)\mathbf{i} + g(x,y)\mathbf{j} + h(z)\mathbf{k}$. What do you notice about its direction?

(*b*) Show, using cartesians, that $\operatorname{curl}(\mathbf{A} \times \mathbf{r}) = 2\mathbf{A}$ if $\mathbf{A} = $ const. and $\mathbf{r} = $ position vector.

A solid body or mass of fluid is rotating about a fixed axis through the origin with uniform angular velocity $\boldsymbol{\Omega}$. Express the velocity $\mathbf{v}$ in terms of the position vector $\mathbf{r}$ and find curl $\mathbf{v}$. Is it what you expected?

10.2 A two-dimensional velocity field has $v_x = ky$, $v_y = kx$, ($k = $ const.). Are its streamlines curved? Does it have vorticity? Consider the motion of various short line elements in the fluid.

*10.3 (*a*) $\vec{PQ}$(dr) is a short line element in the vicinity of a point P in a fluid. If the velocity of Q relative to P is $\mathbf{dv} = (\mathbf{dr} \cdot \text{grad})\mathbf{v}$, with a component $\mathbf{dv}_n$ transverse to dr, show that $\mathbf{dv}_n$ could be attributed to an angular velocity $\mathbf{\Omega} = \mathbf{dr} \times \mathbf{dv}/(\mathbf{dr})^2$ about P. Find $\mathbf{\Omega}$ in terms of the x, y, z derivatives of v's components and l, m, n, the directions cosines of $\vec{PQ}$. (It should be sufficient to record only a representative sample of the 18 terms.)
(*b*) Now consider three mutually perpendicular line elements PQ_1, PQ_2, PQ_3 with direction cosines (l_1, m_1, n_1) etc. respectively and angular velocities, defined as in (*a*), equal to $\mathbf{\Omega}_1$ etc. Show that $\mathbf{\Omega}_1 + \mathbf{\Omega}_2 + \mathbf{\Omega}_3 =$ curl v, noting the result that $l_1^2 + l_2^2 + l_3^2 = 1$, etc. and $l_1 m_1 + l_2 m_2 + l_3 m_3 = 0$, etc. (for consider the base vectors i, j, k referred to PQ_1, PQ_2, PQ_3 as cartesian axes).
(*c*) Interpret (*b*) for the special case of uniform rotation at angular velocity $\mathbf{\Omega}$ about PQ_1, and confirm that curl $\mathbf{v} = 2\mathbf{\Omega}$.
(*d*) Instead of using approach (*b*), show that, if $\vec{PQ}$ is allowed to adopt all possible orientations in three dimensions, the mean value of $\mathbf{\Omega}$ in (*a*) is equal to $\frac{1}{3}$ curl v. (The mean value of l^2, etc. is $\frac{1}{3}$ and of lm, etc. is zero. Can you prove these results? Contrast the two-dimensional equivalents.) Note the contrast with the result that the mean angular velocity of the fluid is $\frac{1}{2}$ curl v.

10.4 (*a*) v is a two-dimensional, solenoidal field, independent of z and with v_z zero, while A is another field, also independent of z, but with A_x and A_y zero. Show that

$$\mathbf{v} \times \text{curl } \mathbf{A} = \text{curl } (\mathbf{v} \times \mathbf{A}).$$

(*b*) Electrically conducting fluid is in hydrostatic equilibrium under $\mathbf{j} \times \mathbf{B}$ forces due to steady currents of intensity j and a magnetic field B. What condition must apply to curl $(\mathbf{j} \times \mathbf{B})$? Using (*a*) investigate whether it is satisfied if B is uniform and in the z-direction and j flows in x, y planes with no z-variation.
(*c*) When a conducting fluid moves steadily under a steady magnetic field B, electric currents are induced only if curl $(\mathbf{v} \times \mathbf{B}) \neq \mathbf{0}$. (Otherwise the e.m.f. $\mathbf{v} \times \mathbf{B}$ can be balanced by the conservative electric field, i.e. $\mathbf{E} + \mathbf{v} \times \mathbf{B} = \mathbf{0}$ with $\mathbf{E} = -$ grad V.) Investigate whether currents are induced in the case of constant-density, two-dimensional motion in x, y planes under a uniform field B in the z-direction. Consider any loop in an x, y plane, moving with this motion. Why is its area constant? Deduce that the rate of change of magnetic flux linked by any such travelling loop is zero (which confirms that no currents are induced).

10.5 A cylindrical tank with a vertical axis contains water, the upper half of which is at rest while the lower half is undergoing axisymmetric vortex motion in horizontal planes. There is a thin vortex sheet or shear layer between the two halves. Considering the two cases:
(*a*) lower half undergoing irrotational circulation (free vortex) about a core of intense vorticity,

(*b*) lower half undergoing 'solid-body' rotation (forced vortex), verify the conservation of the solenoidal vorticity flux at the vortex sheet, given the results:

(i) vorticity flux along the core of a free vortex (for which $v = A/r$) equals $2\pi A$;

(ii) vorticity flux per unit width along a vortex sheet equals the jump in tangential velocity across it. (The vorticity is perpendicular to the relative tangential velocity.)

10.6 (*a*) Combine the steady-flow version of Euler's equation with the result $T \operatorname{grad} s = \operatorname{grad} h - (1/\rho) \operatorname{grad} p$ from thermodynamics (T = temperature, s = entropy, h = enthalpy) to deduce *Crocco's theorem*:

$$\operatorname{grad} h_0 = \mathbf{v} \times \boldsymbol{\omega} + T \operatorname{grad} s,$$

in which $h_0 = h + \frac{1}{2}\mathbf{v}^2$, the stagnation enthalpy. (See also problem 9.12(*f*).) Show that if h_0 = uniform, non-uniform entropy implies vorticity. (This result has implications for the flow behind curved shock waves.)

*(*b*) The simplest design approach to axial-flow turbo-machinery ('*vortex-theory*') is to make the work exchange in the moving blade rows per unit mass flow the same at all radii. Then the stagnation enthalpy h_0 can be independent of radius r at each cross-section. Ideally also the entropy is uniform everywhere. Show that Crocco's theorem is satisfied if at any cross-section between blade rows the axial velocity v_a is uniform, the whirl velocity v_w equals const./r and radial velocities are negligible.

For the annulus lying between radii r and $r + \mathrm{d}r$, relate the torque on a moving blade row (rotating at angular velocity Ω) to the change in whirl momentum across the row, and confirm that the work exchanged per unit mass flowing is independent of radius.

*10.7 Using cartesian coordinates, show that

$$\operatorname{curl}(\mathrm{D}\mathbf{v}/\mathrm{D}t) = \mathrm{D}\boldsymbol{\omega}/\mathrm{D}t + \boldsymbol{\omega} \operatorname{div} \mathbf{v} - (\boldsymbol{\omega} \cdot \operatorname{grad})\mathbf{v}.$$

If the fluid obeys the equation $\rho \mathrm{D}\mathbf{v}/\mathrm{D}t + \operatorname{grad} p = \mathbf{0}$, show that, even if ρ = const. and $\operatorname{div} \mathbf{v} = 0$, the vorticity of a travelling fluid element is not constant if the velocity varies along vorticity lines (as will generally be the case). Consider the application to example 10.5.

10.8 A fluid of viscosity μ and density ρ moves with a velocity v (in the x-direction) which is given by an expression of the form $T \sin ky$ in which T is a function of t and k is constant. Show that the vorticity of any fluid element decays in the ratio e:1 in a time $\rho/\mu k^2$.

10.9 Show that

$$\mathbf{u} = (x - 3y)\mathbf{i} + (y - 3x)\mathbf{j} + 0\mathbf{k}$$

and

$$\mathbf{v} = (x + y)(x - 3y)\mathbf{i} + (x + y)(y - 3x)\mathbf{j} + 0\mathbf{k}$$

share the same field lines. Which field has the property that the closeness of the field lines indicates field intensity and which is conservative? Find its potential.

10.10 (*a*) A differential expression $\epsilon = F\mathrm{d}x + G\mathrm{d}y + H\mathrm{d}z$, where F, G and H are functions of x, y, z is called a *perfect differential* if it equals the increment $\mathrm{d}\phi$ of some other function ϕ of x, y and z. Show that then the curl of the vector field **v** with components (F, G, H) must vanish. Show conversely that if curl **v** vanishes, then ϵ is a perfect differential.
(*b*) What are the two-dimensional equivalents of (*a*)? In thermodynamics $T\mathrm{d}s + v\mathrm{d}p$ is a perfect differential ($\mathrm{d}h$) where T, v and h are functions of s and p. Show that $(\partial T/\partial p)_s = (\partial v/\partial s)_p$, one of Maxwell's thermodynamic equations (not to be confused with his electromagnetic ones).

10.11 (*a*) Show that in *unsteady*, constant-density, irrotational flow, with velocity potential ϕ, the quantity $(\partial\phi/\partial t + \frac{1}{2}\mathbf{v}^2 + p/\rho + gz)$ is uniform in space at each instant, in the absence of viscous forces.
*(*b*) Consider the application of (*a*) to example 10.11.

*10.12 Two-dimensional, irrotational flows at constant density may be found by the usual complex-variable methods for solving Laplace's equation, in which $\Omega = \phi + i\psi = f(z)$, where $z = x + iy$, ϕ and ψ being respectively velocity potential and stream function. Consider the following examples:
(*a*) What flow is represented by $\Omega = Uz$ (U real and constant)?
(*b*) A doublet at the origin (see problem 8.15(*e*)) is specified by $\Omega = UR^2/z$ (R real and constant). Combining this with (*a*) gives $\Omega = Uz + UR^2/z$. Show that then the flow at large distances from the origin is uniform and that the streamline $\psi = 0$ consists of the x-axis together with a circle centred at the origin. What is its radius? The part of this solution external to the circle can be taken as potential flow past a cylinder placed in a uniform stream. Use Bernoulli's equation to find the difference between the maximum and minimum pressures at the surface and to show there is no drag on the cylinder. (It should be appreciated that in real fluids, boundary-layer separation causes this solution to fail on the rear of such a bluff body with the result that there is a high drag in practice. On more 'streamlined' bodies potential flow solutions are very satisfactory although they still ignore the drag associated with viscous shear forces.)
(*c*) What flow is represented by $\Omega = iB \log(z/R)$ (B real and constant)? (Compare section 8.11.) Using the facts that $\int_1^2 \mathbf{v} \cdot \mathrm{d}\mathbf{r} = \phi_2 - \phi_1$ and that ϕ is now multi-valued, show that the circulation round any route enclosing the origin once is non-zero and unique.
(*d*) If the terms in (*b*) and (*c*) are added so that

$$\Omega = Uz + UR^2/z + Bi \log(z/R),$$

show that the streamline $\psi = 0$ still includes the same circle as in (*b*) and that the flow at large distances is still a uniform stream, i.e. the solution external to the circle is a different flow of the same stream past the same cylinder. Show that the circulation Γ by any route once round the cylinder is non-zero and unique. Find the velocity and pressure distributions on the cylinder and show that the drag is still zero but there is a transverse force (*lift*) in the y-direction equal to $\rho U\Gamma$ per unit length of cylinder.

10.13 Show that the circulation $d\Gamma$ of a vector **v** round a small rectangular loop dx by dy in a plane z = const. is given by

$$d\Gamma = \left(\frac{\partial v_y}{\partial x} - \frac{\partial v_x}{\partial y}\right) dx\, dy.$$

10.14 In a fluid Ω_D, the component in a particular direction D of the average angular velocity at any point, could be defined in relation to a small circular area element d**a** of radius r, oriented in the direction D as follows:

(i) Define average peripheral velocity v_t as (circulation $\oint \mathbf{v}\cdot d\mathbf{r}$) $\div 2\pi r$.
(ii) Define Ω_D as v_t/r.

Show that $\Omega_D = \frac{1}{2}$ (component of curl **v** in direction D).

10.15 The variable plane $lx + my + nz = 1$ (where $l^2 + m^2 + n^2 = 1$) cuts the x, y and z axes in the points A, B and C respectively. Show that:
(*a*) the area of the triangle ABC is $1/2lmn$,
(*b*) the circulation of the vector field $\mathbf{v} = \mathbf{i}z + \mathbf{j}x + \mathbf{k}y$ round the triangle in the direction $ABCA$ is equal to $(l + m + n)/2lmn$,
(*c*) the circulation per unit area of triangle is maximised when $l = m = n$ and that then curl **v** is normal to the plane of the triangle, and
(*d*) the circulation is zero when the plane of the triangle is parallel to curl **v**. Why does ABC not have to be small in this problem?

10.16 By means of the argument of problem 4.4(*a*), show that Stokes's theorem implies that curl **v** is solenoidal.

10.17 The following are exercises on Stokes's theorem:
(*a*) Repeat problem 3.20(*a*).
(*b*) Find the flux of curl **v** in example 10.1 through the square whose sides are $x = \pm 1$, $z = \pm 1$ in the plane $y = 0$.
(*c*) By applying Stokes's theorem to a vector field with components $(-Q, P, 0)$ where P and Q are functions of x and y, deduce the same result as was deduced from Gauss's theorem in problem 7.18(*c*).
(*d*) By setting $P = x$, $Q = y$ in (*c*) find a formula for the area of a loop in the x, y plane as a line integral. Change it to polar coordinates with $x = r\cos\theta$ and $y = r\sin\theta$ and deduce that the area $= \frac{1}{2}\oint r^2 d\theta$.
(*e*) In a vortex motion the streamlines are circles in x, y planes, centred on the z-axis and the velocity depends only on distance r from the z-axis. The vorticity is then a z-wise vector of magnitude $(1/r)d(rv)/dr$. Verify that this is consistent with Stokes's theorem applied to a streamline. Also consider the particular cases $v = \Omega r$ (forced vortex) and $v = A/r$ (free vortex).

10.18 (*a*) A two-dimensional solenoidal field **B**, independent of z and with B_z zero, has a stream function $A(x, y)$. Show that if **A** is the vector field $A\mathbf{k}$, then **B** = curl **A**. Why does this confirm that **B** is solenoidal? (**A** is called the *vector potential* of **B**.)
*(*b*) If **v** is another, two-dimensional, solenoidal field, independent of z and with v_z zero, use problems 10.4(*a*) and 10.18(*a*) and Stokes's theorem to show that $\iint \mathbf{v} \times \mathbf{B} \cdot d\mathbf{a}$ vanishes when integrated over a domain in the x, y plane in either of the cases where **v** or **B** is parallel to the boundary of the domain at the boundary. (In the latter case choose $A = 0$ at the boundary.) Compare this with problem 8.3(*b*).

10.19 These problems are applications of the identity for div **u** × **v**.
(*a*) Show that, if **v** = grad ϕ × grad ψ, **v**'s field lines are the intersections of the surfaces ϕ = const. and ψ = const. Wherever either or both of these sets of surfaces approach more closely, grad ϕ and/or grad ψ are larger in magnitude. This suggests that **v** is the kind of field which gets stronger where its field lines are closer. Confirm that this is true because div **v** = 0. Interpret also the effect of changes in the angle between grad ϕ and grad ψ.
(*b*) A constant density, inviscid, steady flow has uniform vorticity. Show that $p_0 = p + \frac{1}{2}\rho v^2$ obeys the Poisson equation

$$\nabla^2 p_0 = \text{const.}$$

*(*c*) If **A** is a vector such that the magnetic field **B** = curl **A** and if curl **H** = **0** everywhere because the magnetic field is due to magnetisation, not to currents, so that the magnetic potential U is single-valued, show that $\iiint \mathbf{H} \cdot \mathbf{B}\, d\tau = 0$, evaluated over the volume enclosed by a magnetic equipotential surface U = const. This confirms that **H** and **B** must be in opposition over part of the volume (which cannot be in non-magnetic or vacuum regions where $\mathbf{B} = \mu_0\mathbf{H}$). Consider the application to figure 6.8, sketching in some equipotentials orthogonal to the **H**-lines.
*(*d*) Currents are induced in a solid electrical conductor of uniform resistivity τ by rotating it at an angular velocity $\boldsymbol{\Omega}$ in a uniform magnetic field **B** inclined at an angle θ to the axis of rotation. Polarisation and the effect of currents on the magnetic field may be neglected. Show that the net charge density q in the conductor obeys the equation

$$\tau \partial q/\partial t + q/\epsilon_0 + 2\mathbf{B} \cdot \boldsymbol{\Omega} = 0.$$

In view of problem 7.15, the first term may be neglected (except over very short time intervals). For what value of θ is q then zero?

Show by using cartesians that curl (**v** × **B**) = $\boldsymbol{\Omega}$ × **B** here, if **v** = velocity. No currents are induced when **v** × **B** is irrotational (compare problem 10.4(*c*)). For what value of θ does this occur? Note that only then does the flux linked by any loop drawn in the moving solid remain constant.

10.20 These problems are applications of the identity for curl s**v**.
(*a*) Heat flows in a non-uniform conductor according to $\mathbf{Q} = -K \operatorname{grad} T$,

where K depends only on temperature T. Show that the heat flow field $\mathbf{Q}$ is irrotational. (Problem 7.3(*a*) is relevant.) Problem 7.12 concerns the potential which must exist for $\mathbf{Q}$.
(*b*) A radial fluid velocity field $\mathbf{v}$ is such that $\mathbf{v} = s\mathbf{r}$ where $s(x, y, z)$ is a scalar field and $\mathbf{r}$ is the position vector. Show that the vorticity lines are the intersections of the surfaces $s =$ const. with spheres centred at the origin.

*10.21 These problems are applications of the identity for $\oint s\,\mathrm{d}\mathbf{r}$.
(*a*) A wire loop of arbitrary shape carries a uniform current J in a uniform magnetic field $\mathbf{B}$. We are now equipped to prove the results of problem 3.19(*b*) more efficiently:
(i) Apply Stokes's theorem to the uniform vector $J\mathbf{I} \times \mathbf{B}$, where $\mathbf{I}$ is an arbitrary, uniform, unit vector, to show that the total force $J \oint \mathrm{d}\mathbf{r} \times \mathbf{B} = \mathbf{0}$.
(ii) Show that the couple on the loop, $J \oint \mathbf{r} \times (\mathrm{d}\mathbf{r} \times \mathbf{B})$, where $\mathbf{r}$ is referred to any origin, is equal to $J \oint (\mathbf{r} \cdot \mathbf{B})\mathrm{d}\mathbf{r}$. Noting that grad $(\mathbf{r} \cdot \mathbf{B}) = \mathbf{B}$, show that the couple $= J\mathbf{A} \times \mathbf{B}$, in which $\mathbf{A} = \iint \mathrm{d}\mathbf{a}$, the vector area of the loop.
(*b*) Show that, for any surface S spanning a loop L, $(n \neq 2)$,

$$\oint_L \frac{\mathrm{d}\mathbf{r}}{r^{n-2}} = (n-2) \iint_S \frac{\mathbf{r} \times \mathrm{d}\mathbf{a}}{r^n} \quad \text{and} \quad \oint_L \log r \,\mathrm{d}\mathbf{r} = - \iint_S \frac{\mathbf{r} \times \mathrm{d}\mathbf{a}}{r^2}.$$

(*c*) Deduce from (*b*) that the vector moment $\mathbf{M}$ about any origin O of a uniform pressure acting on *any* surface spanning a given loop L is unique. (Compare problem 3.22(*b*).) Using the results that

$$(\mathbf{r} \times \mathrm{d}\mathbf{r}) \times \mathbf{r} = \mathbf{r}^2\mathrm{d}\mathbf{r} - \mathbf{r}\mathrm{d}(\tfrac{1}{2}\mathbf{r}^2)$$

and that

$$\mathrm{d}(\mathbf{r}^2\mathbf{r}) = \mathbf{r}^2\mathrm{d}\mathbf{r} + \mathbf{r}\mathrm{d}(\mathbf{r}^2)$$

show that

$$\frac{1}{3}\oint (\mathbf{r} \times \mathrm{d}\mathbf{r}) \times \mathbf{r} = \frac{1}{2}\oint \mathbf{r}^2\mathrm{d}\mathbf{r}$$

and relate $\mathbf{M}$ to the moment of pressures acting on the surface composed of all the radius vectors from O to L.

*10.22 (*a*) Using the identity for curl $s\mathbf{v}$ and the results of problems 7.2(*c*) and 10.1(*b*), show that

$$\text{curl}\,\frac{\mathbf{A} \times \mathbf{r}}{r^n} = \frac{n(\mathbf{r} \cdot \mathbf{A})\mathbf{r}}{r^{n+2}} + \frac{(2-n)\mathbf{A}}{r^n},$$

$\mathbf{A}$ being a constant vector.
(*b*) Hence, using Stokes's theorem, show that

$$\oint_L \frac{\mathbf{r} \times \mathrm{d}\mathbf{r}}{r^n} = n \iint_S \frac{(\mathbf{r} \cdot \mathrm{d}\mathbf{a})\mathbf{r}}{r^{n+2}} + (2-n) \iint_S \frac{\mathrm{d}\mathbf{a}}{r^n},$$

where S is any surface spanning loop L. (The case $n = 0$ duplicates problem 3.19(*a*).)

(*c*) If $\mathbf{da}'$ is an area element on a second surface S' spanning a loop L', composed of line elements $\mathbf{dr}'$, with position vectors $\mathbf{r}'$, show using problem 10.21(*b*) that

$$\iint_{S'} \mathbf{da}' \cdot \oint_L \frac{\mathbf{R} \times \mathbf{dr}}{R^n} = -\oint_L \mathbf{dr} \cdot \iint_{S'} \frac{\mathbf{R} \times \mathbf{da}'}{R^n}$$

$$= \frac{1}{2-n} \oint_L \oint_{L'} \frac{\mathbf{dr} \cdot \mathbf{dr}'}{R^{n-2}},$$

$\mathbf{R}$ being the displacement vector $\mathbf{r} - \mathbf{r}'$ from $\mathbf{dr}'$ to $\mathbf{dr}$ and R its magnitude. If $n = 2$, show that the right-hand integral is replaced by

$$\oint_L \oint_{L'} (\log R) \mathbf{dr} \cdot \mathbf{dr}'.$$

Note that $\mathbf{dR} = \mathbf{dr}$ if the other end of $\mathbf{R}$ is not moving.

(*d*) The theory of diffuse thermal radiation between two surfaces at given temperatures states that the rate at which heat leaves an area element $\mathbf{da}$ of one surface per unit solid angle in a given direction is proportional to the projected area of $\mathbf{da}$ in that direction. If $\mathbf{da}'$ is an element of the second surface, located by a position vector $\mathbf{R}$ relative to $\mathbf{da}$, show that heat passes between $\mathbf{da}$ and $\mathbf{da}'$ at a net rate $k(\mathbf{R} \cdot \mathbf{da})(\mathbf{R} \cdot \mathbf{da}')/R^4$, in which the constant k depends on the surfaces and their temperatures.

With the aid of (*b*) and (*c*) above, show that the net heat exchange rate between the two finite surfaces is equal to

$$\tfrac{1}{2}k \oint_L \oint_{L'} (\log R)\, \mathbf{dr} \cdot \mathbf{dr}',$$

where L and L' are the rims of the two surfaces, i.e. the heat exchange depends only on the rim shape. (Note that four stages of integration have been reduced to two by this result.)

(*e*) Each area element $\mathbf{da}$ of a surface S is made into a magnetic loop dipole $J\,\mathbf{da}$, J being a uniform current. (This is called a 'magnetic shell'.) Using (*b*) above and the result of problem 7.10(*b*), deduce the Biot–Savart law, noting that the currents circulating round the various $\mathbf{da}$ cancel except at the rim – compare the proof of Stokes's theorem.

The mutual inductance between two loops L and L' can be defined as the magnetic flux M linked by one when unit current flows round the other. Using the notation and result of (*c*) above, show that

$$M = \frac{\mu_0}{4\pi} \oint_L \oint_{L'} \frac{\mathbf{dr} \cdot \mathbf{dr}'}{R}.$$

The symmetry of this result shows that the current may flow in either loop, i.e. the inductance is truly *mutual*.

10.23 These problems are exercises on the identity for $\oiint \mathbf{da} \times \mathbf{v}$.

(*a*) By setting $\mathbf{v} = \phi\mathbf{I}$, with $\mathbf{I}$ a uniform unit vector, deduce the corollary to Gauss's theorem.

(*b*) Show that the vector moment about the origin of the forces due to a uniform pressure p acting on a closed surface is zero. Is this consistent with problem 10.21(*c*)?

*10.24 For a finite sphere of radius R and centre O, drawn in a moving fluid of uniform density ρ, a weighted mean vorticity $\boldsymbol{\omega}_2$ which takes more account of the values near O is defined by taking

$$\boldsymbol{\omega}_2 = \iiint (R^2 - r^2) \operatorname{curl} \mathbf{v} \, d\tau / (\tfrac{8}{15} \pi R^5)$$

(in which r = distance from O and the denominator on the right is chosen so as to make $\boldsymbol{\omega}_2 = \operatorname{curl} \mathbf{v}$ if this is uniform). Show that the moment of momentum of the sphere's fluid *contents* equals $\frac{1}{2}\boldsymbol{\omega}_2 I_2$, if I_2 = the moment of inertia of the sphere about any diameter. What happens if the sphere is so small that the variation of curl $\mathbf{v}$ can be ignored in the integral?

11

The curl operator in electromagnetism

11.1 The electromagnetic equations in differential form

There were some indications in the last chapter that the concept of curl, which we had developed to meet the needs of fluid mechanics, would also be relevant and fruitful in electromagnetism. In this respect curl resembles div and grad, whose remarkable versatility we have already exploited. This chapter is devoted to showing what a central role it plays in the proper elucidation and formulation of electromagnetism; the curl concept leads directly to the prediction of several phenomena of the greatest scientific and technological significance.

The aim here is not to give a comprehensive treatment of electromagnetic theory but merely to reveal certain basic phenomena which all engineers should be aware of and to point the way to more detailed and advanced studies. We are at this point finally moving from Zone 2 into Zone 3 in figure 3.2.

Before curl is put to work, it is worth recapitulating the results concerning div and grad in relation to electromagnetism which were encountered in chapter 7 on earlier excursions into Zone 3. They are:

(i) **B** is always solenoidal: $\text{div}\,\mathbf{B} = 0.$

(ii) Gauss's law, relating **E** to its charge sources: $\epsilon_0\,\text{div}\,\mathbf{E} = q,$
in which q includes free and bound charge density.
If q_f refers to free charge, then $\text{div}\,\mathbf{D} = q_f,$
while if q_b is bound charge density, $\text{div}\,\mathbf{P} = q_b.$

(iii) In electrostatics, or wherever $\partial\mathbf{B}/\partial t$ is negligible, **E** is conservative, with a potential V, and then $\mathbf{E} = -\text{grad}\,V.$

(iv) If current and $\epsilon_0\partial\mathbf{E}/\partial t$ are absent or negligible, **H** is conservative with a potential U, and then $\mathbf{H} = -\text{grad}\,U,$
with the proviso that U may be multi-valued unless 'cuts' are inserted.

Now let us apply the integral form of Faraday's law to a small area element **da** and its rim L, the element being so small that $\partial\mathbf{B}/\partial t$ does not vary significantly over it. Then

$$-\frac{\partial\mathbf{B}}{\partial t}\cdot\mathbf{da} = \oint\mathbf{E}\cdot\mathbf{dr} = \text{curl}\,\mathbf{E}\cdot\mathbf{da}$$

in view of section 10.9. Since **da** is arbitrary in direction, and can for instance be oriented successively in the x, y and z-directions, it follows that

$$\text{curl}\,\mathbf{E} = -\frac{\partial \mathbf{B}}{\partial t},$$

the differential form of Faraday's law, expressible only because the curl concept is now to hand. Similarly we may develop the integral form of the Maxwell–Ampère law

$$\oint \mathbf{B} \cdot \mathrm{d}\mathbf{r} = \mu_0 \iint \left(\mathbf{j} + \epsilon_0 \frac{\partial \mathbf{E}}{\partial t}\right) \cdot \mathrm{d}\mathbf{a}$$

(where $\mathbf{j}$ may include polarisation current $\partial \mathbf{P}/\partial t$ and magnetisation is assumed absent) into a differential equivalent

$$\text{curl}\,\mathbf{B} = \mu_0 \left(\mathbf{j} + \epsilon_0 \frac{\partial \mathbf{E}}{\partial t}\right).$$

Example 11.1 Find differential equations in terms of $\partial \mathbf{D}/\partial t$ for a dielectric material and in terms of curl $\mathbf{H}$ for a magnetic material.

If $\mathbf{j}$ consists only of $\partial \mathbf{P}/\partial t$, as in a dielectric insulator, then

$$\text{curl}\,\mathbf{B} = \mu_0 \frac{\partial \mathbf{D}}{\partial t}, \quad \text{for} \quad \mathbf{D} = \mathbf{P} + \epsilon_0 \mathbf{E}.$$

If magnetisation $\mathbf{M}$ is present,

$$\oint \mathbf{B} \cdot \mathrm{d}\mathbf{r} = \mu_0 \iint \left(\mathbf{j} + \epsilon_0 \frac{\partial \mathbf{E}}{\partial t}\right) \cdot \mathrm{d}\mathbf{a} + \mu_0 \oint \mathbf{M} \cdot \mathrm{d}\mathbf{r}$$

and

$$\oint \mathbf{H} \cdot \mathrm{d}\mathbf{r} = \iint \left(\mathbf{j} + \epsilon_0 \frac{\partial \mathbf{E}}{\partial t}\right) \cdot \mathrm{d}\mathbf{a}.$$

These yield the differential equivalents

$$\text{curl}\,\mathbf{B} = \mu_0 \left\{\mathbf{j} + \epsilon_0 \frac{\partial \mathbf{E}}{\partial t} + \text{curl}\,\mathbf{M}\right\} \quad \text{and} \quad \text{curl}\,\mathbf{H} = \mathbf{j} + \epsilon_0 \frac{\partial \mathbf{E}}{\partial t}.$$

(See also problem 11.1(*a*).) curl $\mathbf{M}$ could be termed a 'virtual current'.

Together with the results (i) and (ii) above involving div, these equations involving curl are known as Maxwell's equations. They constitute a complete statement of the laws governing the electromagnetic fields. To complete the description of any phenomenon involving dielectric, magnetic, or conducting media it is also necessary to specify their properties by means of appropriate relations between **E** and **D**, **B** and **H** or **j** and **E** (or e.m.f.), respectively. Maxwell's equations are correct and complete even in relativistic terms, incidentally.

Any reader who thought he previously understood electromagnetism, but at a lower mathematical level, must think again! The curl concept is absolutely essential. It provides *the only way* of making precise, general statements about the mutual interactions of the electric and magnetic fields at a point. At the same time, one must admit that curl and the idea of 'circulation density' are rather difficult concepts on first acquaintanceship, certainly harder to grasp intuitively than div and grad. But curl is so important that it well repays effort spent in coming to terms with it at an intuitive level. The notion that curl **E** can be hunted for with the aid of a search coil, as discussed in section 10.9, is certainly helpful, as are the analogies with fluid velocity fields which help one to picture what kinds of fields are rotational and which are conservative (irrotational).

Example 11.2 Repeat in differential form the analysis and resolution of the crisis in electromagnetism, treated in integral form in section 6.2.

Without Maxwell's correction term, Ampère's law in differential form is curl **B** $= \mu_0 \mathbf{j}$. Then the identity div curl $\equiv 0$ implies that div $\mathbf{j} = 0$ (Kirchhoff's first law) whereas instead the conservation-of-charge equation is div $\mathbf{j} = -\partial q/\partial t \neq 0$ in general. However, $q = \epsilon_0$ div **E** and so

$$\partial q/\partial t = \text{div}\,(\epsilon_0 \partial \mathbf{E}/\partial t),$$

which leads to

$$\text{div}\,(\mathbf{j} + \epsilon_0 \partial \mathbf{E}/\partial t) = 0.$$

It is therefore consistent with div curl $\equiv 0$ to modify Ampère's law into

$$\text{curl}\,\mathbf{B} = \mu_0(\mathbf{j} + \epsilon_0 \partial \mathbf{E}/\partial t).$$

(See also problem 11.1(*b*).)

There is another intriguing point to be made about Maxwell's equations. Newton's law for a particle, $\mathbf{F} = m\,d\mathbf{v}/dt$, might reasonably be said to be a statement of cause and effect. The *cause* is **F** and the *effect* $d\mathbf{v}/dt$ is the term involving the time derivative. Integrating the equation tells us how **v** evolves in time under the influence of cause **F**.

By the same token, in the equation curl $\mathbf{E} = -\partial \mathbf{B}/\partial t$ the rate of change of **B**, i.e. the immediately subsequent evolution of **B**, would appear to be the *effect* of the instantaneous non-uniformity of **E**, as measured by curl **E**, which is the cause. Note that this is a reversal of cause and effect as normally interpreted! However, in most physical problems any distinction between cause and effect is somewhat spurious; phenomena merely co-exist.

Applying Stokes's theorem to Maxwell's curl equations simply recaptures their integral equivalents. For instance

$$\oint \mathbf{E} \cdot d\mathbf{r} = \iint \text{curl}\,\mathbf{E} \cdot d\mathbf{a} = -\iint \frac{\partial \mathbf{B}}{\partial t} \cdot d\mathbf{a}.$$

Example 11.3 In a two-dimensional magnetic problem, all quantities are independent of z and steady currents $\mathbf{j}$ flow in x, y planes, $\epsilon_0 \partial \mathbf{E}/\partial t$ terms and magnetisation being absent. Show that $\mathbf{H}$ is in the z-direction and that H (the magnitude of $\mathbf{H}$) is $\mathbf{j}$'s stream function, constant along current flow lines.

The two-dimensional currents must flow in closed loops; they cannot cross and if they were spirals this would imply current sinks or sources. (Note that this argument fails in three dimensions.) Each family of identical loops in successive x, y planes constitutes an infinite solenoid which produces a magnetic field purely in the z-direction. Hence the total field must also be in the z-direction, with $H_x = H_y = 0$.

The equation curl $\mathbf{H} = \mathbf{j}$, with $\partial/\partial z = 0$, gives

$$j_x = \partial H/\partial y, \quad j_y = -\partial H/\partial x.$$

Comparison with the equations $v_x = \partial \psi/\partial y$, $v_y = -\partial \psi/\partial x$, the relations between $\mathbf{v}$ and its stream function ψ, reveals that H is $\mathbf{j}$'s stream function, which exists because $\mathbf{j}$ is solenoidal and two-dimensional. (Compare problem 10.18(*a*).)

As a further check, consider H's variation along a current flow line:

$$|\mathbf{j}| \text{ times the rate of change of } H \text{ with distance along } \mathbf{j}\text{-lines}$$

$$= \mathbf{j} \cdot \text{grad}\, H = j_x\, \partial H/\partial x + j_y\, \partial H/\partial y = -j_x j_y + j_y j_x = 0,$$

i.e. H is constant along $\mathbf{j}$-lines, as befits a stream function. Problems 11.2 and 11.3 are further exercises in two dimensions.

The incentive to develop field equations in differential rather than integral form was, it will be remembered, to allow solution of more general problems, as integral methods are suitable only for problems with high symmetry or where the sources of the fields (charges, currents, etc.) are known explicitly *ab initio*. We are therefore now able to tackle a wider range of problems with the aid of Maxwell's *differential* field equations.

11.2 Electromagnetic skin-effect; shielding

Figure 11.1(*a*) shows a portion of a coil wound round a long cylindrical block. Raising the current J from zero produces a rising axial magnetic field $\mathbf{B}$ and then 'transformer-action' according to Faraday's law induces a circumferential electric field $\mathbf{E}_i$, as shown. If there were a short-circuited second coil, similarly placed to the first, the electric field would induce a current in it such as would *oppose* the creation of the magnetic field by J. If instead there is no second coil but the block is a conductor, similar concentric opposing currents are induced in the block itself in a distributed fashion. If the conductivity is high enough, sufficient current is induced in the outer layers of the block to stop the electromagnetic disturbance from penetrating into the deep interior of the block, at least for a significant time. This phenomenon is called *electromagnetic shielding.* (It has important applications,

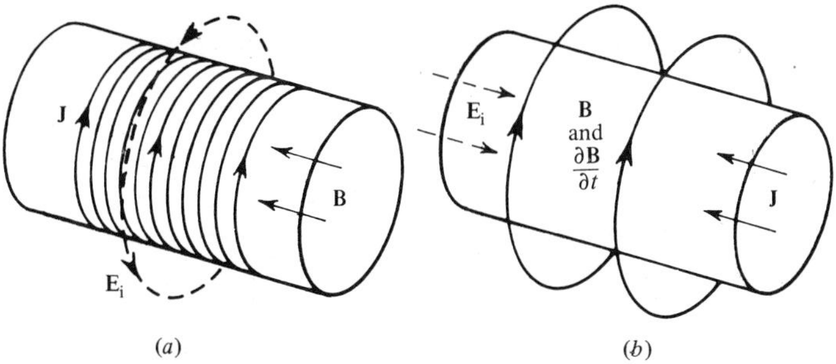

Fig. 11.1

e.g. in the construction of screened rooms for precision electronics work which must be isolated from external electromagnetic disturbances from nearby radio stations, etc.) Since the current and magnetic field penetration in figure 11.1(*a*) are confined to a thin layer or 'skin' on the surface of the block, it is also called the *electromagnetic skin-effect.* If the block is hollow (e.g. a box or screened room) it suffices if the walls are thicker than the skin layers.

Skin-effect also occurs in many other geometrical configurations, including that shown in figure 11.1(*b*), in which the current and magnetic field directions of figure 11.1(*a*) have been interchanged so as to give a rising current J flowing along a straight wire and an associated circular magnetic field **B**. Remembering that the induced electric field $\mathbf{E}_i$ is related to its source $(-\partial\mathbf{B}/\partial t)$ in exactly the same mathematical ways that a magnetic field is related to its source current (e.g. by a law of Biot–Savart form), we see that $\mathbf{E}_i$ is axial, just as **B** is axial in figure 11.1(*a*). Moreover $\mathbf{E}_i$ opposes the current flow, competing against the imposed electric field responsible for J. (Compare example 4.5.) Once again, if the conductivity is high enough, the electromagnetic disturbance in and around the wire finds difficulty in penetrating beyond a limited '*skin depth*' into the deep interior. A common situation is that where J is an alternating current. Then, if the frequency and/or conductivity is high enough, the electromagnetic disturbances achieve only limited penetration in each half-cycle before being superseded by fields of opposite sign, and events are confined to a skin indefinitely. A consequence is that the resistance of a wire to alternating current can appear to be much higher than it is to direct current simply because the former is confined to the outer layers of the wire and is unable to exploit all the metal available.

After this qualitative introduction we turn to mathematical analysis. A relatively simple treatment can be given if we assume that the skin-effect is so severe that the skin depth is small compared with the radius of curvature of the surface of the conductor, or its thickness. Then the surface of the conductor can be treated as virtually plane and the other side of the conductor as virtually at infinity. An adequate model of the situation is then to treat the conductor as occupying the semi-infinite region $x > 0$, say, x being measured into the conductor normally to its surface, with the magnetic field taken in the y-direction and the electric field and current both in the z-direction. This model could apply to both cases shown in figure 11.1 with appropriate choices of axes.

If the skin depth is indeed small, the rate of change of all variables will be much greater in the x-direction than any other and it will be appropriate to take them all as functions only of x and time t.

The conductivity σ is assumed to be good enough for the conduction current, given by Ohm's law, to be far greater than any contribution from polarisation current or the $\epsilon_0 \partial \mathbf{E}/\partial t$ term. This is a very good approximation in metals but can fail in poor conductors such as sea-water, a problem of some importance in connection with submarine telecommunications. (See problem 11.4(*a*).) We also assume that the material is non-magnetic for the present.

Under these conditions, the relevant equations are

$$\mathbf{j} = \sigma \mathbf{E}, \quad \text{curl } \mathbf{B} = \mu_0 \mathbf{j} \quad \text{and} \quad \text{curl } \mathbf{E} = -\partial \mathbf{B}/\partial t,$$

or

$$j_z = \sigma E_z, \quad \partial B_y/\partial x = \mu_0 j_z \quad \text{and} \quad \partial E_z/\partial x = \partial B_y/\partial t,$$

if we use the facts that j_x, j_y, E_x, E_y, B_x and B_z all vanish and there is negligible variation with y or z. The fact that

$$\frac{\partial}{\partial x}\left(\frac{\partial B_y}{\partial t}\right) = \frac{\partial}{\partial t}\left(\frac{\partial B_y}{\partial x}\right)$$

permits elimination of E_z and B_y. This yields an equation for j_z:

$$\frac{\partial j_z}{\partial t} = \frac{1}{\mu_0 \sigma} \frac{\partial^2 j_z}{\partial x^2}.$$

B_y and E_z each also obey the same differential equation. Reference back to section 9.4 reveals that this is a new manifestation of the unsteady diffusion equation in one-dimension. Problem 11.5 concerns two simple cases of this type while problem 11.7(*b*) considers the more general case of three-dimensional diffusion.

The discussion of this equation given in section 9.4 can be immediately adopted for application to the skin-effect phenomenon. The conclusion is that an electromagnetic disturbance penetrates into a conductor a distance that is of the order of $(t/\mu_0\sigma)^{1/2}$ in a time interval t, because a comparison of the differential equations shows that the diffusivity is $1/\mu_0\sigma$. It is important to appreciate that the penetration gets deeper if the conductivity is *reduced*, not increased, for the eddy currents responsible for the shielding effect are then weakened. There is a contrast with thermal diffusion, which is enhanced if the conductivity is *increased*.

The important general idea here is that it is not possible suddenly to change electric or magnetic fields deep inside a conductor; any changes must diffuse in from the surface, just like temperature changes in a thermal conductor. If a magnetic field is required to be manipulated rapidly deep inside a medium, a non-conductor such as ferrite must be used.

11.3 Alternating current skin-effect

A case of great practical interest is that where all quantities depend sinusoidally upon time, with circular frequency ω. The characteristic time for penetration to occur before the fields reverse is then of order $t = 1/\omega$ and therefore the penetration or skin depth should be of order $(\omega\mu_0\sigma)^{-1/2}$. The higher the frequency, the thinner is the skin.

These result are confirmed by the exact analysis of the one-dimensional, x-dependent case, which follows. This adopts the approach, widespread in electrical engineering, of representing sinusoidally fluctuating quantities as the *real part* of complex numbers which contain the factor $e^{i\omega t}$, i being the root of minus one. If we denote 'real part of' by Re, we then put

$$j_z = \text{Re}\{j_z^*\}, \text{ say}, \quad \text{where} \quad j_z^* = X(x)e^{i\omega t},$$

which means $j_z = X \cos \omega t$ if X is real. In general X will be complex and the phase of j_z varies in the usual way in consequence. At the end of the calculation the actual values of physical quantities have to be found by taking real parts only. Inserting the complex form $j_z^* = X(x)e^{i\omega t}$ into the unsteady diffusion equation gives

$$i\omega\mu_0\sigma X e^{i\omega t} = \frac{d^2X}{dx^2} e^{i\omega t} \quad \text{and} \quad i\omega\mu_0\sigma X = \frac{d^2X}{dx^2},$$

an equation whose solutions are of the form $X = Ae^{mx}$ where $m^2 = i\omega\mu_0\sigma$. From the facts that $i = e^{i\pi/2}$ and therefore that $i^{1/2} = e^{i\pi/4} = (1+i)/\sqrt{2}$, it follows that

$$m = \pm(1+i)/\delta,$$

in which δ, dimensionally a length, equals $(2/\omega\mu_0\sigma)^{1/2}$. Thus the full solution is

$$j_z^* = Ae^{i(\omega t + x/\delta) + x/\delta} + Be^{i(\omega t - x/\delta) - x/\delta}.$$

The conducting medium is regarded as occupying the semi-infinite region $x > 0$ and responding to disturbances at the surface $x = 0$. As the deep interior of the medium at large values of x is devoid of electromagnetic fields and currents, we require that $j_z^* \to 0$ as $x \to \infty$. This implies that A must vanish and only the second solution survives. Taking its real part, with B assumed real, gives

$$j_z = Be^{-x/\delta} \cos(\omega t - x/\delta).$$

At the surface $x = 0$ and $j_z = B \cos \omega t$. The applied electric field at the surface is

$$E_z = (B/\sigma) \cos \omega t,$$

from Ohm's law.

The penetration of current is seen to take the form of a wave-like oscillation (the cosine factor) which has a wave-speed $\omega\delta$ into the conductor and a wavelength equal to $2\pi\delta$, while the wave's amplitude decays according to the exponential factor. The amplitude decays in the ratio $1:e^{2\pi}$ or $1:533$ in one wavelength, i.e. the decay of the wave with depth is very rapid. The length δ is clearly a measure of the penetration depth, which is seen to be indeed of order $(\omega\mu_0\sigma)^{-1/2}$ as expected.

Example 11.4 In view of the importance of the change in impedance of a wire due to the concentration of current in a skin, calculate the magnitude and phase of the skin current, and the impedance per unit surface area.

Per unit length of perimeter measured perpendicular to the current, the total current J equals the real part of

$$\int_0^\infty B\mathrm{e}^{i(\omega t-x/\delta)-x/\delta}\,\mathrm{d}x = B\mathrm{e}^{i\omega t}\int_0^x \mathrm{e}^{(-x/\delta)(1+i)}\,\mathrm{d}x = \frac{B\delta\mathrm{e}^{i\omega t}}{1+i} = \frac{B\mathrm{e}^{i(\omega t-\pi/4)}}{(\omega\mu_0\sigma)^{1/2}}$$

if we insert the value of δ and put $1+i=(\sqrt{2})\mathrm{e}^{i\pi/4}$. The real part is

$$\frac{B}{(\omega\mu_0\sigma)^{1/2}}\cos(\omega t-\pi/4).$$

The most interesting aspect of this result is the phase relationship between this total current and the axial electric field $(B/\sigma)\cos\omega t$ which drives it. The current lags by the very unusual phase angle of $\pi/4$. (This is easily demonstrated experimentally with a wire at a frequency high enough for the skin depth to be small compared with the radius.) The amplitude ratio of E_z to J, namely $(\omega\mu_0/\sigma)^{1/2}$, equals the impedance per unit length (parallel to the current) and per unit periphery (perpendicular to the current).

In the example above the impedance rises with frequency according to the half power. Raising the resistivity $1/\sigma$ raises the impedance, but again only as the half power. Problem 11.4(*b*) considers the modification which occurs when the conducting medium is linearly magnetisable, with **M** and **H** proportional to **B**. The result is that μ_0 in the preceding formulae is replaced by μ, where μ/μ_0 may be of the order of 1000 for ferromagnetic materials. It should be remembered that the linear model is only very approximate for real ferromagnetic media, however.

Typical values for the skin depth δ at the mains frequency of 50 Hz are 10 mm for copper and 0.7 mm for soft iron. Both these values have important technological implications. There is little point in making copper conductors thicker than 10 mm, as otherwise the interior will carry very little current at 50 Hz and so (for this and other reasons) busbars are made of flat strips thinner than 10 mm. Alternatively the interior of a thicker circular conductor can safely be made of stronger material, less able to conduct electricity, in order to bear the tension in long conductors slung between pylons. The result for soft iron is important in connection with transformers, whose cores have to be laminated to stop eddy skin currents from shielding the inside of each core element against penetration by magnetic field. This is achieved by making the laminations considerably thinner than the 0.7 mm skin depth that would prevail in a thicker member. Then the field penetrates the full thickness with little inhibition and precious volume is not wasted.

11.4 Magnetic pressure and levitation

Whether the skin currents vary sinusoidally in time or not, there is a normal force exerted on the surface layer of the conductor due to the $\mathbf{j} \times \mathbf{B}$ forces there. The total force per unit area of surface, directed into the conducting medium in the x-direction, is

$$-\int_0^\infty j_z B_y \,\mathrm{d}x = -\frac{1}{\mu_0}\int_0^\infty B_y \frac{\partial B_y}{\partial x}\,\mathrm{d}x = -\frac{1}{2\mu_0}\int_0^\infty \frac{\partial}{\partial x}(B_y^2)\,\mathrm{d}x$$

$$= \frac{B^2}{2\mu_0}, \qquad (B = \text{field at surface}),$$

in which we have used the facts that $\mu_0 j_z = \partial B_y/\partial x$ and that $B_y \to 0$ deep inside the conductor. In other words there is an effective *magnetic pressure* equal to $B^2/2\mu_0$ wherever a tangential magnetic field B is being excluded from deep penetration into a conductor by skin currents. For a typical magnetic field strength of 1 Tesla, this pressure is roughly 4 atmospheres. Applied over large areas, very great forces can obviously be generated in this way.

There is considerable interest nowadays in *magnetic levitation*, one form of which exploits the effective magnetic pressure on a surface associated with skin currents. An early application was in metallurgy where an a.c. coil can be made to levitate and then melt samples of metals and alloys *in vacuo* under conditions where contact of the melt with a crucible would be unacceptable. Another intriguing possibility which is receiving considerable attention is the levitation of high-speed vehicles, for which a magnetic support system has many advantages over the wheel or air cushion.

Figure 11.2 shows schematically a typical system in which a magnetic field emanating from a coil on the underside of a vehicle is prevented by skin currents from penetrating beyond a certain skin depth into the flat metallic track beneath in the time that it takes for the magnet to pass a given point in the track. Problem 11.6 considers some simple order-of-magnitude calculations for such a situation and includes the interesting result that the magnetic drag force *falls* as the vehicle speed inceases, a very attractive characteristic that compensates in some degree for the rapidly rising aerodynamic drag.

One view of this process is to say that a moving conductor behaves like a *diamagnetic* material, for it excludes and is repelled by a magnetic field as is a superconductor.

Another application of magnetic pressure is in *magnetic forming*, whereby metal parts are shaped by intense, pulsed magnetic fields which rise too rapidly to penetrate the metal significantly, and magnetic pressure is made to confine plasma in some schemes for controlled thermonuclear fusion.

11.5 Electromagnetic waves

In chapter 6 it was asserted that Maxwell's insertion of the $\epsilon_0 \partial \mathbf{E}/\partial t$ term into the equation relating magnetic fields to their sources totally transformed the nature of

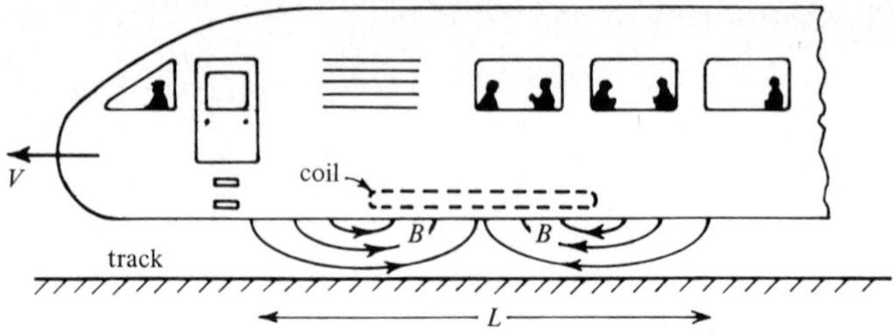

Fig. 11.2. Magnetically-levitated high-speed vehicle.

electromagnetism because it predicted the existence of electromagnetic waves. The development of the curl operator, because it has permitted the precise statement of the laws of electromagnetism in the form of differential equations, now opens the way for a theoretical demonstration of the waves. One implication of their existence is that an electromagnetic disturbance can only penetrate even *empty space* at a finite (though very high) rate, whereas the previous sections revealed that a disturbance penetrates comparatively sluggishly into a *conductor*.

In contrast to the previous sections dealing with skin-effect, we now retain the vital $\epsilon_0 \partial \mathbf{E}/\partial t$ term but for simplicity assume that ohmic conduction current is absent instead. We also exclude dielectric or magnetic media, and confine attention to one-dimensional phenomena in which all quantities depend only upon one space coordinate, x say, and upon time t. They are independent of the transverse coordinates y and z. (Problem 11.7(c) refers to the more complex question of three-dimensional electromagnetic wave phenomena, a subject of great importance in connection with radio antennae and microwave systems. A full treatment would be far beyond the scope of this book.)

Another name for a one-dimensional wave is a *plane wave*, for in each plane $x = \text{const.}$, perpendicular to the direction of propagation, conditions are uniform at each instant in time.

In the absence of conduction, polarisation and magnetisation, Maxwell's equations

$$\text{curl}\,\mathbf{E} = -\frac{\partial \mathbf{B}}{\partial t} \quad \text{and} \quad \text{curl}\,\mathbf{B} = \mu_0 \epsilon_0 \frac{\partial \mathbf{E}}{\partial t}$$

under one-dimensional conditions have y and z-components as follows:

$$\left.\begin{aligned} -\frac{\partial E_z}{\partial x} &= -\frac{\partial B_y}{\partial t} \\ \frac{\partial E_y}{\partial x} &= -\frac{\partial B_z}{\partial t} \end{aligned}\right\} \qquad \begin{aligned} -\frac{\partial B_z}{\partial x} &= \mu_0 \epsilon_0 \frac{\partial E_y}{\partial t} \\ \frac{\partial B_y}{\partial x} &= \mu_0 \epsilon_0 \frac{\partial E_z}{\partial t} \end{aligned}$$

Notice that E_z and B_y interact just with each other, as do E_y and B_z. The two interactions can be regarded as independent phenomena which can co-exist and be superposed if necessary. From the fact that

$$\partial^2 E_z/\partial x \partial t = \partial^2 E_z/\partial t \partial x$$

it follows that

$$\frac{1}{\mu_0 \epsilon_0} \frac{\partial^2 B_y}{\partial x^2} = \frac{\partial^2 B_y}{\partial t^2} .$$

Similarly

$$\frac{1}{\mu_0 \epsilon_0} \frac{\partial^2 E_z}{\partial x^2} = \frac{\partial^2 E_z}{\partial t^2} ,$$

if we concentrate solely on the E_z/B_y interaction. Both equations are recognisable as examples of the one-dimensional wave equation, discussed in section 9.7. (Notice, however, that the electromagnetic equations are naturally linear and the wave equation prevails for all amplitudes, in contrast to the case of sound waves, where the linearisation process demands small amplitudes.) From the known properties of the wave equation it follows that the one-dimensional E_z/B_y distribution propagates as waves along the x-axis. Any E_y/B_z distribution would propagate independently but at the same speed.

At last it is revealed how Maxwell's correction of Ampère's law predicts electromagnetic wave behaviour. The wave speed, as usual the square root of the coefficient of the second x-derivative in the wave equation, is seen to be $c = (\mu_0 \epsilon_0)^{-1/2}$, which takes the approximate value 3×10^8 m/s. It is called the *velocity of light*, even though light waves occupy only a short frequency band in the electromagnetic spectrum. The same speed c applies to all of this spectrum, from radio to γ-rays.

Problem 11.8(*b*) reveals that the velocity of the waves in a linear dielectric medium is reduced below the value c, and this accounts for refraction phenomena. Often the wave speed depends on frequency. (This is the phenomenon of *dispersion* which accounts for the rainbow, amongst other things.) A similar effect occurs in plasma at frequencies so high that the inertia of the electrons must be allowed for. (See problem 11.12.)

It is appropriate to pause and review our earlier assertions that **E** and **B** were abstractions concocted to systematise observations on charged particles. This view might still appear adequate for electromagnetic waves in the vicinity of charges, residing perhaps on nearby conductors. However, the fact is that electromagnetic waves can exist far out in intergalactic space, billions of miles from the charges whose agitation originally gave them birth in stars which may have long been annihilated! It begins to look as though **E** and **B**, non-material though they are, do have an independent existence as they help each other along through space. They are 'there' in the sense that a suitable detector could demonstrate their influence at any point in their paths. So we are forced to abandon our original view that **E** and **B** are mere abstractions, invented by *Homo sapiens*.

On later pages we do discuss situations where the waves are close to and associated with changing charge distributions on conductors, however.

The earlier discussion of the one-dimensional wave equation in section 9.7 is mostly applicable here and will not be repeated. We shall merely highlight the ways in which electromagnetic waves are different from sound waves.

Consider the simple mode that involves just E_z and B_y, E_y and B_z being absent.

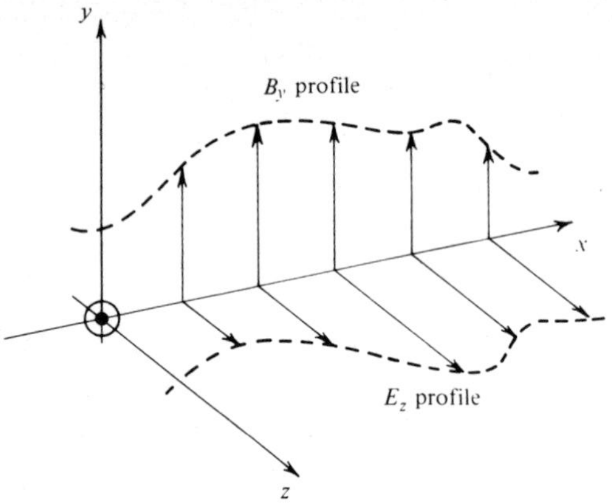

Fig. 11.3. Plane-polarised, plane electromagnetic wave.

In this wave the vector variables **E** and **B** are normal to each other and to the direction of propagation. The wave is therefore called a *transverse wave* (in contrast to the *longitudinal* sound wave, in which the vector variable **v** lies in the direction of propagation). In this case **E** is always parallel to the same direction, as is **B**, and such a wave is called *plane-polarised.* Such waves are universal in television transmission, for instance. The transverse spikes on a TV aerial must be oriented in the **E**-direction for successful reception. Figure 11.3 represents a plane-polarised wave diagrammatically.

The established terminology is confusing. The epithet 'plane' can refer both to the existence of planes x = const. over which all quantities are uniform and to the planes into which **E** and **B** may be polarised. On top of this, the term 'polarisation' can refer both to the orientation of the **E** and **B** vectors and to the behaviour of a dielectric, if present. It is therefore possible to have plane-polarised, plane waves in a polarisable medium. In this phrase each epithet is used in both of its senses!

Example 11.5 Show that $|\mathbf{E}|$ is proportional to $|\mathbf{B}|$ in a unidirectional plane-polarised wave.

Let the wave be travelling towards positive x. Then the solutions of the wave equation take the form

$$B_y = f(w) \quad \text{and} \quad E_z = g(w) \quad \text{where} \quad w = x - ct,$$

and

$$\frac{\mathrm{d}g}{\mathrm{d}w} = \frac{\mathrm{d}g}{\mathrm{d}w}\frac{\partial w}{\partial x} = \frac{\partial E_z}{\partial x} = \frac{\partial B_y}{\partial t} = \frac{\mathrm{d}f}{\mathrm{d}w}\frac{\partial w}{\partial t} = -c\frac{\mathrm{d}f}{\mathrm{d}w}.$$

So

$$g = -cf,$$

if we ignore the constant of integration. (If retained this would imply merely a uniform (and uninteresting) magnetic and/or electric field superposed on the wave fields.) Hence $E_z = -cB_y$, i.e. $|\mathbf{E}|$ is proportional to $|\mathbf{B}|$. For a wave travelling the opposite way, however, $E_z = +cB_y$.

In contrast to the example above, figure 11.3 shows a wave with dissimilar E_z and B_y profiles. (See also problem 11.9(*a*).) In this case there has to be propagation in both directions. Remember that the linear wave equation allows waves to propagate in both positive and negative x-directions. This occurs for example when an incident wave reflects off a suitable plane wall normal to the x-axis. A sheet of highly conducting metal makes a good reflector because it imposes the boundary condition that the tangential electric field must be virtually zero. (From example 11.4 it is apparent that the impedance of the surface layers in the skin-effect tends to zero like $1/\sqrt{\sigma}$ as $\sigma \to \infty$.) As with sound waves, a condition such as this is met by the creation of a suitably tailored reflected wave train. (See problem 11.9(*b*).) Standing waves, resonant cavities, etc. then become obvious possibilities. (See example 11.7 below.) Skin-effect occurs in the conducting walls because there have to be tangential currents flowing in the metal surfaces to account for the confinement of the magnetic field and then there is the usual magnetic pressure exerted on the surfaces, also known in this manifestation as *radiation pressure.*

11.6 Sinusoidal waves

Reference has already been made to the electromagnetic spectrum. This apparently simple idea rests on the more sophisticated fact that any physical distribution can be expressed as the superposition of a series of sinusoidal wave-forms by means of fourier series (or fourier integrals, if all frequencies are present), a process known as *harmonic analysis.* The basic building block in electromagnetism is a plane-polarised unidirectional sinusoidal wave of a single frequency, which is called a *monochromatic* wave in the case of light waves. In such a wave we could have

$$B_y = A \cos \{2\pi(x-ct)/\lambda\}$$

on which A = amplitude and λ = wavelength. Then

$$E_z = -cA \cos \{2\pi(x-ct)/\lambda\}.$$

At a fixed point, x = const. and B_y and E_z oscillate in time at frequency c/λ. The circular or angular frequency is $2\pi c/\lambda$. Problem 11.10 explores the encounter of such a wave with a dielectric body.

Some interesting compound waves are possible if two different plane-polarised, plane, monochromatic waves of the same frequency are superposed.

Example 11.6 Construct a compound wave in which **E** and **B** rotate but have constant magnitude.

Consider the combination of the B_y/E_z wave above with the independent wave in which

$$B_z = A \sin \{2\pi(x-ct)/\lambda\} \quad \text{and} \quad E_y = cA \sin \{2\pi(x-ct)/\lambda\},$$

which satisfy the appropriate differential equations. Figure 11.4, which is a view of this combined wave, looking in the positive x or propagation direction, expresses

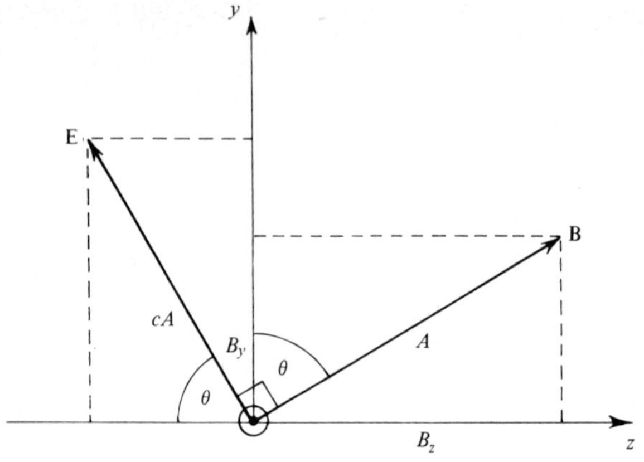

Fig. 11.4. Transverse section of a circularly-polarised wave.

the equations for E_y, E_z, B_y and B_z graphically, with $\theta = 2\pi(x - ct)/\lambda$. It reveals how **B** and **E** have constant magnitudes (A and cA respectively), are still perpendicular to each other and to the propagation direction, and now rotate in each transverse plane x = const. as time passes, instead of oscillating as in a plane-polarised wave. At a given instant of time the **E** and **B** vectors take on a corkscrew or helical distribution in terms of position. Such a wave is called *a circularly polarised* wave. One feature of such a wave is that the **E** (or **B**) field is parallel or anti-parallel to its own curl field. (Compare example 10.1.)

Example 11.7 Two electromagnetic waves in which

$$E_z = C \sin\{2\pi(x - ct)/\lambda\} \quad \text{and} \quad E_z = C \sin\{2\pi(x + ct)/\lambda\}$$

co-exist *in vacuo*. Show that a standing wave results. Where could two, highly conducting, plane boundaries be placed so as to provide a resonant cavity for these waves? Given these boundaries, what *other* resonant frequencies would be possible?

Here

$$\begin{aligned} E_z &= C[\sin\{2\pi(x - ct)/\lambda\} + \sin\{2\pi(x + ct)/\lambda\}] \\ &= 2C \sin(2\pi x/\lambda) \cos(2\pi ct/\lambda) \end{aligned}$$

i.e. a standing wave-form, with a profile $\sin(2\pi x/\lambda)$ multiplied by a variable amplitude factor $2C \cos(2\pi ct/\lambda)$.

Conducting boundaries in which $E = 0$ could be placed wherever $\sin(2\pi x/\lambda) = 0$, e.g. at $x = 0$ and x = any multiple of $\lambda/2$, the semi-wavelength. (Compare the acoustic cases in section 9.8.)

If boundaries occur at $x = 0$ and $n\lambda/2$ (n integral), another wave of wavelength λ' resonates in this cavity provided

$$\sin(2\pi x/\lambda') = 0 \quad \text{at} \quad x = 0 \text{ and } n\lambda/2,$$

i.e. if $2\pi(n\lambda/2)/\lambda' =$ any multiple of $\pi(k\pi$, say) or $\lambda' = n\lambda/k$. The resonant frequencies are $c/\lambda' = (k/n)(c/\lambda)$, which of course include the original wave, for which $k = n$.

Problem 11.11 invites study of another compound wave, made out of two plane-polarised waves of the same frequency but travelling in directions that are inclined to each other. The interference patterns which they produce reveal how such a wave system could occur when there is repeated oblique reflection off opposite sides of a rectangular pipe with highly conducting walls. Such a system is known as a wave-guide. This pattern is one of the simplest of many alternative ways (or *modes*) in which high-frequency, short-wavelength electromagnetic waves, known as *microwaves*, can be piped along ducts of various shapes. The next section discusses what might be regarded as the most primitive wave-guide of all.

11.7 Electromagnetic waves and charge motions in circuits

It is time to consider the problem that should have been worrying the reader ever since he began the study of electricity: 'How fast does electricity travel?' If the reader refers back to example 2.3 concerning the speed of conduction electrons in copper and observes just how small it is, he will be more worried still, because obviously electrical signals travel very rapidly in practice. A typical specific question that deserves an answer is: 'How long does it take a d.c. circuit to reach its steady state after a switch is closed, this delay being associated with the building-up of the electrostatic charge distributions needed to provide the electric fields that drive the currents along in the steady state?'

We shall approach this kind of problem by investigating a particularly simple configuration which is also of some technological interest in its own right. Consider a parallel plate capacitor: how rapidly can it be charged or discharged from one edge of the gap?

One technological significance of this is that the rapid discharge of capacitors is the best method of supplying very high powers for very short times in a reasonably cheap way. It finds application for instance in producing intense, brief light-sources or in energising experiments on controlled thermonuclear fusion. It is of obvious interest to know what limits there are on the speed with which charging or discharging can be accomplished. (In practice capacitors are often wound up into a 'swiss roll' configuration to save space but this does not materially affect their behaviour.)

Figure 11.5 shows a cross-section of the parallel plate capacitor, viewed parallel to the edges A and B at which charging or discharging is to occur. For simplicity we treat the plates as stretching indefinitely to the right and in both directions perpendicular to the paper and assume that there is no dielectric or magnetic material in the gap. Suppose that initially the capacitor is uncharged. Suddenly a source of current and voltage is switched on across AB, right along these edges and the capacitor begins to charge. Points such as CD, deep inside the gap, cannot instantaneously charge up, for this would imply the immediate appearance of an electric field there,

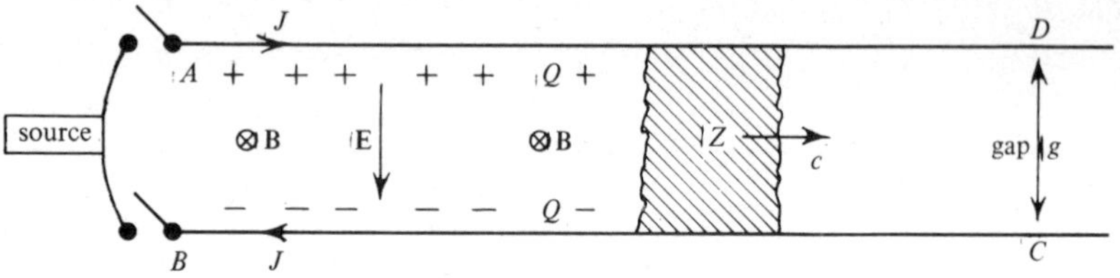

Fig. 11.5. Charging a parallel-plate capacitor from one edge.

whereas we now know that changes in the electric field must take a finite time to travel in the form of an electromagnetic wave to any other point. It is evident that some kind of 'charging zone' Z rushes across the gap at a velocity c. (Its form and width depend on the details of the switching-on process.) Before it arrives at a point, nothing happens, and where it has passed (if the voltage across AB is held steady) the capacitor is left uniformly charged with charges $\pm Q$ per unit area on the upper and lower plates, respectively, and with a vertical electrical field E. This charging process involves a supply and extraction of current at a rate J per unit length measured perpendicular to the paper. But this current flow produces a magnetic field. Actually this is produced by a combination of J and $\epsilon_0 \partial E/\partial t$, for at any point where Z is passing $\epsilon_0 \partial E/\partial t$ is in vigorous action. The part of the capacitor between AB and Z can be regarded as a long solenoid of variable width, round which flows a current J per unit length of solenoid. In Z, the $\epsilon_0 \partial E/\partial t$ effects replace J. The resulting magnetic field B is of course perpendicular to the paper. The four quantities E, Q, J and B are all zero in the virgin territory ahead of the zone Z. Their values behind the zone are easily related:

(i) *Charge conservation*: The zone sweeps over plate area at a rate c per unit distance measured normal to the paper, depositing charge at a rate Qc on the upper plate. Thus $J = Qc$.

(ii) *Electric field in a parallel plate capacitor*: The standard result $\epsilon_0 E = Q$ applies since the gap contains no dielectric.

(iii) *Magnetic field in a long solenoid of arbitrary cross-section*: The standard result $B = \mu_0 J$ applies, since the gap contains no magnetic material. (See problem 4.10(a).)

(iv) *Faraday's law applied round the loop ABCD.*

We are tacitly assuming that the plates are so highly conducting that negligible electric field is needed in the direction of J to drive it. In other words, we are assuming that ohmic dissipation is not causing the wave to attenuate as it travels. The electric field along CD is also zero and so the only non-zero contribution to $\oint \mathbf{E} \cdot \mathbf{dr}$ evaluated round $ABCD$ comes from the stretch AB and equals the 'input voltage' Eg, if we take the capacitor gap width to be g. But the rate of change of magnetic flux linked by $ABCD$ is Bcg, because the zone Z is sweeping area in figure 11.5 at a rate cg, populating it with flux at an intensity B. Faraday's law therefore gives $E = cB$. The reader should check the signs for consistency.

Suggestions in chapter 4 that a capacitor sometimes behaves like an inductor are now amply confirmed. Obviously a lumped-circuit view is quite inadequate in a problem such as this. The passage in between the plates has now turned into a *wave-guide.*

Combining (i), (ii), (iii) and (iv) reveals that

$$c^2 = 1/\mu_0\epsilon_0$$

and we have found the formula for the velocity of electromagnetic waves by a very simple method that involves no differential equations.

As the wave arrives at each section, the abundant conduction electrons nearby spring into slow motion to extend the stream of current J. They vastly outnumber the relatively few electrons needed to create the surface charges $\pm Q$. The conduction electrons do not move far along the plates.

One interesting feature of this problem is that, as seen from the source on the left, the capacitor looks like a pure resistance, even though there is no dissipation. This constitutes a very simple example of the concept of *radiation resistance*, important in antenna theory. The charging process can be looked upon as if the source were 'shining' a beam of electromagnetic radiation into the aperture AB. Across a square of aperture, g times g (or indeed of *any* size), the apparent radiation resistance is $Eg \div Jg$ which equals $(\mu_0/\epsilon_0)^{1/2}$ or $1/c\epsilon_0$, which takes the value 120π ohms. This important magnitude also has significance in connection with transmission line theory. For instance, if we were to terminate the passage between the parallel plates on the right of figure 11.5 by a resistive connection of impedance 120π ohms per square of end face, the incident wave would find this indistinguishable from more wave-guide beyond and the wave would not reflect. The connection would correctly meet the relation between E and J set up by the incident wave. This impedance is called the *characteristic impedance* of the wave guide or transmission line.

Any other termination would set up an end condition which could only be satisfied by having reflected waves as well. When a capacitor of finite dimensions is charged or discharged rapidly by means of wave action, the waves may reflect back and forth several times, giving a pulsating effect at the terminals (a phenomenon known as *ringing*) until dissipative effects finally cause the waves to die away. Notice how differently high-speed charging behaves from the familiar low-speed exponential charging processes that are typical of lumped-circuit theory.

Problem 11.13 considers an analogous situation where an inductor suddenly finds itself behaving like a capacitor. Problem 11.14 explores first the discharging of a capacitor through a short-circuit and secondly the way in which energy is best extracted from a capacitor by means of a suitably matched load equal to the characteristic impedance. Note that in the first case, because the load is a short-circuit, no energy is extracted from the capacitor. The discharging wave merely changes all the stored electrical energy into stored magnetic energy. Later waves repeatedly reverse the process.

Example 11.8 A coaxial cable consists of two concentric conductors of negligible resistance and radii a and b, separated by a non-magnetic dielectric of constant k. A constant voltage V is suddenly applied between the conductors at one end. Examine the form of the initial and reflected waves if at the other end the conductors are joined by a resistance R. What if $R = 0$? When is there no reflection?

The analysis given above for plane conductors is easily modified. The magnetic field lines are concentric circles and E and B vary with radius (see figure 11.6). Let J and Q now be total current and charge per unit *length* of cable, respectively.

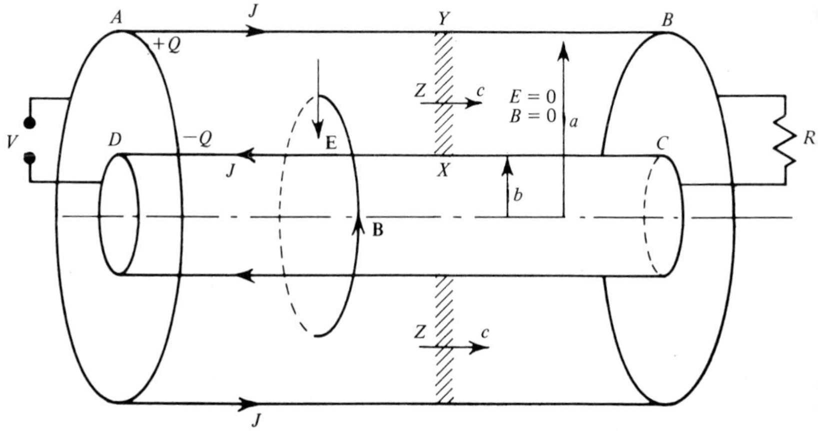

Fig. 11.6. Electromagnetic step-wave in a coaxial cable.

Initial wave. Charge conservation again requires $J = Qc$ but now $k\epsilon_0 E = Q/2\pi r$, as usual for a concentric cylindrical capacitor. Also $B = \mu_0 J/2\pi r$ from the application of Ampère's law to a field line. Applying Faraday's law to loop $ABCD$ gives

$$V = \int_D^A E\mathrm{d}r = \frac{Q}{2\pi k\epsilon_0}\log\frac{a}{b} = \oint \mathbf{E}\cdot\mathrm{d}\mathbf{r} = c\int_X^Y B\mathrm{d}r = \frac{c\mu_0 J}{2\pi}\log\frac{a}{b},$$

if we allow for the rate at which the wave front XY fills new space with non-uniform B. Hence

$$Q/k\epsilon_0 = c\mu_0 J = c^2\mu_0 Q \quad \text{and} \quad c = (k\mu_0\epsilon_0)^{-1/2},$$

the reduced velocity of light in a dielectric.

Reflected wave. This is created at the instant when the initial wave reaches the end. Let dashes denote the changes it produces in J, Q etc. Because of its reversed velocity $J' = -cQ'$ but $k\epsilon_0 E' = Q'/2\pi r$, still. After the reflection, the *total* conditions near the right-hand end are given by

$$J_{\text{tot}} = cQ - cQ' \quad \text{and} \quad E_{\text{tot}} = (Q + Q')/2\pi k\epsilon_0 r.$$

These are related by the resistance R, for

$$\text{p.d.} = \int_C^B E_{\text{tot}}\mathrm{d}r = \frac{Q + Q'}{2\pi k\epsilon_0}\log\frac{a}{b} = RJ_{\text{tot}} = Rc(Q - Q').$$

This equation determines the ratio of Q and Q', i.e. of the 'strengths' of the initial and reflected waves. If $R = 0$, $Q = -Q'$ and the reflected wave reduces the total charge to zero, as current pours out through R. For no reflection to occur,

$$Q' = 0 \quad \text{and} \quad R = \log(a/b)/2\pi k c \epsilon_0,$$

the characteristic impedance of the cable.

Returning to the original query that started this section, we begin to see what happens when signals are sent from a switch or other initiating agency into any electrical system. The circuit or system, whatever its geometric form, behaves essentially like a wave-guide; electromagnetic waves rattle about between the conductors for some time, finally setting up appropriate charge distributions and also, in general, radiating some energy right away from the system if it is exposed. This stage lasts a time which is a low multiple of the quantity (maximum dimension of system) ÷ (velocity of light) and which is typically about 10^{-9} seconds in bench-top systems. It is in this sense that we can now answer the original question 'How fast does electricity travel?' by saying that it travels at the speed of light. The individual electrons may only creep along, but the changes in their pattern of behaviour can travel at 3×10^8 m/s. It should also be clear now why, for all phenomena occurring at frequencies well below 10^9 Hz (for bench-top systems), electromagnetic wave propagation can be regarded as virtually instantaneous. Then the simple pre-Maxwellian view adopted in most electrotechnology becomes valid. Mathematically it is equivalent to putting $c = \infty$, $\epsilon_0 = 0$ and neglecting all $\epsilon_0 \partial \mathbf{E}/\partial t$ effects. It is this simple model which enables us to use the concept of inductance, for instance; the magnetic field everywhere can be treated as proportional to the current, *without any delays*, and so flux-linked-per-unit-current (i.e. inductance) becomes meaningful. Problem 11.25 goes further into these and related issues.

Treating electromagnetic propagation as virtually instantaneous has a parallel in fluid mechanics in the case of 'incompressible' flow, where the compressibility is neglected, the bulk modulus being so large compared with typical pressure changes that the speed of sound is virtually infinite and pressure disturbances propagate virtually instantaneously throughout the field of flow.

There has been space in this book for only a brief glimpse of the vast subject of electromagnetic wave theory, which forms the basis for telecommunications technology and for a good many natural phenomena besides. The theory becomes rapidly too mathematical for us to pursue here. It is necessary to devise a mathematical formulation in terms of *retarded potentials* to describe how the fields change so as to allow for the delays that ensue owing to the finite travel time of the waves. One of the intriguing aspects of the situation is that, despite the delays which now relate cause and effect at different places, Maxwell's equations in integral form are still correct. For example

$$\oint \mathbf{E} \cdot \mathbf{dr} = -\iint (\partial \mathbf{B}/\partial t) \cdot \mathbf{da}$$

is an equation that appears to indicate that $\mathbf{E}$ in one place is determined by $\partial \mathbf{B}/\partial t$

at other places *at the same instant.* Yet these same equations in differential form are the very thing that predicts electromagnetic waves, with their concomitant delays! Another point is that Newton's law concerning the equality of action and reaction needs modification. The reactions between two interacting particles can no longer be instantaneously equal if one has not yet received news of what the other is up to! In the theory, this is allowed for by associating momentum with electromagnetic waves. (See problem 11.9(*c*).)

*11.8 Energy in electromagnetism

It is possible that for several years the reader has been exposed to proofs and calculations concerning electromagnetic phenomena which invoked ideas of energy conservation. But energy is to some extent a legitimate abstraction invented by man to assist the systematic description and prediction of physical events. Whenever the energy book-keeping fails, one invents a new kind of energy to save the situation, if possible. Energy arguments are therefore not really acceptable until a relevant energy conservation theorem or law has been demonstrated. Such a demonstration is clearly necessary in the context of electromagnetism. We shall find that the conservation of electromagnetic energy is rather more subtle than expected.

Electromagnetism concerns the interaction of material media (whose energy behaviour is described by the subject of thermodynamics) with the electromagnetic field, characterised by the vectors **E** and **B**. Simple examples such as the charging of a capacitor *in vacuo*, where energy appears to be supplied but the thermodynamic state of the only material present (the plates) does not change, make it evident that the sustaining of energy conservation principles will demand the invention of new kinds of energy 'stored in the electromagnetic field', distinct from material energy.

There are two ways in which energy can be fed into the material and the space it occupies by the **E** and **B** fields.

I: The Lorentz force $Q(\mathbf{E} + \mathbf{v} \times \mathbf{B})$ can do work on any charge Q moving at velocity $\mathbf{v}$ in the material. The rate of working is $Q\mathbf{E} \cdot \mathbf{v}$ because the $\mathbf{v} \times \mathbf{B}$ part is normal to the motion and does no work. By summing $Q\mathbf{E} \cdot \mathbf{v}$ terms over all the charges in unit volume, we see that

$$\text{total input power density} = \mathbf{E} \cdot \sum Q\mathbf{v} = \mathbf{E} \cdot \mathbf{j}.$$

(As usual **E** is a local, blurred-out mean field, as discussed in chapter 2.) Notice that **j** may include conduction current, convection current and polarisation current $\partial\mathbf{P}/\partial t$. Thus the rate of power input to a dielectric is $\mathbf{E} \cdot \partial\mathbf{P}/\partial t$ per unit volume and its equilibrium thermodynamics is governed by such relations as

$$T\mathrm{d}S = \mathrm{d}U - \mathbf{E} \cdot \mathrm{d}\mathbf{P}.$$

If the medium is an ohmic conductor, for which $\mathbf{E} = \tau\mathbf{j}$, the input power density is the familiar joule dissipation term $\tau\mathbf{j}^2$, unless the conductor is moving in a magnetic field, in which case $\mathbf{E} \cdot \mathbf{j}$ includes the mechanical work exchanges associated with the magnetic force $\mathbf{j} \times \mathbf{B}$. (See problem 11.15.) There may also be other effects such as thermoelectricity. We shall use $\mathbf{j}$ to denote the sum of all three kinds of current in what follows. It does not include the Maxwell effect $\epsilon_0 \partial \mathbf{E}/\partial t$, however. This operates *in vacuo*, and does not relate to material. (Nor does $\mathbf{j}$ include the virtual current curl $\mathbf{M}$ in magnetic material. See example 11.1.) From elementary studies of capacitors, the reader will be acquainted with the idea of an energy density $\frac{1}{2}\epsilon_0 \mathbf{E}^2$ associated with an electric field $\mathbf{E}$ even *in vacuo*. When this field changes there must be an input power density

$$\partial/\partial t(\tfrac{1}{2}\epsilon_0 \mathbf{E}^2) = \mathbf{E} \cdot \epsilon_0 \partial \mathbf{E}/\partial t,$$

which can now be associated with the $\epsilon_0 \partial \mathbf{E}/\partial t$ term. The total input power associated with $\mathbf{E}$ to both the space and its material contents is seen to be

$$\mathbf{E} \cdot \mathbf{j} + \mathbf{E} \cdot \epsilon_0 \partial \mathbf{E}/\partial t = \mathbf{E} \cdot (\mathbf{j} + \epsilon_0 \partial \mathbf{E}/\partial t) = \mathbf{E} \cdot \text{curl}\, \mathbf{H}.$$

II: When a magnetic field is created by currents in a winding, the total input power density is $\mathbf{H} \cdot \partial \mathbf{B}/\partial t$ (see example 6.2), which can also be written as

$$-\mathbf{H} \cdot \text{curl}\, \mathbf{E}.$$

If there is no magnetisable material present $\mathbf{B} = \mu_0 \mathbf{H}$ and the input is $\mathbf{H} \cdot \partial(\mu_0 \mathbf{H})/\partial t$ or $\partial/\partial t(\frac{1}{2}\mu_0 \mathbf{H}^2)$, so that $\frac{1}{2}\mu_0 \mathbf{H}^2$ can be regarded as the energy density stored in the space due to the field. If there is material magnetised to an intensity $\mathbf{M}$ present, where $\mathbf{B} = \mu_0(\mathbf{H} + \mathbf{M})$, the extra energy input density is

$$\mathbf{H} \cdot \partial \mathbf{B}/\partial t - \mathbf{H} \cdot \partial(\mu_0 \mathbf{H})/\partial t = \mu_0 \mathbf{H} \cdot \partial \mathbf{M}/\partial t,$$

which explains why the equilibrium thermodynamics of magnetic material is governed by such relations as $TdS = dU - \mu_0 \mathbf{H} \cdot d\mathbf{M}$. Combining I and II we see that the total input power density $\mathcal{E}$ takes the terse form

$$\mathcal{E} = (\mathbf{E} \cdot \text{curl}\, \mathbf{H} - \mathbf{H} \cdot \text{curl}\, \mathbf{E}).$$

We now face the question: 'Is this quantity zero – i.e. is there *local* energy conservation?' Local conservation would mean, for instance, that any electromechanical work was extracted from electric or magnetic energy stored in the material or the space at the same spot.

We can make progress with the aid of the purely mathematical result:

$$-\text{div}\,(\mathbf{E} \times \mathbf{H}) \equiv \mathbf{E} \cdot \text{curl}\, \mathbf{H} - \mathbf{H} \cdot \text{curl}\, \mathbf{E}.$$

(See section 10.11.)

We are evidently forced to take an interest in the new and rather mysterious vector

$$\mathbf{S} = \mathbf{E} \times \mathbf{H}$$

(called the *Poynting vector*) because of the fact that

$$\mathcal{E} = -\text{div}\, \mathbf{S}.$$

Local energy conservation would imply that $\mathcal{E} = 0$ and $\mathbf{S}$ is solenoidal, but this is not in general true. The quantity $-\text{div}\, \mathbf{S}$ is the net influx (not efflux, because of the minus sign) of the vector field $\mathbf{S}$ into any volume element per unit volume.

Evidently **S** must represent an energy flux intensity (in watts/metre^2) if $-\operatorname{div}\mathbf{S}$ represents an energy input power density (in watts/metre^3). The energy conservation statement is called *Poynting's theorem*:

$$-\operatorname{div}\mathbf{S} = \text{power density into the material } (\mathbf{E}\cdot\mathbf{j} + \mu_0\mathbf{H}\cdot\partial\mathbf{M}/\partial t)$$

$$+ \text{ power density into the space } \frac{\partial}{\partial t}(\tfrac{1}{2}\epsilon_0\mathbf{E}^2 + \tfrac{1}{2}\mu_0\mathbf{H}^2),$$

if on the right-hand side we split $\mathcal{E}$ into its various constituents. This result implies that energy is not conserved locally but appears to flow, appearing here, disappearing there, according to the Poynting vector field **S**.

Integrated over a finite volume with the aid of Gauss's theorem, Poynting's theorem states that the total rate of supply of energy inside the volume equals the total Poynting flux inwards through the surface enclosing the volume. This is another very good example of a relationship between a surface integral and a volume integral which enables an interesting quantity, affecting all parts of the interior of a volume, to be evaluated merely from knowledge of conditions at its surface, which involves much less effort and information.

For a finite volume, the total material and field energy content is conserved within that volume only when there is no net Poynting flux through the surface of the volume, e.g. in the case where **E** or **H** is zero at the surface, or normal to it (for then $\mathbf{E}\times\mathbf{H}$ is tangential). This result makes it clear that energy conservation statements in electromagnetism should be made with great care, with particular attention to the choice of surface enclosing the region under discussion. Some examples follow. Problems 11.15 and 11.16 give other exercises on Poynting's theorem.

*11.9 Applications of Poynting flux

The first example is a destructive one. Consider a 'still life' composition consisting of a stationary permanent magnet and a stationary body that is electrically charged. Although it is quite clear that nothing whatever is happening at the macroscopic level, the locality is permeated by **E** and **H** fields and therefore by an **S** field as well. Is there or is there not energy constantly in transit here?

The lesson from this example is that **S** itself does not have any physical meaning; only the *divergence* of **S** is physically significant. In the non-event just described we would in fact find that $\operatorname{div}\mathbf{S} = 0$ everywhere. Moreover, any arbitrary solenoidal field can be added on to **S** to give a new field $\mathbf{S}'$ which would be just as acceptable in Poynting's theorem. (Problem 11.17 exhibits a case in point.) Thus **S** is a somewhat arbitrary and paradoxical abstraction.

Next consider a simple configuration consisting of a long, straight, resistive wire, carrying a steady axial current under the influence of an axial electric field which prevails both in the wire and the space around it. There is a circumferential magnetic field inside the wire and in the space nearby, and **S** is therefore directed radially inwards. (Problem 11.16(*a*) goes into details.) This picture of radial energy flow conflicts totally with the widely prevailing notion that electrical power comes *along* the wires from the power station, through the electricity meter and into our equipment. Which way *does* it flow?

The answer is that either view of the situation is acceptable because energy and its flux are largely our own inventions (within certain natural constraints), to be exploited as we choose. Problem 11.17 shows how to convert **S** into an energy flux which *is* along the wire. But it is a perfectly good alternative view to think of the energy as flowing through the space between the wires and of the meter as being mounted across an 'aperture' at entry to the waveguide formed by the domestic circuitry.

For comparison, consider a coaxial cable, with concentric inner and outer cylindrical conductors of zero resistance, delivering d.c. power from a source to a resistive end-load. The potential difference between the conductors is constant and the electric field is now radial between them, the magnetic field still being circumferential. Evidently **S** is now axial and the cable appears to be acting as an annular pipe, guiding energy flux from source to load.

Finally, figure 11.7 shows a combined case, a short coaxial cable with resistive conductors. The orthogonal **E** and **S** field lines are now curved because of the axial components of **E** that are necessary to overcome resistance, and the spread of the energy along **S**-lines to its various destinations is clearly displayed. The increasing curvature of the **E**-lines occurs because the radial electric fields become smaller towards the load end but the axial electric fields stay constant.

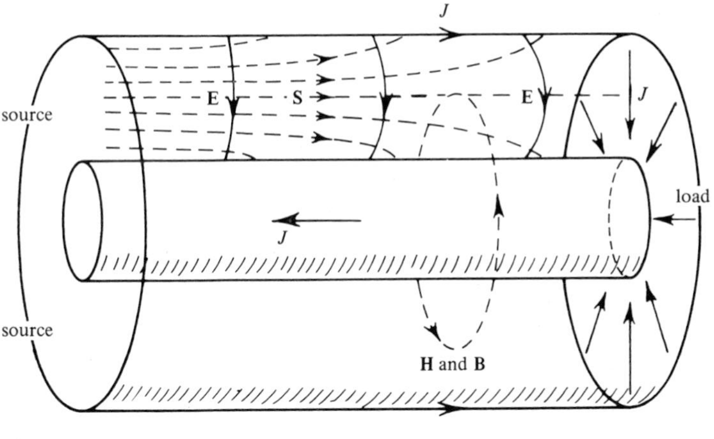

Fig. 11.7. Energy flow in a short coaxial cable.

A squirrel-cage induction motor provides a good application of Poynting's theorem. The cylindrical rotor runs inside a cylindrical stator, both being made of laminated iron. Electrical a.c. power is fed into a winding set in the cylindrical face of the stator, producing a magnetic field across the gap to the rotor, such that the magnetic field pattern rotates about the axis of the machine. This induces axial electric fields in the gap and in the axial copper conductors (the squirrel-cage) which are buried in the rotor near its surface and are joined together at the ends of the machine. Mechanical power is delivered by the rotor shaft and so electromagnetic power must somehow be crossing the gap from the stator in the form of a Poynting flux $\mathbf{E} \times \mathbf{H}$. If $\mathbf{H}$ were radial (as is normally supposed), $\mathbf{E} \times \mathbf{H}$ would have no radial component across the gap. An induction motor only functions because $\mathbf{H}$ is not quite radial.

Though in all the examples cited so far, entirely adequate methods other than Poynting's theorem are available for analysing the energetics of the situation, there are problems in which the Poynting flux offers a particularly convenient way of expressing energy conservation and transport. This is especially true of plane electromagnetic waves. Referring again to figure 11.3 reveals that $\mathbf{S}$ (which is $\mathbf{E} \times \mathbf{B}/\mu_0$ for waves in non-magnetic material) is oriented in the direction of propagation. For once, the Poynting vector is actually pointing instead of disappointing! In plane-polarised, unidirectional waves it fluctuates in intensity, but in circularly polarised, sinusoidal waves it obviously stays absolutely steady in space and time, directed normally to the paper in figure 11.4, while $\mathbf{E}$ and $\mathbf{B}$ are rotating. Problem 11.18 explores the prospects for power transmission in waveguides.

Example 11.9 For the case considered in example 11.7 calculate the H_y distribution and account for the energy flow between the forms $\frac{1}{2}\epsilon_0\mathbf{E}^2$ and $\frac{1}{2}\mu_0\mathbf{H}^2$ via the Poynting flux.

We have

$$\frac{\partial B_y}{\partial t} = \mu_0 \frac{\partial H_y}{\partial t} = \frac{\partial E_z}{\partial x} = \frac{4\pi C}{\lambda} \cos\frac{2\pi x}{\lambda} \cos\frac{2\pi ct}{\lambda}$$

and

$$H_y = \frac{2C}{c\mu_0} \cos\frac{2\pi x}{\lambda} \sin\frac{2\pi ct}{\lambda},$$

if we integrate for t, rejecting all but the standing wave solution. We also have the x-wise (and only) component of the Poynting vector,

$$S_x = -E_z H_y = -\frac{C^2}{c\mu_0} \sin\frac{4\pi x}{\lambda} \sin\frac{4\pi ct}{\lambda},$$

in view of the result $\sin 2\phi = 2 \sin\phi \cos\phi$. Meanwhile

$$\tfrac{1}{2}\epsilon_0\mathbf{E}^2 = \frac{2C^2}{c^2\mu_0} \sin^2\frac{2\pi x}{\lambda} \cos^2\frac{2\pi ct}{\lambda}$$

and

$$\tfrac{1}{2}\mu_0\mathbf{H}^2 = \frac{2C^2}{c^2\mu_0} \cos^2\frac{2\pi x}{\lambda} \sin^2\frac{2\pi ct}{\lambda}.$$

Figure 11.8 shows these two energy densities at the instants $t = 0$ and $t = \lambda/4c$ (one-quarter of a period later) for the shortest possible resonant cavity (between boundaries $x = 0$ and $\lambda/2$). The energy evidently surges from the middle to the ends, or vice versa, every quarter period. (The total energy is constant – see problem 11.16(*c*).) Per unit cross-section, the total energy to the left of a particular position x is given by

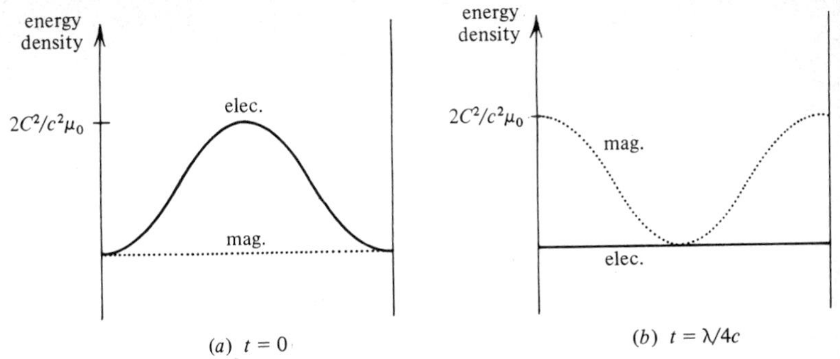

Fig. 11.8. Energy fluctuations in a standing wave: elec. $= \frac{1}{2}\epsilon_0 \mathbf{E}^2$; mag. $= \frac{1}{2}\mu_0 \mathbf{H}^2$.

$$\frac{2C^2}{c^2\mu_0}\int_0^x \left(\sin^2\frac{2\pi x}{\lambda}\cos^2\frac{2\pi ct}{\lambda} + \cos^2\frac{2\pi x}{\lambda}\sin^2\frac{2\pi ct}{\lambda}\right)\mathrm{d}x$$

$$= \frac{C^2}{c^2\mu_0}\left\{x - \frac{\lambda}{4\pi}\sin\frac{4\pi x}{\lambda}\cos\frac{4\pi ct}{\lambda}\right\}.$$

Its rate of rise is

$$\frac{C^2}{c\mu_0}\sin\frac{4\pi x}{\lambda}\sin\frac{4\pi ct}{\lambda},$$

i.e. $-S_x$, the rate of energy flow from right to left, as it should be.

---

*11.10 Some more mathematical points

We already know that if curl $\mathbf{v} = \mathbf{0}$, the vector $\mathbf{v}$ has a *scalar potential* ϕ such that $\mathbf{v} = \text{grad}\,\phi$. This is consistent with the identity curl grad $= \mathbf{0}$.

If div $\mathbf{v} = 0$, the identity div curl $= 0$ suggests that there should exist a vector $\mathbf{w}$ such that $\mathbf{v} = \text{curl}\,\mathbf{w}$. This is in fact the case, and $\mathbf{w}$ is called $\mathbf{v}$'s *vector potential*. For example the magnetic field $\mathbf{B}$, for which div $\mathbf{B} = 0$, has a *magnetic vector potential* $\mathbf{A}$, such that $\mathbf{B} = \text{curl}\,\mathbf{A}$. $\mathbf{A}$ plays a big part in advanced electromagnetic theory. To define $\mathbf{A}$ completely, another condition such as div $\mathbf{A} = 0$ is necessary. If the vector $\mathbf{v}$ is neither solenoidal nor irrotational, it is possible to use both kinds of potential and express $\mathbf{v}$ as

$$\mathbf{v} = \text{grad}\,\phi + \text{curl}\,\mathbf{w},$$

the two parts being respectively irrotational and solenoidal. This split is particularly significant in the case of $\mathbf{E}$, the electric field, the sum of $\mathbf{E}_s$, the electrostatic part (for which $\mathbf{E}_s = -\,\text{grad}\,V$) and $\mathbf{E}_i$, the induction part, which obeys

$$\text{curl}\,\mathbf{E}_i = -\,\partial\mathbf{B}/\partial t = -(\partial/\partial t)\,\text{curl}\,\mathbf{A}.$$

It is therefore possible to have $\mathbf{E}_i = -\,\partial\mathbf{A}/\partial t$, consistently with div $\mathbf{E}_i = 0$ and div $\mathbf{A} = 0$. The divergence of $\mathbf{E}$ is provided wholly by div $\mathbf{E}_s$. The split of $\mathbf{E}$ can therefore be written

$$\mathbf{E} = -\text{grad } V - \partial\mathbf{A}/\partial t.$$

Problem 10.18(*a*) reveals the relation between stream function and vector potential in two-dimensions, while problems 11.17 and 11.19 take the idea of vector potential further.

Another important, related, general idea is that a vector field $\mathbf{v}$ is determinate once its 'sources' are defined over the region of interest (together with appropriate boundary conditions). There are two basic kinds of sources:

(*a*) those which define div $\mathbf{v}$, and

(*b*) those which define curl $\mathbf{v}$.

Once again $\mathbf{E}$ provides a convenient example. The charge distributions present define div $\mathbf{E}$ (= div $\mathbf{E}_s$) and the changing magnetic field defines curl $\mathbf{E}$ (= curl $\mathbf{E}_i$). $\mathbf{E}_s$ is defined by the known values of div $\mathbf{E}_s$ and curl $\mathbf{E}_s$ (zero) while $\mathbf{E}_i$ is defined by the known values of curl $\mathbf{E}_i$ and div $\mathbf{E}_i$ (zero), and $\mathbf{E}$ is the sum of $\mathbf{E}_s$ and $\mathbf{E}_i$.

Problem 11.20 considers the application of similar ideas to the magnetic fields of magnetised material and considers the forces on magnets.

Problems 10.19(*c*), 10.22(*e*), 11.21 to 11.25 take the reader further into electromagnetic theory. Attention is particularly drawn to the last two for the light they throw upon the reasons for the success of pre-Maxwellian, pre-relativistic electrotechnology.

Apart from some references to Maxwell's stresses in chapter 12, this is the last we shall hear of electromagnetism in this book. Most of the basics of the subject have been mentioned. The main omission has been relativistic electrodynamics, which develops the theory in a fully consistent form, recognising the fact that electromagnetic waves *in vacuo* travel at the same speed relative to all observers. The interested student who goes further with the subject will find that there are many more fascinating aspects to master in detail, whether in plasma electrodynamics, wave propagation and interaction theory, geomagnetism, telecommunication and microwave technology, controlled fusion technology or superconductor technology, to name some of the most interesting areas.

Problems 11

11.1 (*a*) Is the last result of example 11.1 consistent with the result $H = -\text{grad } U$ in the absence of current or $\epsilon_0 \partial\mathbf{E}/\partial t$?
(*b*) Is the solenoidal property of $\mathbf{B}$ consistent with the divergence of Faraday's law in differential form?

11.2 (*a*) Express the circular magnetic field at a distance r from the centre line of a long, straight wire of radius a and carrying a current of uniform intensity j (see example 4.1) in terms of x, y components in a transverse plane. Verify that curl $\mathbf{H} = \mathbf{0}$ $(r > a)$ but curl $\mathbf{H} = \mathbf{j}$ $(r < a)$.
(*b*) A long, straight, circular solenoid carries NJ ampere-turns per unit length, with J rising in time at a rate low enough for $\epsilon_0 \partial\mathbf{E}/\partial t$ terms to be

negligible. By using Faraday's law in integral form, find the circular electric field inside the solenoid as a function of r (distance from the axis), and then in terms of x, y components in a transverse plane. Verify that curl $\mathbf{E} = -\partial\mathbf{B}/\partial t$.

(*c*) A vacuum magnetic field has a uniform, steady magnitude B, but is rotating at a steady angular velocity $\boldsymbol{\Omega}$ (in the z-direction), inducing electric fields in the z-direction, as occurs in an induction motor. The symmetry is such that $\mathbf{E} = \mathbf{0}$ on the z-axis and no field varies with z. Consider a stationary rectangular loop with one side of unit length on the z-axis and the opposite side cutting x, y planes at the point (x, y). Taking $\mathbf{B}$ as inclined at an angle Ωt to the x-axis, write down the flux linking the loop and deduce $E_z(x, y)$ from Faraday's law in integral form. Verify that curl $\mathbf{E} = -\partial\mathbf{B}/\partial t$.

11.3 (*a*) Generalise problem 10.4(*b*) to the case where $\mathbf{B}$ is non-uniform, obeying $\mu_0\mathbf{j} = \text{curl } \mathbf{B}$, by applying the identity $(\mathbf{v} \cdot \text{grad})\mathbf{v} \equiv (\text{curl } \mathbf{v}) \times \mathbf{v} + \text{grad } (\frac{1}{2}\mathbf{v}^2)$. Show that $p + \mathbf{B}^2/2\mu_0 =$ const. and consider how and why p does or does not vary along current flow lines. (Refer also to example 11.3.)

*(*b*) Account for the fact that $p + B^2/\mu_0 =$ const. in example 7.4 (rather than $p + B^2/2\mu_0$).

11.4 (*a*) Using the solution given in section 11.3, which neglects $\epsilon_0\partial\mathbf{E}/\partial t$ and $\partial\mathbf{P}/\partial t$, find the maximum amplitudes of these quantities at frequencies of 10^4 and 10^9 Hz in sea-water, taking the values $\sigma = 2.5$ mho/m and $\mathbf{P} = 79\epsilon_0\mathbf{E}$, and compare them with that of $\mathbf{j}$. Comment on the approximation in these cases. What is the skin depth?

(*b*) Adapt the discussion in section 11.3 to the case where the medium is linearly magnetisable with $\mathbf{B} = \mu\mathbf{H}$ ($\mu =$ const.) and $\mathbf{j} = \text{curl } \mathbf{H}$.

11.5 (*a*) A slab of uniform electrical conductivity σ contains a magnetic field $\mathbf{B}$ such that $\mathbf{B} = \mathbf{A} \sin px$, where $p =$ const. and $\mathbf{A}$ is a vector in the y-direction whose magnitude A depends only upon time t. Neglecting $\partial\mathbf{D}/\partial t$ and $\mathbf{M}$, find a relation between A and $\mathrm{d}A/\mathrm{d}t$ and show that the magnetic field at each point decays in the ratio e:1 in a time $\mu_0\sigma/p^2$. (Compare problem 9.7(*a*).)

(*b*) If instead the slab has a plane face at $x = 0$ and a rising tangential electric field $E_z = E_0\mathrm{e}^{mt}$ is applied there, find a solution of form $E(x)\mathrm{e}^{mt}$ describing the diffusion of the fields into the slab.

11.6 Consider the magnetically-levitated vehicle shown in figure 11.2, travelling at a speed V. It is supported by an average field B and magnetic pressure $B^2/2\mu_0$ exerted over an area of length L and width W. The skin-effect current flowing in the track has roughly to equal B/μ_0 per unit distance measured perpendicular to the current in order to exclude the field from deep penetration. (Do you see why?) In the time L/V that it takes the field to pass any given point in the track, the penetration distance is of order $(L/V\mu_0\sigma)^{1/2}$. By equating (drag × speed) to total ohmic dissipation in the skin, show that the lift/drag ratio of the vehicle is of order $\frac{1}{2}(\mu_0\sigma VL)^{1/2}$. (The inclusion of the factor $\frac{1}{2}$ is somewhat arbitrary as this is only an order of magnitude calculation.) Evaluate this ratio if $\sigma = 4 \times 10^7$ mho/m, $V = 100$ m/s, $L = 4$ m.

If the magnetic field is provided by a horizontal, constant-flux, superconducting coil, why is the suspension stable (in the sense that lowering the vehicle raises the lift force)?

11.7 (*a*) The operator ∇^2 can be generalised to apply to a vector by defining $\nabla^2\mathbf{v} = \mathbf{i}\nabla^2 v_x + \mathbf{j}\nabla^2 v_y + \mathbf{k}\nabla^2 v_z$. Demonstrate the vector identity grad (div $\mathbf{v}$) $\equiv$ curl (curl $\mathbf{v}$) + $\nabla^2\mathbf{v}$, using cartesians. (N.B. The operator ∇^2 applied to a vector needs special treatment whenever curvilinear coordinates are used.)

(*b*) For a medium of conductivity σ deduce the three-dimensional unsteady diffusion equation $(\mu_0\sigma)^{-1}\nabla^2\mathbf{B} = \partial\mathbf{B}/\partial t$. Neglect $\partial\mathbf{D}/\partial t$ and $\mathbf{M}$.

(*c*) For electromagnetic waves *in vacuo*, deduce the three-dimensional wave equation

$$c^2\nabla^2(\mathbf{B} \text{ or } \mathbf{E}) = (\partial^2/\partial t^2)(\mathbf{B} \text{ or } \mathbf{E}),$$

noting that the order of curl and $\partial/\partial t$ operators is immaterial.

11.8 (*a*) In section 11.5 the equations for div **B** and div **E** and the x-components of Maxwell's curl equations are ignored. Deduce from them that B_x and E_x are constant in space and time, if present at all, when variables depend only on x and t.

(*b*) Develop the theory of one-dimensional electromagnetic waves in a linear dielectric medium for which $\mathbf{D} = k\epsilon_0\mathbf{E}$ and show that the wave speed $= ck^{-1/2}$.

11.9 (*a*) Plane-polarised, periodic waves of wave length λ travelling in the $\pm x$-directions are such that at time $t = 0$ the total profile of B_y is as shown in figure 11.9(*a*), the value of E_z being instantaneously zero. By graphical techniques find the E_z and B_y profiles of the two oppositely-travelling waves which must be present and demonstrate that at time $t = \frac{1}{4}\lambda/c$, the total B_y profile is as shown in figure 11.9(*b*). (Note that the two E_z profiles must cancel out at $t = 0$.) Also study the behaviour of the total E_z profile.

(*b*) A plane, highly-conducting wall lies at $x = 0$. A plane, plane-polarised electromagnetic wave travelling x-wards *in vacuo* arrives from negative x. It consists of a square wave of length cT in which B_y has a constant value B_0 for a duration T seconds, being otherwise zero. Find what reflected wave is created to match the condition $E_{\text{tangential}} = 0$ at the wall and calculate the magnetic pressure on the wall and its duration.

*(*c*) What is the momentum of a solid body which, bouncing elastically, would impart the same impulse to the wall per unit area. If this momentum were to be attributed to the incident wave, what would be its

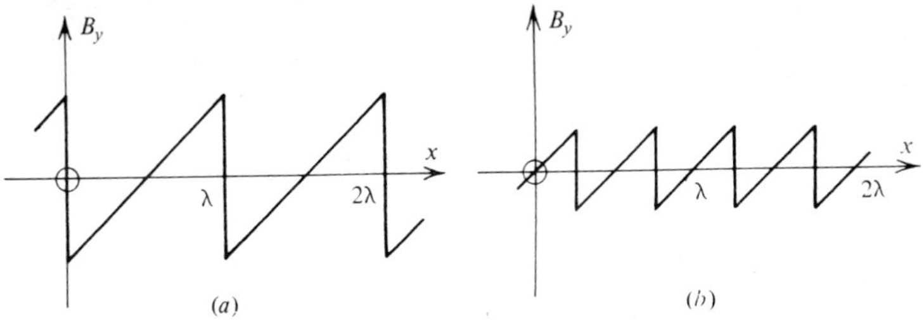

Fig. 11.9

momentum per unit volume? (This is a particular case of a general result that electromagnetic momentum density is $\mathbf{E} \times \mathbf{H}/c^2$.)

*11.10 A plane-polarised, plane wave *in vacuo* in which $E_z = A_i \cos \omega(x/c - t)$ arrives at the plane face ($x = 0$) of a dielectric medium in which the reduced velocity of light is $c_0 = ck^{-1/2}$. (See problem 11.8(*b*).) The wave carries on with altered amplitude and speed but the same frequency such that $E_z = A_0 \cos \omega(x/c_0 - t)$ and there is also a reflected wave in which $E_z = A_r \cos \omega(x/c + t)$. At $x = 0$, E_z and B_y must be continuous. (Why?) Find the ratio A_0/A_i.

*11.11 Consider the plane-polarised, sinusoidal B_y/E_z wave propagating in the x-direction, discussed in section 11.6. Refer it to new x' and y'-axes but the same z-axis. Let the new x'-axis be inclined at an angle α to the old x-axis, so that x equals $(x' \cos \alpha - y' \sin \alpha)$. Then

$$E_z = -cA \cos \{2\pi(x' \cos \alpha - y' \sin \alpha - ct)/\lambda\}$$

Now superpose another plane-polarised wave of the same frequency and amplitude and with its electric field in the same direction but propagating in a direction in the x, y plane that is also inclined at an angle α to the x'-axis but lies on the other side of it. Changing the sign of α for this wave gives

$$E_z = -cA \cos \{2\pi(x' \cos \alpha + y' \sin \alpha - ct)/\lambda\}.$$

Show that, in the combined wave, E_z is zero where $y' = \pm \lambda/(4 \sin \alpha)$ for all values of x' and t. Since also $E_{y'} = 0$ everywhere, the combined solution could represent a wave-mode inside a rectangular wave-guide, extending in the x'-direction, with highly-conducting walls at $z = 0$ and $z =$ (a const.) and at $y' = \pm \lambda/(4 \sin \alpha)$. (The boundary condition on $\mathbf{E}$ is merely that its tangential component vanishes at the walls.) If the guide's y'-wise width is w, show that the frequency n must exceed $c/2w$, known as the *cut-off frequency*.

Show that the wave pattern propagates with a speed (known as the *phase velocity*) equal to $c \sec \alpha$ (greater than c!) in the x'-direction. But if the wave-guide is suddenly activated from one end, show that, because the constituent waves are travelling obliquely and reflecting repeatedly, the wave front travels in the x'-direction at a speed (known as the *group velocity*) equal to $c \cos \alpha$ (less than c). (Is this group velocity consistent with the value derived from the dispersion relation for a given value of w? This assumes acquaintance with general group velocity theory.)

Verify that the magnetic field boundary condition (that **B** has no normal component) is satisfied at the walls, where skin currents exclude the field.

*11.12 When a one-dimensional wave in which $E_z = E \sin (kx - \omega t)$ passes through a plasma, the ions may be treated as stationary, the magnetic force on the electrons may be neglected and the random collisions and thermal motions of the electrons ignored. Show that the current density due to the transverse simple harmonic oscillation of the electrons in the electric field is given by $j_z = (ne^2E/m\omega) \cos (kx - \omega t)$, where m = electron mass, $-e$ = electron charge and there are n electrons per unit volume. Since the oscillations are *transverse*, the charge motions produce no departure from neutrality (as $\operatorname{div} \mathbf{j} = 0$).

The magnetic field B_y of the wave is affected by j_z. Show that the speed ω/k of the waves $= c(1 - \omega_p^2/\omega^2)^{-1/2}$ in which $\omega_p = (ne^2/m\epsilon_0)^{1/2}$, the plasma frequency. (See section 5.6.) Note that wave propagation fails if $\omega < \omega_p$, which is therefore a *cut-off frequency*. (This is why radio waves of sufficiently low frequency reflect off the ionosphere.)

Plasma oscillation is the case where $\omega = \omega_p$ when j_z and $\epsilon_0 \partial E_z/\partial t$ cancel and there is no magnetic field and no dependence on x (for $k = 0$).

11.13 A steady current is flowing in a short-circuited inductor of negligible resistance. The inductor takes the form of a long flat tube of rectangular cross-section (width a, depth b, with $a \gg b$) and the current flows uniformly round each cross-section at an intensity J per unit length of tube.

What happens if the tube suddenly splits slightly all along one of its narrow faces? What voltage appears across the gap, if no current can cross it? How long is it before all current flow first ceases and what is the state of the system at that instant and subsequently?

11.14 (*a*) A parallel plate capacitor with an air gap is initially uniformly charged. A short-circuit is suddenly connected across one edge and the capacitor discharges by means of an electromagnetic wave, which in due course reflects simultaneously all along the far edge. (The reflection occurs because the end condition is that the current in the plates must be zero.) Effects at the other two edges of the plates should be ignored. Show that after a certain time t_0 all the capacitor's stored electric energy has become magnetic energy and that after time $2t_0$ the capacitor returns instantaneously to its original state, but with the charges reversed in sign.
(*b*) If the short circuit is replaced by a matched resistive load $g(\mu_0/\epsilon_0)^{1/2}$ ohms per unit length of edge (the gap being g) show that after time $2t_0$ the capacitor is completely discharged and all action ceases. (*Hint*: The discharging wave is now half as 'strong' as in (*a*).)

*11.15 For a moving conductor show that $-\text{div}\,(\mathbf{E}\times\mathbf{H})$ equals the sum of the ohmic dissipation and the electromechanical work $\mathbf{v}\cdot\mathbf{j}\times\mathbf{B}$, if $\partial/\partial t\{\frac{1}{2}\epsilon_0\mathbf{E}^2+\frac{1}{2}\mu_0\mathbf{H}^2\}$ is negligible and there is no polarisation or magnetisation. Is this result altered by the Hall effect? (See problem 2.5(*d*).)

*11.16 (*a*) A long, straight, cylindrical, resistive wire of radius R carries a uniform axial current of intensity j. Show that the radial Poynting flux into a concentric cylinder of unit length and radius r equals the ohmic dissipation within it. Consider both the cases $r>R$ and $r<R$.
(*b*) Explore the Poynting flux in problem 6.3(*a*) and relate it to stored electric and magnetic energy.
(*c*) For the case considered in examples 11.7 and 11.9 calculate the magnetic pressure on the plane boundaries. At what frequency does it and the Poynting vector fluctuate? Verify that the total energy is constant. When is it uniformly distributed? Re-examine the energy flow and storage if there is a linear dielectric medium present.
(*d*) For a circularly polarised, plane wave *in vacuo*, show that $\mathbf{B}^2$ at a point is constant in time and therefore that $\mathbf{B}$ and $\partial\mathbf{B}/\partial t$ are mutually perpendicular and $\mathbf{B}\cdot\text{curl}\,\mathbf{E}=0$. Prove similar results for $\mathbf{E}$. Deduce that the Poynting vector has no divergence (i.e. the energy flux is maintained throughout).

*11.17 A vector $\mathbf{S}'$ equal to $\mathbf{E}\times\mathbf{H}$ plus any solenoidal vector is equally satisfactory for use in Poynting's theorem. Consider the choice

$$\mathbf{S}' = \mathbf{E}\times\mathbf{H}+\text{curl}\,(V\mathbf{H})$$

where V is electrostatic potential. If there are changing magnetic fields present $\mathbf{E}=-\text{grad}\,V-\partial\mathbf{A}/\partial t$. (See section 11.10.) Show that

$$\mathbf{S}' = V\mathbf{j}+(\mathbf{H}\times\partial\mathbf{A}/\partial t)$$

if $\partial\mathbf{D}/\partial t$ is negligible. Consider the application of this result to problem 11.16(*a*). Note that the energy flux now appears to be along the wire, as is normally supposed.

*11.18 A wave-guide transmits electromagnetic waves in which the maximum electric field strength is E and the maximum magnetic field strength at the wall is B, where $E\approx cB$ in order of magnitude. The gas and its pressure inside the waveguide are such that E must be limited below 10^6 V/m, to avoid breakdown (electric gas discharge) and the strength of the walls is such that the magnetic pressure must not exceed 10^6 N/m^2 to avoid deformation and possible bursting. Which condition governs and what is the maximum possible Poynting flux intensity? What area of cross-section would be needed to transmit a gigawatt (10^9) of power, if the average axial Poynting flux is $\frac{1}{10}$ of this maximum? (Other limitations are the wall cooling problem and the difficulty of generating high power microwaves.)

*11.19 (*a*) What is the vector potential of $\mathbf{j}$ when it is solenoidal?

(*b*) Show that the magnetic vector potential **A** obeys the equation

$$\nabla^2 \mathbf{A} = -\mu_0 \mathbf{j}$$

in the absence of $\partial\mathbf{D}/\partial t$ and **M**. (Refer to problem 11.7.)

11.20 (*a*) Neglecting $\partial\mathbf{D}/\partial t$, show that in and around a non-conducting magnetic body

$$\text{curl}\,\mathbf{B} = \mu_0\,\text{curl}\,\mathbf{M} \qquad \text{div}\,\mathbf{B} = 0$$

$$\text{curl}\,\mathbf{H} = 0 \qquad \text{and} \qquad \text{div}\,\mathbf{H} = -\,\text{div}\,\mathbf{M}.$$

These pairs of equations reveal how the 'sources' of **B** and **H** can be regarded as residing in the appropriate non-uniformities of **M** and how the 'sources' are respectively of the 'curl' and 'div' type.

The **B**-field stems from a 'virtual current' curl **M**. For instance in figure 6.8 this is only present near the flanks *FF* where curl **M** is very large as **M** falls rapidly to zero in a thin layer analogous to a vortex sheet. The virtual current behaves like a solenoid. Is curl $\mathbf{M} \neq \mathbf{0}$ deep *inside* a magnet also possible?

The **H**-field has sources wherever div $\mathbf{M} \neq 0$, e.g. at the ends *EE* in figure 6.8, where **M** falls rapidly to zero in a thin layer and is obviously not solenoidal. Figure 6.8(*c*) shows clearly how **H** 'flows' from its sources at the right end (where $-$ div **M** is positive) to its sinks at the left end. These surface sources and sinks are known as the 'North' and 'South' *poles* of the magnet, respectively. (Is div $\mathbf{M} \neq 0$ deep *inside* a magnet also possible?) Note that some flux emerges from the flanks as well as the 'pole faces' in this case.

*(*b*) Figure 11.10 shows the cross-section of a long, magnetised tube of inner and outer radii a and $2a$, with a cut across the diameter *AD* all along the tube. **M** lies in the cross-section along concentric circles and its magnitude depends only on r, the distance from the axis. There are no electric currents. Show that **H** is zero everywhere.

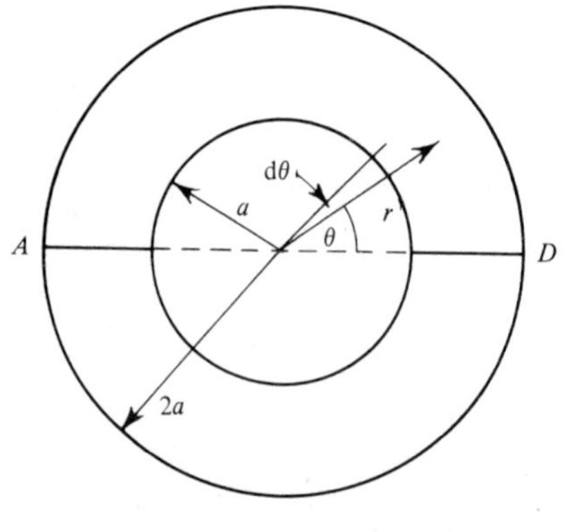

Fig. 11.10. The two halves of a circular magnet.

The force on magnetic material per unit volume may be calculated as (curl **M**) × **B**, the virtual current curl **M** taking the place of actual **j**. Note that allowance must be made for virtual current sheets at a surface whenever the tangential value of **M** drops to zero outside the material. There is an analogy with the magnetic pressure due to skin-effect, discussed in section 11.4, except that here the force is *outwards* from the material. By integrating with respect to polar coordinates r, θ calculate the total force (perpendicular to AD, by symmetry) on the upper half of the tube per unit length in the cases:

(i) M proportional to $1/r$ and (ii) M uniform, and verify that in each case it is equivalent to a tensile stress $\mathbf{B}^2/2\mu_0$ distributed over the faces AB and CD, attracting the two halves of the magnet together. Note where the magnetic force is actually exerted on the material, however. (Chapter 12 relates this result to 'Maxwell's stresses'.)

11.21 This problem takes further the discussions of problems 2.3, 2.4 and 3.10(*d*).

(*a*) In a *non-uniform* electrostatic field **E**, the net force **F** on an electric dipole **p** consisting of two charges $\pm q$ at the ends of a short line element d**r** (where $\mathbf{p} = q\mathrm{d}\mathbf{r}$) is obviously $q\mathrm{d}\mathbf{E}$, where d**E** is the change along d**r** of **E**, which equals $(\mathrm{d}\mathbf{r} \cdot \mathrm{grad})\mathbf{E}$. Hence $\mathbf{F} = (\mathbf{p} \cdot \mathrm{grad})\mathbf{E}$. Show that, since curl $\mathbf{E} = \mathbf{0}$ and **p** is a constant vector,

$$\mathbf{F} = \mathrm{grad}\,(\mathbf{p} \cdot \mathbf{E}).$$

*(*b*) By analogy we should expect a force $\mathbf{F} = \mathrm{grad}\,(J\mathrm{d}\mathbf{a} \cdot \mathbf{B})$ on a small magnetic dipole loop $J\mathrm{d}\mathbf{a}$ located in a *non-uniform* magnetic field **B**. Prove this result by noting that $\mathbf{F} = J \oint \mathrm{d}\mathbf{r} \times \mathbf{B}$ so that if **I** is an arbitrary unit vector, $\mathbf{F} \cdot \mathbf{I} = J \iint \mathrm{curl}\,(\mathbf{B} \times \mathbf{I}) \cdot \mathrm{d}\mathbf{a}$ (by Stokes's theorem) $= J\mathrm{d}\mathbf{a} \cdot \mathrm{curl}\,(\mathbf{B} \times \mathbf{I})$ (since the loop is small) $= \mathbf{I} \cdot \mathrm{grad}\,(J\mathrm{d}\mathbf{a} \cdot \mathbf{B})$ using the facts that div $\mathbf{B} = 0$ and **I** and Jd**a** are constant vectors. Recalculate the force on the ring in problem 4.8 by using this result.

*11.22 A common stratagem in electrotechnology is to use windings etc. to produce a wave-like travelling magnetic field pattern. Note that this travels much slower than an electromagnetic wave and $\epsilon_0 \partial \mathbf{E}/\partial t$ is negligible. Such a 'wave' travelling at constant velocity **b** (components b_x, b_y, b_z) may be written as a function of three variables $\mathbf{B} = \mathbf{B}(u, v, w)$, where $u = x - b_x t$, $v = y - b_y t$, $w = z - b_z t$. (This is a generalisation of the one-dimensional wave travelling purely x-wards.) Deduce that then

$$\partial \mathbf{B}/\partial t + (\mathbf{b} \cdot \mathrm{grad})\mathbf{B} = \mathbf{0},$$

i.e. an observer travelling with the wave velocity sees no change in **B**.

Now consider a travelling loop L (e.g. a loop drawn in the armature of a linear induction motor). Show that $-\mathrm{D}\phi/\mathrm{D}t$, the e.m.f. round L, equals $\oint (\mathbf{v} - \mathbf{b}) \times \mathbf{B} \cdot \mathrm{d}\mathbf{r}$, where **v** is the velocity of each point on the loop. It will be necessary to verify and use the fact that $\mathrm{curl}\,(\mathbf{b} \times \mathbf{B}) = -(\mathbf{b} \cdot \mathrm{grad})\mathbf{B}$ when **b** is uniform and div $\mathbf{B} = 0$.

This result demonstrates the validity and usefulness of the idea of *slip velocity* $(\mathbf{v}-\mathbf{b})$, the velocity of the conductor relative to the magnetic field 'wave'.

11.23 (*a*) For electrically conducting fluid, the equation of motion is

$$\rho\, \mathrm{D}\mathbf{v}/\mathrm{D}t + \operatorname{grad} p = \mathbf{j}\times\mathbf{B}.$$

Show that this equation is satisfied by *any* steady $\mathbf{v}$ and $\mathbf{B}$ distribution such that $\mathbf{B} = (\mu_0\rho)^{1/2}\mathbf{v}$ and $\operatorname{div}\mathbf{B} = 0$, provided ρ is constant and the pressure is given by $p + \frac{1}{2}\rho\mathbf{v}^2 = \text{const}$. The currents affect the magnetic field according to the equation $\mu_0\mathbf{j} = \operatorname{curl}\mathbf{B}$, $\partial\mathbf{D}/\partial t$ and $\mathbf{M}$ being absent. From Ohm's law for a moving conductor show that here $\operatorname{curl}(\tau\mathbf{j})$ must vanish (τ = resistivity). Since $\operatorname{curl}\mathbf{j} \neq \mathbf{0}$ in general this requires that τ be vanishingly small.

*(*b*) An electrically conducting fluid of constant density ρ and negligible resistivity is permeated by a uniform magnetic field B_0 in the x-direction. Show that the equations of fluid motion and electromagnetism, with magnetisation and $\partial\mathbf{D}/\partial t$ neglected, can be satisfied by a transverse velocity field v_y and a transverse magnetic field B_y (additional to B_0) which depend only upon x and t, there being currents and electric fields in the z-direction. The only pressure gradients are in the x-direction. Show that v_y and B_y obey the one-dimensional wave equation, with a wave speed of $B_0/\sqrt{(\mu_0\rho)}$. These waves are known as *Alfvén waves*. Consider their relationship to problem 11.23(*a*).

11.24 A straight metal rod of radius a and uniform resistivity is moving axially with velocity v and carries an axial current of uniform density j which produces a circular magnetic field. To balance the radial e.m.f. $\mathbf{v}\times\mathbf{B}$ there must be a radial electric field whose non-uniformity demands a charge density q in the metal. From the divergence of Ohm's law

$$\tau\mathbf{j} = \mathbf{E} + \mathbf{v}\times\mathbf{B}$$

show that $q = vj/c^2$. Show that the contribution of the convection current qv to j is negligible if (as is normally the case) $v \ll c$. Show also that the radial component of the electric force density $q\mathbf{E}$ is negligible compared with the radial magnetic force density $\mathbf{j}\times\mathbf{B}$.

This result illustrates why electrostatic effects are legitimately neglected in electrotechnological problems generally even though the charge distributions are physically vital for maintaining the prevailing electric fields.

Notice also that, relative to an observer travelling with the rod, q vanishes (a relativistic effect).

11.25 (*a*) If $\mathbf{K}$ is a constant vector and $\mathbf{v}$ is a solenoidal, irrotational vector field with a potential ϕ, show that

$$\operatorname{div}(\phi\mathbf{K}) = \mathbf{K}\cdot\mathbf{v}, \quad \operatorname{curl}(\phi\mathbf{K}) = -\mathbf{K}\times\mathbf{v} \quad \text{and} \quad \nabla^2(\phi\mathbf{K}) = 0.$$

By using the result of problem 11.7(*a*), show that

$$\text{curl}\,(\mathbf{K}\times\mathbf{v}) = -\,\text{grad}\,(\mathbf{K}\cdot\mathbf{v}).$$

*(*b*) Apply (*a*) to the solenoidal, irrotational vector field $\mathbf{v} = \mathbf{r}/r^3$ with $\mathbf{K} = J\mathrm{d}\mathbf{a}/4\pi$, $J\mathrm{d}\mathbf{a}$ being a magnetic loop dipole at the origin whose magnetic *scalar* potential is $\mathbf{K}\cdot\mathbf{v}$. (See problem 4.3.) Deduce that $\mathbf{A} = \mu_0\mathbf{K}\times\mathbf{v}$ is magnetic *vector* potential for the dipole field, noting that div $\mathbf{A} = 0$.

Since any finite steady current loop can be broken into an array of small dipole loops (cf. the proof of Stokes's theorem) its potentials can be found by suitable integrations.

*(*c*) Next apply (*a*) with $\mathbf{v} = \mathbf{r}/r^3$ again but $\mathbf{K} = \mu_0 J\mathrm{d}\mathbf{r}/4\pi$, where $J\mathrm{d}\mathbf{r}$ is an isolated steady Biot–Savart current element at the origin, which must also be an electric dipole of strength $Q\mathrm{d}\mathbf{r}$ that is varying at a constant rate with $J = \mathrm{d}Q/\mathrm{d}t$. As the magnetic field $\mathbf{B} = \mu_0 J\mathrm{d}\mathbf{r}\times\mathbf{r}/4\pi r^3$ is steady, curl $\mathbf{E} = \mathbf{0}$ and there is an electric potential V equal to $Q\mathrm{d}\mathbf{r}\cdot\mathbf{r}/4\pi\epsilon_0 r^3$ such that $\mathbf{E} = -\,\text{grad}\,V$. (See problem 3.10(*d*).) Show from (*a*) that $\epsilon_0\partial\mathbf{E}/\partial t =$ curl $\mathbf{B}/\mu_0$ and hence that Maxwell's equations are satisfied.

This problem shows that it is valid to view a steady Biot–Savart current element as being also a steadily changing dipole, as discussed in section 6.4. But note that if the current J took different values at some time t earlier, then the associated $\mathbf{E}$ and $\mathbf{B}$ fields will be propagating outwards as electromagnetic waves and curl $\mathbf{E}$ will not vanish except within a sphere of radius ct. If J is an alternating current of frequency n, which can only be treated as quasi-steady for periods short compared with $1/n$, then the Biot–Savart law correctly gives the electromagnetic field of the varying dipole within a sphere of radius small compared with c/n. This is called the *near-field of the dipole.* Its *far-field* is given by a solution of the electromagnetic wave equations and is quite different in its properties. Its intensity falls off more slowly with distance than does the near-field. The analysis is beyond the scope of this book.

More generally, this example shows why at frequencies n low enough for c/n to be large compared with the scale of the apparatus, a 'Biot–Savart view', in which magnetic fields are simply proportional to the currents responsible for them, without any delays, is acceptable. On this rests the whole of ordinary low-frequency electrotechnology.

12

Integrals in fluid mechanics

12.1 Line, surface and volume integrals

To complete the course that we set ourselves in figure 3.2 we turn finally to Zone 4 and explore the possibilities of integral formulations in fluid mechanics. Hitherto the full possibilities have only been briefly glimpsed, e.g. in connection with pressure forces on submerged bodies, evaluated by means of the *surface integrals* discussed in chapter 3.

The revelation through Stokes's theorem in chapter 10 that rotationality and circulation are connected suggests strongly that the *line integral*, velocity circulation $\oint \mathbf{v} \cdot d\mathbf{r}$, might have significance in fluid mechanics, however unnatural that may seem. We shall in fact find that a study of this circulation leads to the Kelvin and Helmholtz theorems which are the main subjects of this chapter and which are the most general and profound statements to be made about the essential nature of fluid mechanics.

Volume integrals play their part too, we shall find, particularly in connection with momentum conservation statements. (Their relevance to mass conservation in fluid mechanics was remarked upon in chapter 5.) Circulation integrals and momentum integrals prove to be interrelated in an interesting and useful way in the case of two-dimensional flow.

12.2 Rotation in fluids

It will be recalled that chapter 10 began with an endeavour to break out of the impasse presented by the fact that Euler's equation

$$\rho \mathrm{D}\mathbf{v}/\mathrm{D}t + \operatorname{grad} p = \mathbf{0}$$

contained the further unknown p, the pressure. The endeavour took the form of concentrating on the mean angular velocity of the fluid and led to the invention of the curl operator. We found in particular that under certain circumstances fluid that was not initially rotating (with curl $\mathbf{v} = \mathbf{0}$) continued to be in that state downstream.

We noted also that curl grad $\equiv \mathbf{0}$. It is therefore tempting to eliminate the embarrassing p-term from Euler's equation by merely taking the curl of the whole equation. If ρ is constant, this gives

$$\operatorname{curl}(\mathrm{D}\mathbf{v}/\mathrm{D}t) = \mathbf{0}.$$

This is certainly a kinetic statement which does not involve pressure. Unfortunately it is not equivalent to the simple statement

$$D/Dt\,(\text{curl}\,\mathbf{v}) = D\boldsymbol{\omega}/Dt = \mathbf{0},$$

which is *not* true, in general. There are complications owing to the fact that $D\mathbf{v}/Dt$ is quadratic in $\mathbf{v}$. (Problem 10.7 explores the details.) It is just *not* the case that a fluid particle conserves its mean angular velocity or vorticity ω as it travels along. The way in which a fluid conserves its state of rotation is rather more subtle. It turns out that *circulation* is the key to the matter, as the following example suggests.

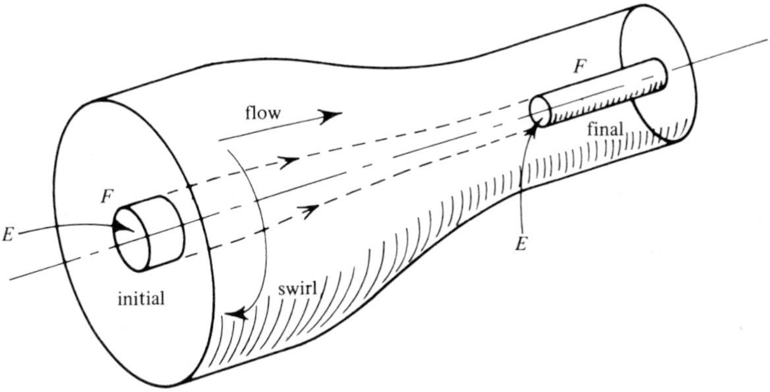

Fig. 12.1. Swirling flow in a converging pipe.

Example 12.1 Figure 12.1 shows a steady inviscid swirling flow along a pipe of contracting cross-section. The small cylindrical fluid element F has a radius r initially. Later it has a radius $r/2$. Show that the circulation round the element is conserved.

As the element is acted on only by normal pressure forces, its angular momentum about the axis is conserved, but its moment of inertia has fallen to a quarter of its first value. (Its radius of gyration is halved.) Its angular velocity and vorticity have therefore quadrupled and its peripheral speed (radius × ang. vel.) has doubled. Hence the *circulation* (peripheral speed × circumference) round the end face E of the element is the *same* in both its initial and final positions. Notice that it is not necessary for the fluid to have constant density here.

Another way of demonstrating the result is to invoke Stokes's theorem and to note that the circulation round E is the product of the *vorticity* (normal to E) and the *area* of E, which have respectively changed by factors of 4 and $\frac{1}{4}$, leaving the circulation unchanged.

12.3 Circulation round a travelling loop

It is obviously necessary to investigate the more general behaviour of the circulation Γ round any loop of travelling fluid particles. We denote by $D\Gamma/Dt$ the rate of change of Γ in time as the loop travels. Figure 12.2 shows a short line element

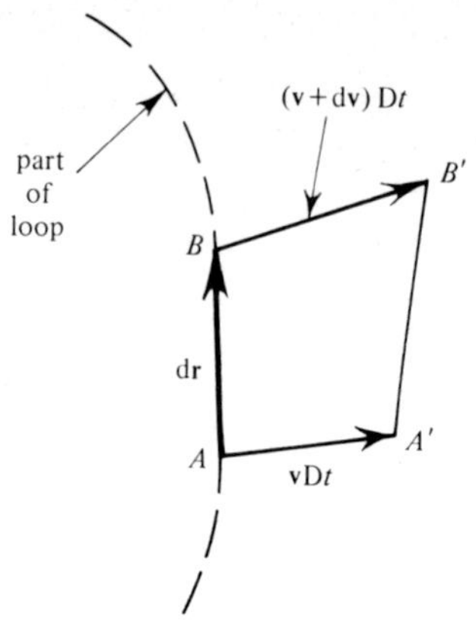

Fig. 12.2. Part of a line of moving fluid particles.

composed of fluid particles, forming part of such a loop, in its positions AB and $A'B'$ respectively before and after some short time interval Dt. The velocities at A and B are $\mathbf{v}$ and $\mathbf{v} + d\mathbf{v}$, respectively, where $d\mathbf{v}$ represents the change in $\mathbf{v}$ consequent upon a change of position $d\mathbf{r}$ or $\overrightarrow{AB}$. The displacement vectors $\overrightarrow{AA'}$ and $\overrightarrow{BB'}$ that occur during Dt are respectively $\mathbf{v}Dt$ and $(\mathbf{v} + d\mathbf{v})\,Dt$, if we ignore higher order small quantities that arise from the change in time of $\mathbf{v}$. The vector $\overrightarrow{A'B'}$ is such that

$$\overrightarrow{A'B'} = \overrightarrow{AB} + \overrightarrow{BB'} - \overrightarrow{AA'} = d\mathbf{r} + d\mathbf{v}Dt.$$

The velocity prevailing at $A'B'$ after the time interval Dt is $\mathbf{v} + D\mathbf{v}$ and so, during Dt, the contribution to the circulation from AB changes from

$$\mathbf{v}\cdot d\mathbf{r} \quad \text{to} \quad (\mathbf{v} + D\mathbf{v})\cdot(d\mathbf{r} + d\mathbf{v}Dt),$$

i.e. it rises by $D\mathbf{v}\cdot d\mathbf{r} + \mathbf{v}\cdot d\mathbf{v}Dt$ + higher order terms, and is rising at a *rate* $(D\mathbf{v}/Dt)\cdot d\mathbf{r} + \mathbf{v}\cdot d\mathbf{v}$. Hence the whole circulation is rising at a rate

$$\frac{D\Gamma}{Dt} = \oint \left\{\frac{D\mathbf{v}}{Dt}\cdot d\mathbf{r} + \mathbf{v}\cdot d\mathbf{v}\right\} = \oint \frac{D\mathbf{v}}{Dt}\cdot d\mathbf{r},$$

the circulation of the *acceleration* field, because $\mathbf{v}\cdot d\mathbf{v}$ integrates to $\frac{1}{2}\mathbf{v}^2$, which does not change round the loop.

If the fluid is acted on by a force field $\mathbf{F}$ per unit volume, which will in general include $-\operatorname{grad} p$,

$$D\mathbf{v}/Dt = \mathbf{F}/\rho \quad \text{and} \quad \frac{D\Gamma}{Dt} = \oint \frac{\mathbf{F}}{\rho}\cdot d\mathbf{r}.$$

The first conclusion from this result is that Γ for a travelling loop does not change if $\mathbf{F}/\rho$, the force field *per unit mass*, has no circulation round that *particular* loop, even though other forces such as viscous forces capable of affecting circulation may act in *other* parts of the flow field, away from that loop. Notice also that gravity *on its own* could never produce changes of circulation, for gravity contributes a force such that $\mathbf{F}/\rho = \mathbf{g}$, which is conservative and has no circulation round a loop.

More generally, the force field may be such that $\oint(\mathbf{F}/\rho)\cdot\mathbf{dr}$ vanishes round *all* loops, i.e. $\mathbf{F}/\rho$ is conservative and curl $(\mathbf{F}/\rho) = \mathbf{0}$. Then the circulation Γ round *any* travelling loop stays constant. In particular when all loops originate in an upstream region devoid of rotation, or when an unsteady flow is initiated from a state devoid of rotation, then $\Gamma = 0$ for all loops both initially and for all subsequent times. If $\Gamma = 0$ for all loops then curl $\mathbf{v} = 0$ everywhere, always. This is a more general demonstration of the possibility of irrotational (potential) flow than could be presented in chapter 10. Potential flow can therefore occur under quite general conditions, such as unsteady or variable density flow, provided that curl $\mathbf{F}/\rho = \mathbf{0}$.

12.4 Kelvin's theorem in constant-density flow

We shall concentrate first on the simpler case where the fluid density does not vary significantly. Then

$$\rho \mathrm{D}\Gamma/\mathrm{D}t = \oint \mathbf{F}\cdot\mathbf{dr}.$$

This vanishes for all loops if curl $\mathbf{F} = \mathbf{0}$, which is true when the only force acting is $\mathbf{F} = -\text{grad } p$, the pressure gradient, for curl grad $\equiv \mathbf{0}$. The presence of gravity adds a term $\rho\mathbf{g}$ to $\mathbf{F}$ and curl $\mathbf{F}$ still vanishes, for $\mathbf{g}$ is irrotational and ρ uniform. Any other irrotational $\mathbf{F}$ is equally acceptable.

This result is *Kelvin's theorem* for a fluid of constant density:

The circulation round any travelling loop of fluid particles does not change when any force field acting on the fluid (in addition to pressure gradient) is irrotational.

In contrast, fluid acted on by viscous forces, which are usually rotational, does not obey Kelvin's theorem. Example 10.8 and problem 10.8 exhibited cases where rotational forces alter a fluid's state of rotation.

Kelvin's theorem (when it applies) confirms that potential flow will occur whenever all fluid elements start from a state devoid of rotation. This confirms the conclusions of section 10.7, which was however confined to steady flow. On the other hand, fluid that is initially rotating will continue to rotate in whatever way allows circulation to be conserved. For example, Kelvin's theorem confirms the rotary behaviour of the swirling flow exhibited in figure 12.1. Problems 12.1 and 12.2 give other exercises on Kelvin's theorem.

These ideas enable us to explain a phenomenon which often puzzles thinking persons blessed with curiosity. Though the phenomenon occurs in all sorts of guises and is of technological importance, the problem may be posed in very homely terms: Why can one *blow* out a candle, yet why cannot one *suck* it out? The question essentially concerns the asymmetry that occurs on the two sides of an orifice through which fluid is passing, as in figure 12.3. On the downstream side there is a jet, separated from the relatively still air outside the jet by edge regions *CD* in which the fluid is rotating as shown by the arrowed circles. The fluid there acquired this rotation as it came along *BC* through viscous action near the wall, and so Kelvin's theorem is not infringed. The reasons as to why the boundary layer *BC* separates from the wall at *C* rather than flowing back up the other side of the wall are too complicated to go into here.

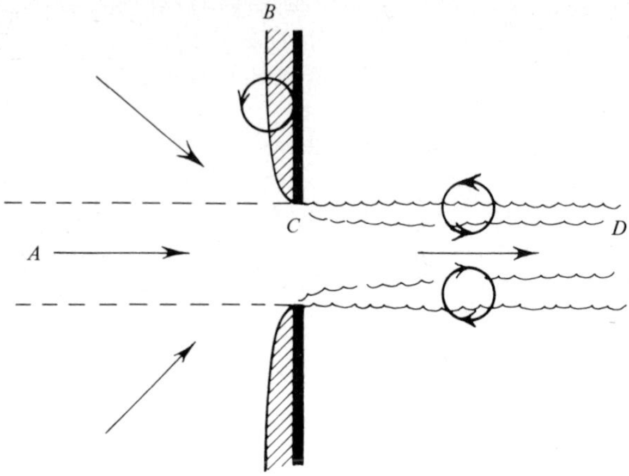

Fig. 12.3. Vorticity in a jet.

A jet cannot occur on the upstream side, however, because this would involve the appearance of rotation on a line such as AC and the approaching fluid there does not suffer significant viscous or other forces such as could generate rotation. The upstream flow must therefore be irrotational (except at the walls) and as a result it approaches the orifice from all directions. Its approach velocity falls off more or less proportionally to $(\text{distance})^{-2}$, i.e. very rapidly. The consequence is that to suck out a candle, one nearly has to swallow the flame!

Another obvious conclusion from Kelvin's theorem is that fluid cannot flow through a vortex sheet, or shear layer of intense vorticity. To do so, fluid particles would suddenly have to acquire – and as suddenly lose – a high level of vorticity.

*12.5 The case of variable density

If we exclude viscous and gravity or other body forces, then $\mathbf{F}/\rho = (-\text{grad}\, p)/\rho$. We shall approach the question as to whether such a force field causes change of circulation round a travelling loop when ρ is variable first by the *differential* approach and then secondly by the *integral* approach, both of which are instructive.

The *differential* approach involves the criterion: does curl $\{(\text{grad}\, p)/\rho\}$ vanish? The curl of a scalar times a vector can be expanded (see section 10.11) as follows:

$$\begin{aligned} \text{curl}\,\{(\text{grad}\, p)/\rho\} &= \rho^{-1}\,\text{curl}\,(\text{grad}\, p) + \text{grad}\,(\rho^{-1}) \times \text{grad}\, p \\ &= (\text{grad}\, p) \times (\text{grad}\, \rho)/\rho^2, \end{aligned}$$

since curl grad $\equiv \mathbf{0}$. For the product grad $p \times$ grad ρ to vanish, in general, grad p and grad ρ must be parallel and the constant p and constant ρ surfaces must coincide, i.e. there is some single relationship $\rho = f(\rho)$ between the variables. (See problem 7.3(*a*).) This then is the condition for $\mathbf{F}/\rho$ to be irrotational and $\mathrm{D}\Gamma/\mathrm{D}t$ to vanish.

It is easy to understand this result physically. Figure 12.4 shows a small sphere of fluid in the two cases where grad p and grad ρ are (*a*) not parallel, and (*b*) parallel.

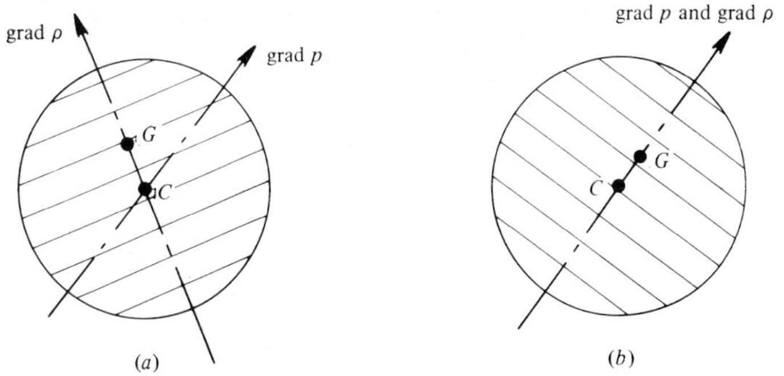

Fig. 12.4. Accelerating spheres of non-uniform fluid.

The sets of parallel lines in the figure represent portions of constant-density surfaces, which are virtually plane if the sphere is small. As regards the mass distribution of the sphere there is symmetry about the diameter parallel to grad ρ. The centre of mass G will therefore lie on this diameter, but will be offset as shown from C, the geometrical centre of the sphere, in the direction of grad ρ, which points to where the mass density is greater.

The force ($-$grad p) per unit volume acting on the sphere produces acceleration in its own direction and acts through C, as do all the individual pressure forces over the sphere.

In case (a) the force exerts a turning moment about G and the fluid element therefore tends to change its state of rotation as it accelerates away. In (b), however, the centre of mass is not offset from the line of acceleration and the turning moment is absent. The crucial role played by the relationship between the pressure and density distributions in determining whether or not fluid elements change their rotation is therefore made clear in convincing physical terms.

The *integral* approach, on the other hand, involves the question whether $\oint(\mathbf{F}/\rho)\cdot d\mathbf{r}$, i.e. $-\oint(\rho^{-1}\,\text{grad}\,p)\cdot d\mathbf{r}$, vanishes or not, round a loop of finite extent. We can rewrite $(\text{grad}\,p)\cdot d\mathbf{r}$ as dp, the change in pressure along $d\mathbf{r}$ and then the integral becomes $-\oint dp/\rho$, a form which should be familiar to the reader from thermodynamics, where $1/\rho$ = specific volume.

A circuit round each loop in the fluid at a particular instant reveals variations in p and ρ which can be plotted on a pressure/specific volume diagram as in figure 12.5. In general, this plot forms closed loops such as L if p and ρ vary freely as the physical loop is traversed. But it is possible for p and ρ everywhere to be related by some law $\rho = f(p)$ in which case the plot becomes a curve such as C. But notice that $\pm\oint dp/\rho$ is equal to the area of the loop on the pressure/volume diagram and vanishes for all such loops only in the case where the loops collapse on to a single curve $\rho = f(p)$. In this case the positive and negative contributions to $\oint dp/\rho$ exactly cancel each other. As a result $D\Gamma/Dt$ vanishes when $\rho = f(p)$ but not if each value of p can correspond to a range of different values of ρ and vice versa. The conclusions are therefore entirely consistent with those of the differential approach.

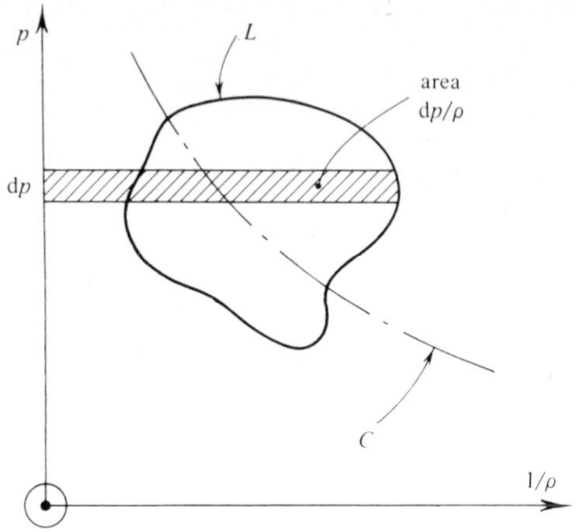

Fig. 12.5

The form of the pressure–density relation depends on thermodynamics. For example the fluid may have started uniform and then undergone only reversible and adiabatic processes, in which case it will be in a state of uniform entropy, with the result that there is a unique p, ρ relationship. (In the case of a perfect gas this would take the form $p/\rho^{\gamma} =$ const. with $\gamma =$ specific heat ratio.) Then $\mathrm{D}\Gamma/\mathrm{D}t$ would vanish. If instead the gas had gone through a non-uniform shock wave, its entropy would be non-uniform and the circulation round a travelling loop could be changing. (See also problem 10.6.)

Another case is that where the fluid has attained a state of uniform temperature. For a perfect gas p/ρ is then constant.

The conclusions of this section are not essentially changed if there is a conservative force field (per unit mass) acting as well as a pressure gradient. The gravity field $\mathbf{g}$ is the most familiar example. With $\mathbf{g}$ present, $\mathbf{F}/\rho$ becomes $(-\mathrm{grad}\ p)/\rho + \mathbf{g}$ and its curl is unchanged by the irrotational $\mathbf{g}$ term. The argument via the differential approach is therefore unaffected, as is the discussion of the behaviour of the small spheres in figure 12.4; the gravity force acts through G and does not alter the turning moment.

Under the integral approach, $\oint (\mathbf{F}/\rho) \cdot \mathrm{d}\mathbf{r}$ gains a term $\oint \mathbf{g} \cdot \mathrm{d}\mathbf{r}$ which vanishes because $\mathbf{g}$ is conservative. It is instructive to consider the interplay of pressure and gravity in the familiar case of natural convection. Figure 12.6 shows a typical example: a 'radiator' and a cold window cause temperature inequalities across the room. Round a loop such as L, $\oint \mathrm{d}p/\rho$ does not vanish and so circulation (using the term in both its technical and layman's senses) is generated progressively as shown by the arrows until limited by viscous or turbulent drag or inability of the heat transfer processes to keep pace. Problem 12.3 provides scope for further thought in this area. Notice that the lay usage of 'circulation' would refer to a closed particle path whereas the technical use refers to *any* loop, not in general a particle path or streamline.

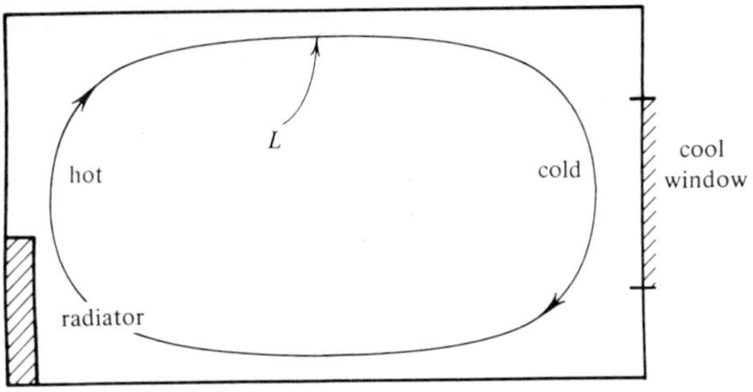

Fig. 12.6. Natural convection.

We can sum up the conclusions of this section by restating *Kelvin's theorem* for the case where density varies:

The circulation round any travelling loop of fluid particles does not change when: (*a*) all points in the fluid obey a single pressure/density relation (not forming a loop on a diagram of pressure against density or specific volume) and (*b*) any other force field (per unit *mass*) acting on the fluid is irrotational.

12.6 Helmholtz's theorem

Kelvin's theorem followed from a consideration of the circulation round a loop defined by a closed line of travelling fluid particles, identified perhaps by a line of dye, smoke or bubbles. Helmholtz's theorem relates to a travelling line of fluid particles in the case where initially they all lie on the same vorticity line (which may be a line vortex – see section 10.4). It is assumed that the conditions which make Kelvin's theorem valid are prevailing.

Let PQ be an element of the line of travelling fluid particles which at some instant t_0 coincides with a vorticity line. With a view to *reductio ad absurdum*, let us postulate that at some later time PQ is *not* part of a vorticity line, i.e. PQ is inclined to the local vorticity. It will then be possible to choose a small area element which contains PQ and through which there is a non-zero vorticity flux. By Stokes's theorem, the circulation $d\Gamma$ round the element is therefore non-zero.

The area element consists of a set of fluid particles. Consider the area element which these particles defined at the earlier instant t_0. By Kelvin's theorem, the circulation round this element is also $d\Gamma$, which is non-zero. But the element contains the particles of PQ, which is part of a vorticity line at instant t_0, and so the vorticity flux through this element and therefore the circulation round this element is zero. This clear conflict means that our postulate is inconsistent with Kelvin's theorem. The only possible conclusion is *Helmholtz's theorem*: under conditions where Kelvin's theorem holds, the fluid particles which initially lie on a vorticity line continue to do so subsequently.

This important and illuminating result means that, instead of vorticity lines being merely instantaneous features of the velocity field, they can be regarded as having a continuing existence from one instant to the next, moving with the fluid as would dye-lines, provided of course that Kelvin's theorem conditions prevail. Kelvin's and Helmholtz's theorems together provide powerful aids to the qualitative understanding of fluid motions in which the influence of viscosity etc. is not too important, even in the many cases which are beyond the reach of detailed mathematical analysis.

Helmholtz's theorem can be regarded as establishing the fact of *vorticity convection*, which is explored further in problem 12.4, where it is revealed that vorticity, being a *vector*, obeys a slightly more subtle convection law than does a convected *scalar* quantity, such as temperature (and that magnetic field can behave similarly sometimes).

12.7 Vorticity convection

It is possible now to understand when and when not the vorticity of a fluid particle will be conserved as it moves along. The situations where it *is* conserved are most readily identified by contrast with those situations where it is *not* conserved. There are basically two mechanisms whereby vorticity conservation is disrupted:

(*a*) stretching of vorticity-lines,

(*b*) tilting of vorticity-lines.

It is now legitimate, in the wake of Helmholtz's theorem, to talk thus of events happening in time to vorticity lines as they travel with the fluid.

The phenomenon of vorticity-stretching is most simply seen in the case of incompressible flow. Figure 12.1 revealed a simple case of it, for it is only another view of the angular momentum conservation phenomenon discussed in section 12.2. In figure 12.1, the symmetry is such that the centre-line is a vorticity line. There the portion of that line which lies within the cylindrical fluid element is stretched as it moves from its initial to its final position at the same time that the vorticity becomes more intense. If the fluid is incompressible so that the element's volume is constant, both vorticity and length rise by the same factor (four). It is also clear that compressibility would complicate the relation between vorticity-enhancement and line-stretching.

Figure 12.7 shows a more general situation, with curl **v** inclined at an arbitrary angle to **v**. The short segment B of a field tube of the *vorticity* field deforms and stretches as it is imagined convected to a later position C. In the incompressible case, the increase in length is accompanied by a decrease in the area A of the cross-section, round which the circulation is preserved (by Kelvin's theorem). Thus the vorticity flux along the field tube is preserved as it moves (by Stokes's theorem) and as A decreases, the vorticity intensity rises, just as in the simpler example. Conversely the velocity field may be such as would shorten vorticity lines, in which case the vorticity level will drop. The crucial consideration is clearly the question as to how the velocity component in the direction of vorticity changes as one goes along a vorticity-line at an instant. Problems 12.5 and 12.6 explore this question. Vorticity-stretching is the key mechanism in maintaining the turbulence which is such a common feature of fluid motion in practice. The example below shows how vorticity *shortening* may be exploited.

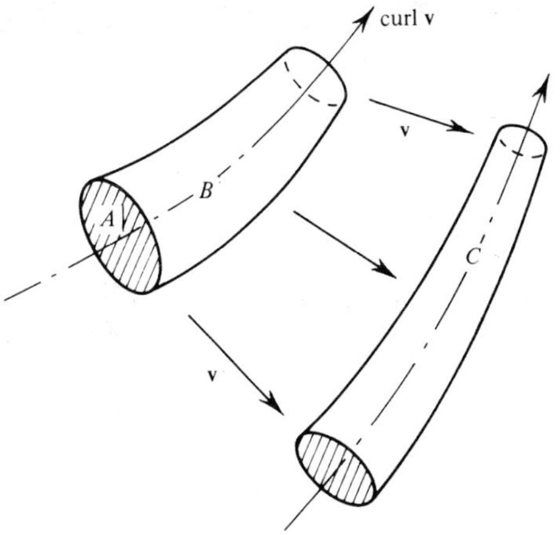

Fig. 12.7. Part of a moving vorticity tube.

Example 12.2 Incompressible fluid is flowing along a duct of square cross-section of side a, with a rectilinear velocity profile $v_1 = Ae^{y_1/a}$, where y_1 is perpendicular distance from one side of the cross-section. The duct then contracts smoothly to a square cross-section of side $a/2$. Estimate the velocity profile v_2 downstream when rectilinear flow has been re-established. (Ignore viscosity, turbulence, boundary layers, etc.)

Consider the streamlines which start upstream at the level y_1 and finish downstream at the level y_2, measured from the side of the duct downstream. The vorticity lines are transverse at each point and the vorticities are related by the relation

$$2\mathrm{d}v_2/\mathrm{d}y_2 = \mathrm{d}v_1/\mathrm{d}y_1 = (A/a)\mathrm{e}^{y_1/a},$$

because each vorticity line has shortened in the ratio 2 : 1. Also the total volumetric flow between 0 and y_1 equals the total flow between 0 and y_2, i.e.

$$\frac{a}{2}\int_0^{y_2} v_2 \mathrm{d}y_2 = a\int_0^{y_1} v_1 \mathrm{d}y_1 = a^2(A\mathrm{e}^{y_1/a} - A) = a^2\left(2a\frac{\mathrm{d}v_2}{\mathrm{d}y_2} - A\right).$$

Differentiating gives

$$\tfrac{1}{2}v_2 = 2a^2\mathrm{d}^2v_2/\mathrm{d}y_2^2,$$

which has solutions

$$v_2 = B\mathrm{e}^{y_2/2a} + C\mathrm{e}^{-y_2/2a}.$$

The vorticity condition requires that

$$B\mathrm{e}^{y_2/2a} - C\mathrm{e}^{-y_2/2a} = A\mathrm{e}^{y_1/a}.$$

Putting $y_1 = y_2 = 0$ gives $B - C = A$ while putting $y_1 = a, y_2 = \frac{1}{2}a$ gives $B\mathrm{e}^{1/4} - C\mathrm{e}^{-1/4} = A\mathrm{e}$. Hence

$$v_2 = \frac{A}{e^{1/2}-1}\{(e^{5/4}-1)e^{y_2/2a} + (e^{5/4}-e^{1/2})e^{-y_2/2a}\}.$$

The range of variation of velocity has dropped from 1:2.718 upstream to 1:1.069 downstream. This result reveals why wind-tunnel contractions are used to produce uniform streams. Problem 12.7 takes the matter further.

Figure 12.7 also shows the other aspect of convection: vorticity-line tilting. In general the velocity field will be causing each element of a vorticity line to be changing its orientation as it is carried along. For this the crucial question is how the velocity components transverse to the vorticity line vary as one goes along the vorticity line, i.e. does the line itself have an angular velocity? Problem 12.5 takes up this question also. (See also problem 12.7.)

12.8 Secondary flow

An important example of this tilting mechanism is provided by the phenomenon of *secondary flow*. Figure 12.8 shows how it occurs, for instance, in a bend in a pipe. The fluid at entry will normally be travelling faster at A, the centre of the pipe, than at B, nearer the wall. In consequence, in the fluid between A and B there will be transverse vorticity curl **v** as shown. (In other parts of the pipe curl **v** will still be transverse but oriented differently.)

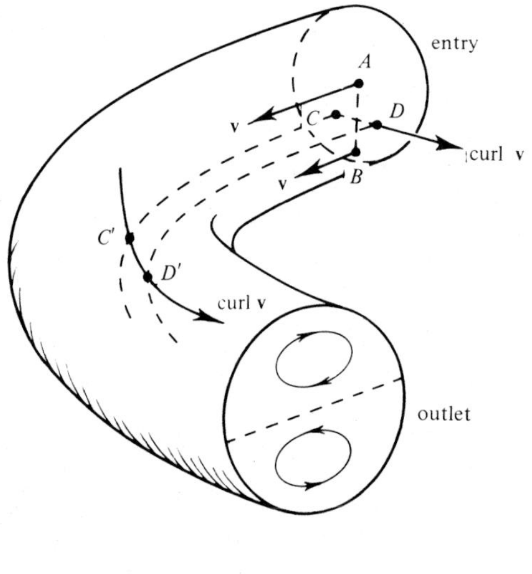

Fig. 12.8. Flow in a pipe-bend.

As the vorticity line element CD is convected round the bend to $C'D'$, say, it will cease to be purely transverse, as the figure shows. Nor will it stay parallel to its original state. This vorticity-line tilting occurs for two reasons: the route round the bend on the inner arc DD' is shorter; and in the bend D travels faster than C because centrifugal effects make the pressure lower at D' than at C' and the fluid on DD' speeds up in comparison with that on CC' according to Bernoulli's equation. The key point to note is that the vorticity-line element $C'D'$ now has a component along the pipe and the fluid must be swirling helically, complicating the motion along CC' and DD' still further. The opposite effect occurs in the top half of the pipe, with the result that the flow at outlet contains the two opposite swirling motions shown in the figure. Note that the circulation shown in the lower half at outlet is consistent with the streamwise vorticity component according to the right-hand screw convention. Vorticity oriented in the main streaming direction is called *secondary vorticity*, and the associated flow which has to be superposed on the primary stream is called *secondary flow*. The phenomenon is highly non-linear and so simple superposition is not accurate except when the secondary effects are very weak. The mathematical difficulties of three-dimensional rotational flow are such that the treatment given in this book must be confined merely to a few simple qualitative ideas, but problem 12.8 involves a mathematical result.

In the case of flow round the bend it is possible to take an alternative view: the fluid traversing the bend can be treated as though exposed to a centrifugal 'force' field which acts more strongly on faster fluid near the centre of the pipe than fluid elsewhere. The swirling motion at outlet then results. Of course if the bend is long enough the secondary flow carries this faster fluid away from the centre and the picture becomes more complicated.

In pipe bends this phenomenon is responsible for the high dissipation or 'hydraulic losses' which occur, because the transfer of high-speed fluid nearer to the walls greatly enhances wall friction effects. It is also important in problems of river-bed erosion, e.g. on meanders or near bridge piers, whenever a flow containing vorticity is appropriately diverted, and in flow through turbo-machinery.

Helmholtz's and Kelvin's theorems apply exactly only in the absence of viscous and other effects. Their influence can still be very strong however, even in the presence of these other forces, although the pure vorticity-convection phenomena are then modified or diluted by vorticity diffusion due to viscosity. (Problem 12.9 provides food for thought.) The viscous modifications are relatively weak whenever there is a high value for the Reynolds numbers, which as usual measures the relative importance of inertia effects and viscous effects.

12.9 Two-dimensional, constant-density flows

The convection of vorticity becomes much simpler in two-dimensional flows where the fluid moves only in x, y planes, say, there being no variation in the z-direction. In such motions the vorticity is in the z-direction and the vorticity lines always have the same length and orientation, i.e. the stretching and tilting mechanisms are inactive. The result is that each fluid particle travels with constant vorticity. This makes rotational flow in two-dimensions rather more tractable, although it is still in general a non-linear problem, markedly more difficult to solve than irrotational (potential) flow.

It is easy to see direct from Kelvin's theorem why fluid particles conserve their vorticity in this case. Consider a small travelling area element in an x, y plane, regarded as the cross-section of a travelling prism of fluid of unit length in the z-direction. As the prism's volume is conserved (if the flow is incompressible) and its length is constant, the area element retains a constant magnitude. Kelvin's theorem requires that the circulation round it is preserved and so, by Stokes's theorem, the vorticity intensity through it is also constant. (It is also equally clear that compressibility could allow the vorticity to change.)

The kinetic statement that determines such motions, with the unknown pressure eliminated, is therefore

$$\mathrm{D}/\mathrm{D}t\,(\mathrm{curl}\,\mathbf{v}) = \mathbf{0} = \{\partial/\partial t + (\mathbf{v}\cdot\mathrm{grad})\}\,\mathrm{curl}\,\mathbf{v}. \qquad (\alpha)$$

It will be noted that this is non-linear, i.e. quadratic, in terms of $\mathbf{v}$.

Although the velocity potential ϕ does not exist when curl $\mathbf{v} \neq \mathbf{0}$, we may make use of the stream function ψ in two-dimensional incompressible flows with div $\mathbf{v} = 0$, even when they are unsteady. Then $v_x = \partial\psi/\partial y$, $v_y = -\partial\psi/\partial x$ as usual and

$$|\,\mathrm{curl}\,\mathbf{v}\,| = \frac{\partial v_y}{\partial x} - \frac{\partial v_x}{\partial y} = -\nabla^2\psi.$$

The equation (α) can clearly be turned into a non-linear equation for ψ. However, some problems permit a simpler approach. For example it is possible for the vorticity to be uniform at all points and all times, for this certainly satisfies (α). Then

$$\nabla^2\psi = \text{const.},$$

a simple case of Poisson's equation, discussed in section 9.2. The problem has become linear again. Problems 12.10 and 12.11 explore examples, and another follows.

***Example 12.3** A cubical tank of side $2a$ containing fluid of low viscosity is rotated at an angular velocity Ω about an axis parallel to an edge through its centre until viscosity has caused the fluid too to be rotating uniformly at Ω. The tank is then suddenly stopped. Find the subsequent motion, before viscosity has had time to act.

By Kelvin's theorem the uniform vorticity (2Ω) is instantaneously unchanged, as the whole process is two-dimensional. In the new steady state a stream function ψ can be used, with $\nabla^2\psi = -2\Omega$, a Poisson equation. ∇^2 is $\partial^2/\partial x^2 + \partial^2/\partial y^2$, if we take x, y axes normal to the axis of rotation. Let the walls be $x = \pm a, y = \pm a$, at which $\psi = \text{const.} = 0$, say, for the boundary is a streamline. The problem is now virtually identical to example 9.2, from which it follows that the solution here is

$$\psi = \frac{32a^2\Omega}{\pi^3}\sum\frac{(-1)^{(n-1)/2}}{n^3}\cos\frac{n\pi y}{2a}\left\{1 - \frac{\cosh(n\pi x/2a)}{\cosh(n\pi/2)}\right\}, \quad (n \text{ odd}),$$

which is a specification of the motion.

A more general problem concerns steady flows, in which the streamlines ψ = const. are also particle paths, along which the vorticity $-\nabla^2\psi$ must therefore be constant. The equation $\nabla^2\psi = f(\psi)$ summarises the situation. If $f(\psi)$ is proportional to ψ the equation is again linear and of the Helmholtz type discussed in section 9.3. Problem 12.12 investigates some simple cases and reveals how unlike potential flow these rotational flows can be.

12.10 The momentum flux integral

So far this chapter has concentrated on the important properties of the *line integral* (circulation) in fluid mechanics. We turn now to a study of *volume* and *surface integrals* in relation to momentum statements. The reader has probably met these already, at least through simple examples in relation to the force on a pipe bend, the theory of the Pelton wheel, the thrust of a rocket, etc. Many phenomena and devices depend on the ability of changes in fluid momentum to generate pressure forces, reacting on a system in a useful way. The bee and the helicopter hover by accelerating air downwards.

If we confine attention to steady flow, it is possible to relate the pressure and other forces acting on the contents of a control volume (a fixed region of space through which fluid is passing) to the net flux of fluid momentum through the control surface, which encloses the volume. In the analysis below we shall ignore other forces such as the weight of the fluid but these are easily added. (See problem 12.16.)

The momentum flux concept is a *vector* generalisation of the *scalar* fluxes that have figured frequently on earlier pages. Through an element $\mathbf{da}$ of the surface, the mass flow rate outwards is the scalar $\rho\mathbf{v}\cdot\mathbf{da}$ and we define the associated momentum flux as $(\rho\mathbf{v}\cdot\mathbf{da})\mathbf{v}$, a vector. The net momentum flux out of the whole surface is therefore

$$\oint\!\!\oint (\rho\mathbf{v}\cdot\mathbf{da})\,\mathbf{v}.$$

Consider the x-component of this vector, $\oint\!\!\oint (\rho\mathbf{v}\cdot\mathbf{da})\,v_x$, which can be rewritten as $\oint\!\!\oint v_x\rho\mathbf{v}\cdot\mathbf{da}$ because v_x is a scalar. Gauss's theorem transforms this into the volume integral

$$\iiint \operatorname{div}(v_x\rho\mathbf{v})\,\mathrm{d}\tau = \iiint \rho(\mathbf{v}\cdot\operatorname{grad})\,v_x\,\mathrm{d}\tau,$$

for

$$\operatorname{div}(v_x\rho\mathbf{v}) = v_x \operatorname{div}\rho\mathbf{v} + \rho\mathbf{v}\cdot\operatorname{grad} v_x, \quad \text{and} \quad \operatorname{div}\rho\mathbf{v} = 0,$$

the equation of mass conservation in steady flow. The y and z components of the momentum flux can be treated similarly. Putting them all together gives the vector statement

$$\oint\!\!\oint (\rho\mathbf{v}\cdot\mathbf{da})\,\mathbf{v} = \iiint \rho(\mathbf{v}\cdot\operatorname{grad})\,\mathbf{v}\,\mathrm{d}\tau.$$

(Problem 12.13 studies a comparable result in electromagnetism.) This relation states that the net momentum flux equals the integral of the masses $\rho\,\mathrm{d}\tau$ of the elements contained in the volume, each multiplied by its steady-flow acceleration $(\mathbf{v}\cdot\text{grad})\mathbf{v}$. By Newton's law, this integral is obviously equal to the total force acting on the contents of the control surface. In particular, if we insert the fact that $\rho(\mathbf{v}\cdot\text{grad})\mathbf{v} = -\text{grad}\ p$, in the absence of other forces, the volume integral becomes $-\iiint \text{grad}\ p\,d\tau$, which is equal to the surface integral $-\oiint p \cdot d\mathbf{a}$, by Gauss's corollary, the net pressure force exerted on the fluid inside the control surface by the fluid outside, or by adjoining solid suraces. We thus have a formal relationship between pressure force and momentum flux at the control surface:

$$\oiint (\rho\mathbf{v}\cdot d\mathbf{a})\mathbf{v} + \oiint p\,d\mathbf{a} = 0.$$

This vector equation is in general category, remarked upon in section 5.3, where a statement about the contents of a closed surface becomes expressible in terms purely of conditions at the surface. This is particularly attractive because it involves both less effort and less information than would evaluation of the corresponding volume integrals. Notice however that the inertia effects inside the volume can be represented by a surface integral only in the case of steady flow. Problem 12.14 explores an alternative proof, while problem 12.15 concerns some other surface integrals relevant only to potential flow. Some simple examples of the momentum integral follow.

Example 12.4 A plane normal shock wave is a thin, stationary zone, a few mean-free-paths thick, in which high speed gas suddenly slows down with a violent pressure and density change. Show that $p + \rho v^2 = \text{const.}$ across the wave.

In the wave the fluid changes from a uniform state characterised by properties with the suffix 1, to another uniform state, with the suffix 2. Consider a control surface, straddling the wave and with unit cross-section as seen in the flow direction. The streamwise components of $\oiint (\rho\mathbf{v}\cdot d\mathbf{a})\mathbf{v}$ and $\oiint p\,d\mathbf{a}$ are respectively

$$-\rho_1 v_1^2 + \rho_2 v_2^2 \quad \text{and} \quad -p_1 + p_2$$

if we observe the convention that $d\mathbf{a}$ points outwards from the control surface. Hence

$$p_1 + \rho_1 v_1^2 = p_2 + \rho_2 v_2^2,$$

one of the Rankine–Hugoniot relations for a shock wave. In passing one may note the contrast with the result $p + \frac{1}{2}\rho v^2 = \text{const.}$, which is Bernoulli's equation for situations where the density does *not* change and where the dissipative processes that occur in a shock wave are absent.

Example 12.5 Incompressible fluid flows normally through a stationary wire gauze or perforated sheet in which the holes occupy a fraction $1/R$ of the area. Estimate the loss coefficient, defined as $\Delta p/(\frac{1}{2}\rho v^2)$, where v = velocity of stream, ρ = density, and Δp = overall pressure loss.

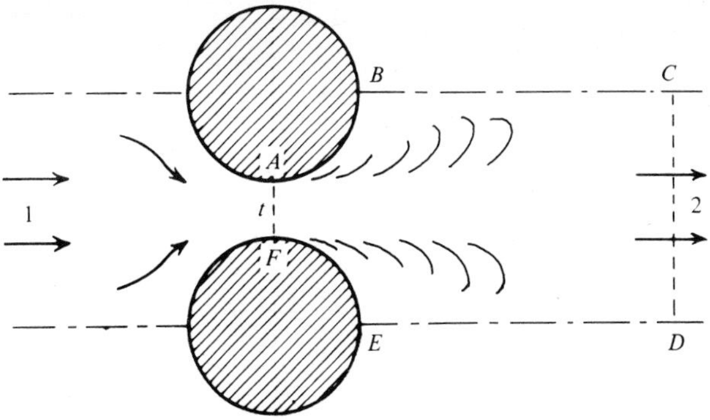

Fig. 12.9. Streamtube passing through one orifice in a gauze or perforated plate.

Figure 12.9 shows the streamtube (of upstream cross-section A) associated with one orifice. Let the suffices 1, t and 2 refer to upstream, the throat and downstream (where the stream has become virtually uniform again). If we treat conditions as uniform over the throat, we have

$$v = v_1 = v_2 = v_t/R,$$

the density being constant. Bernoulli's equation, applied from 1 to t, gives

$$p_1 - p_t = \tfrac{1}{2}\rho(v_t^2 - v_1^2) = \tfrac{1}{2}\rho v^2(R^2 - 1).$$

Bernoulli's equation is not applicable from t to 2 because eddying flow would occur. To make progress we assume that pressure variations are negligible in the relatively stagnant regions near AB and FE, for the jet breaks away at A and F (compare figure 12.3), i.e. we assume that p_t prevails all along $BAFE$. Applying the streamwise momentum flux integral to the control surface $ABCDEF$ then gives

$$(p_t - p_2)A = \rho v A(v_2 - v_t) = \rho v^2 A(1 - R).$$

Eliminating p_t, we find that $\Delta p/(\tfrac{1}{2}\rho v^2) = (R-1)^2$.

Example 12.6 Figure 12.10 shows two-dimensional flow through a blade row (e.g. in a turbine). Upstream and downstream the flow is uniform, with the velocity components shown in the figure. Neglecting viscosity and compressibility, show that the resultant force R on each blade per unit length is perpendicular to $\mathbf{v}_m$, the mean of the upstream and downstream velocities, and equals $\rho v_m \Gamma$, in which v_m is the magnitude of $\mathbf{v}_m$ and Γ is the circulation round a blade.

As the density is constant, mass conservation requires that $u_1 = u_2 = u$, say. Let the pressures upstream and downstream be p_1 and p_2 respectively. Consider a control surface $ABCD$ between the streamlines AB and CD, of unit extent perpendicular to the paper. If AD and BC both equal the blade spacing b, AB and CD have the same shape, pressure and velocity distribution. If R is inclined at an angle α to the row, the two components of the momentum equation are:

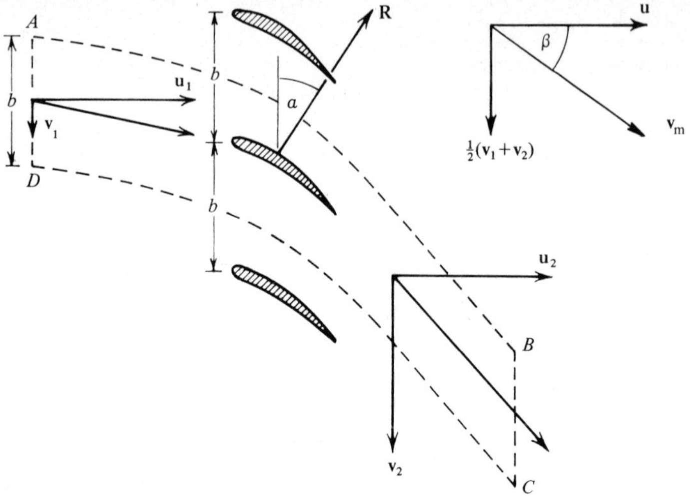

Fig. 12.10. Flow through a blade-row.

$$(p_1-p_2)\,b-R\sin\alpha = ub\rho(u_2-u_1) = 0$$

and

$$R\cos\alpha = ub\rho(v_2-v_1).$$

Note that the force on the *fluid* is the reverse of R and that the pressure forces on AB and CD cancel mutually. Bernoulli's equation gives

$$p_1-p_2 = \tfrac{1}{2}\rho(u_2^2+v_2^2-u_1^2-v_1^2) = \tfrac{1}{2}\rho(v_2^2-v_1^2).$$

Hence

$$R\sin\alpha = \tfrac{1}{2}(v_2+v_1)\,b\rho(v_2-v_1) \quad \text{and} \quad \tan\alpha = \tfrac{1}{2}(v_2+v_1)/u.$$

But the mean velocity $\mathbf{v}_m$ has components u and $\frac{1}{2}(v_2+v_1)$ and is inclined to the u-direction at an angle β such that $\tan\beta = \frac{1}{2}(v_2+v_1)/u$. Evidently $\alpha=\beta$ and R is perpendicular to v_m. Eliminating α gives

$$R = b\rho(v_2-v_1)\sqrt{(u^2+\tfrac{1}{4}(v_2-v_1)^2)} = b\rho(v_2-v_1)\,v_m = \rho v_m\Gamma,$$

because the circulation round the loop $ABCD$ is $b(v_2-v_1)$, for the contributions from AB and CD are equal and opposite.

Problems 12.16 and 12.17 give further exercises on the use of the momentum equation. Another important application follows.

12.11 Circulation in irrotational flow

The circulation and momentum integrals come together rather unexpectedly in the case of two-dimensional irrotational steady flow past an immersed body. This is a case of great interest in aviation for it is the first step towards understanding the behaviour of lifting aerofoils (the wings of aircraft).

The first point that needs explaining is how an irrotational flow can contain

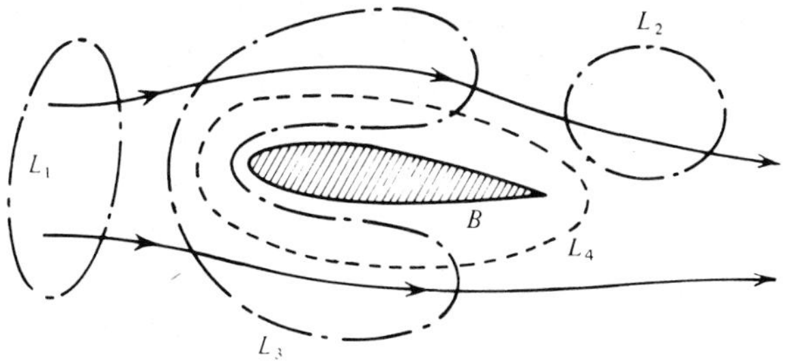

Fig. 12.11

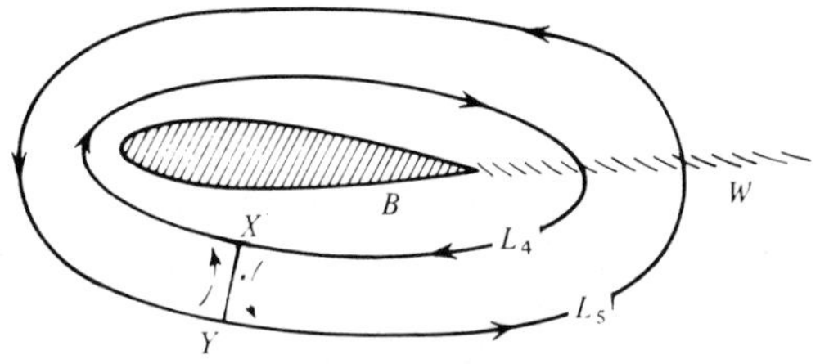

Fig. 12.12

any circulation, in view of Stokes's theorem. Figure 12.11 shows a two-dimensional steady flow from an upstream region of uniform (and so irrotational) flow, known as the *free stream*, past an immersed body B. An upstream loop, such as L_1, devoid of circulation, can be convected into shapes such as L_2 or L_3, round which there can never be circulation, in view of Kelvin's theorem. However, convection cannot turn L_1 into a loop such as L_4, which encloses the body. Kelvin's theorem therefore does not appear to preclude circulation round such a loop, particularly when one reflects that L_4 encloses not only irrotational fluid and the solid body B (which being at rest contains no vorticity) but also the boundary layers on the body in which the vorticity will be intense and will probably not add up to zero total vorticity flux. The mainly clockwise vorticity in the boundary layer above the body may not exactly cancel the mainly anticlockwise vorticity in the lower boundary layer. The boundary layers may be modelled mathematically as vortex sheets. So non-zero circulation round L_4 need not infringe Stokes's theorem.

It is easy to show that the circulations round any two loops L_4 and L_5, both of which enclose the body and which lie entirely in the irrotational flow outside the boundary layer, are identical. In figure 12.12 the two loops are bridged by any line XY. Applying Stokes's theorem to the compound loop that consists of L_4, L_5 and XY traversed twice in opposite directions as shown by the arrows reveals that there is no net circulation round the compound loop, for the shaded area which it encloses carries no vorticity flux. The contributions to the circulation from XY cancel out and we see that:

$$\begin{aligned}\text{clockwise circulation round } L_4 &= -\text{ anticlockwise circulation round } L_5\\ &= \text{clockwise circulation round } L_5.\end{aligned}$$

Thus the clockwise circulation round *any* loop enclosing the body outside the boundary layer takes on a unique value.

The uniqueness of the circulation is explicable in terms of a multi-valued velocity potential, which changes by the same amount each time a circuit is made round the body, as was discussed in sections 4.3 and 10.8, which provides an alternative approach. Problem 10.12(*d*) explores the particular case of potential flow past a circular cylinder with circulation, a rather academic case because in real flows past bluff bodies the potential flow downstream is replaced by an eddying wake containing vorticity.

The fact that there is usually a thin shear layer or streamwise *wake W* behind the body even when it is 'streamlined' does not affect the argument, because the pressures on each side of the wake are equal and Bernoulli's equation (which prevails throughout the irrotational part of the flow) indicates equality of velocity on each side of the wake. The wake is therefore a shear layer or vortex sheet of zero *net* vorticity content. (See problem 10.5, last part.)

We have still to explain how the circulation round the body is generated and its level fixed in the first place. Problem 10.12 appears to suggest that, in the case of a circular cylindrical body at least, a variety of values for the circulation are equally probable in principle. In fact, for an aerofoil section with a sharp trailing edge, the value of the circulation is determined during the starting process.

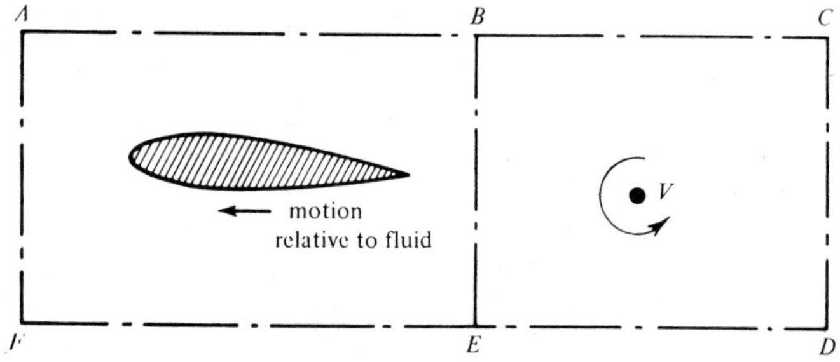

Fig. 12.13

When the relative motion of the aerofoil and the fluid commences, there is no way in which circulation round a loop such as *ACDF* in figure 12.13 can be created, for it is well clear of any viscous forces which may arise near the body. (As was remarked in section 12.3, Kelvin's theorem for a particular loop holds when there are no rotational forces on the loop, irrespective of viscous forces elsewhere.) Nevertheless it is possible for a clockwise circulation to be established round the aerofoil in the potential flow which ensues. If the wing is unsymmetrical about the direction of advance relative to the undisturbed fluid ahead, and if it has a sharp trailing edge, some of the vorticity which is created in the boundary layers on the aerofoil during starting is left behind and may agglomerate in the anticlockwise *starting vortex V*. The strength of *V* (measured by the unique circulation round it) and the circulation established round the aerofoil are equal and opposite. In

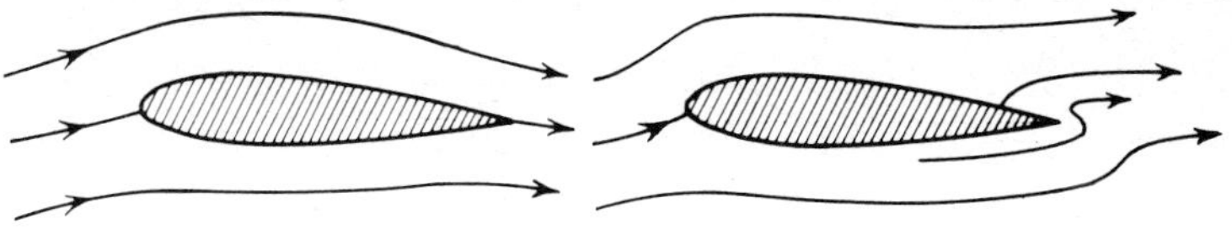

Fig. 12.14

particular the circulations round *ABEF* and *BCDE* are equal and opposite so that the circulation round *ACDF* can indeed remain zero. (There can be circulation round *ABEF* because *BE* was interrupted by the aerofoil for a period during starting.)

The existence of the starting vortex is easily demonstrated with a teaspoon in a teacup, for the surface dimple reveals the vortex.

The sharp trailing edge is crucial. By complex processes in the boundary layer it fixes the circulation round the aerofoil by selecting the flow shown in (*a*) in figure 12.14 rather than one such as (*b*) which would involve unrealistic fluid behaviour at the trailing edge. That the streamlines leave the trailing edge smoothly as shown in (*a*) is known as the *Zhukovskii condition*.

The presence of clockwise circulation means that the fluid (outside the boundary layer) passes over the top of the aerofoil at higher speed on the whole than it does beneath the aerofoil. As Bernoulli's equation applies throughout the irrotational flow, the pressure is therefore generally lower above the aerofoil than below it. Thus circulation implies *lift*, i.e. a force on the aerofoil transverse to the direction of the free stream. Essentially the same processes enable the blades of a turbine, fan or propeller to exert a force on a fluid stream.

12.12 The circulation-lift theorem

We shall now invoke the momentum integral theorem to demonstrate that in fact the lift is simply proportional to the circulation. (The proof for flow past a circular cylinder figured in problem 10.12(*d*).)

Consider two-dimensional, steady, irrotational flow past an aerofoil with circulation Γ, where far from the aerofoil the undisturbed stream has a uniform velocity $\mathbf{U}$ and uniform pressure P. We wish to relate Γ to $\mathbf{F}$, the force on the aerofoil due to fluid pressure, per unit length of aerofoil measured perpendicular to the planes of flow. Consider the momentum integral for the control volume of fluid lying between the aerofoil surface and a distant loop L, having unit extent perpendicular to the flow planes. There is no contribution to the momentum integral from the plane end faces of this control volume and the pressure forces on these faces are mutually cancelling. Figure 12.15 shows an element $\mathrm{d}\mathbf{r}$ of the loop L, which also constitutes the edge view of an area element $\mathrm{d}\mathbf{a}$ of unit length measured perpendicular to the paper. Note that the vectors $\mathrm{d}\mathbf{r}$ and $\mathrm{d}\mathbf{a}$ are perpendicular. As the surface of the control volume includes both the surface of the aerofoil and the surface at L, the

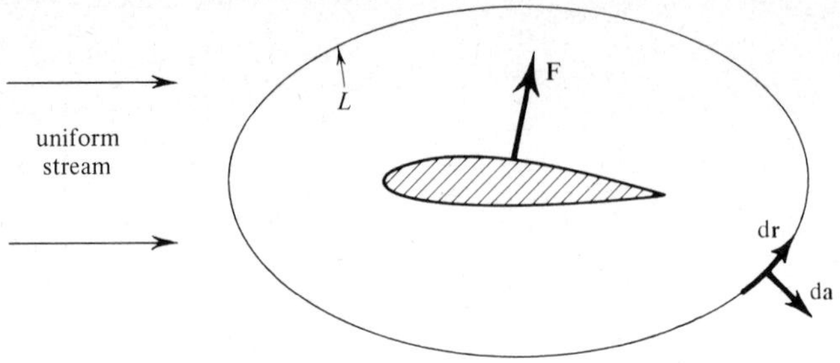

Fig. 12.15

total pressure force on the fluid contents of the control volume consists of the two parts $-\mathbf{F}$ and $-{}_L\oint p\mathbf{da}$ ($+\mathbf{F}$ is the force *on the aerofoil*). The momentum statement for the control volume is therefore

$$\rho \underset{L}{\oint} \mathbf{v}(\mathbf{v}\cdot\mathbf{da}) + \underset{L}{\oint} p\mathbf{da} + \mathbf{F} = \mathbf{0},$$

since ρ is constant. Note that there is no contribution to the momentum flux at the surface of the aerofoil, for there is no mass flux through that part of the control surface. Let us write $\mathbf{v} = \mathbf{U} + \mathbf{v}'$ so that $\mathbf{v}'$ tends to zero far from the body. The quadratic terms in $\mathbf{v}'$ then become negligible in comparison with the first-order terms when we let L become very distant. (See problem 12.19.) Bernoulli's equation states that

$$P + \tfrac{1}{2}\rho\mathbf{U}^2 = p + \tfrac{1}{2}\rho\mathbf{v}^2 = p + \tfrac{1}{2}\rho(\mathbf{U}^2 + 2\mathbf{U}\cdot\mathbf{v}' + \mathbf{v}'^2),$$

or

$$p + \rho\mathbf{U}\cdot\mathbf{v}' = P = \text{const.},$$

if we ignore the second-order term. Then

$$\oint p\mathbf{da} = -\rho\oint(\mathbf{U}\cdot\mathbf{v}')\mathbf{da}, \quad \text{for} \quad \oiint P\mathbf{da} = 0.$$

Meanwhile

$$\mathbf{v}(\mathbf{v}\cdot\mathbf{da}) = \mathbf{U}(\mathbf{v}\cdot\mathbf{da}) + \mathbf{v}'(\mathbf{v}\cdot\mathbf{da}) = \mathbf{U}(\mathbf{v}\cdot\mathbf{da}) + \mathbf{v}'(\mathbf{U}\cdot\mathbf{da})$$

if we ignore a second-order term in $\mathbf{v}'$. But

$$\oint \mathbf{U}(\mathbf{v}\cdot\mathbf{da}) = \mathbf{U}\oint \mathbf{v}\cdot\mathbf{da} = \mathbf{0},$$

since $\operatorname{div}\mathbf{v} = 0$ in incompressible flow. Hence

$$\rho\oint \mathbf{v}(\mathbf{v}\cdot\mathbf{da}) = \rho\oint \mathbf{v}'(\mathbf{U}\cdot\mathbf{da}).$$

The momentum statement can now be rewritten as

$$\mathbf{F} = \rho\oint\{(\mathbf{U}\cdot\mathbf{v}')\mathbf{da} - \mathbf{v}'(\mathbf{U}\cdot\mathbf{da})\} = \rho\oint \mathbf{U}\times(\mathbf{da}\times\mathbf{v}')$$

(by a standard identity in vector algebra)

$$= \rho \mathbf{U} \times \oint d\mathbf{a} \times \mathbf{v}'.$$

The first point to note is that $\mathbf{F}$ is perpendicular to the free-stream $\mathbf{U}$, i.e. potential flow theory predicts a pure lift force (but no drag, since viscous effects have been ignored). The vector $\oint d\mathbf{a} \times \mathbf{v}'$ is perpendicular to the paper and of magnitude $\oint \mathbf{v}' \cdot d\mathbf{r}$ (see section 5.8), which equals $\oint \mathbf{v} \cdot d\mathbf{r}$ or Γ since $\oint \mathbf{v} \cdot d\mathbf{r} - \oint \mathbf{v}' \cdot d\mathbf{r} = \oint \mathbf{U} \cdot d\mathbf{r} = 0$, $\mathbf{U}$ being uniform. We see that the lift force equals $\rho U \Gamma$ or $\rho \mathbf{U} \times \boldsymbol{\Gamma}$, where $\boldsymbol{\Gamma}$ is a vector perpendicular to the paper. (Note the analogy with the $\mathbf{J} \times \mathbf{B}$ force.) This is *Zhukovskii's circulation/lift theorem*, which reveals the two quantities as simply proportional. In practice, provided the flow does not break away into eddies (the phenomenon of *stalling*), the theorem predicts the lift on a two-dimensional aerofoil very accurately. For a small angle of incidence the Zhukovskii condition has the result that the lift and circulation are proportional to this angle, measured from the position of zero lift.

The absence of drag might appear to be paradoxical, particularly when the two-dimensional case is compared with three-dimensional flow past a wing of finite span in the next section. Also the existence of lift, which must be associated with net downwards momentum being created, appears to be at odds with the notion that the velocity tends to $\mathbf{U}$ at great distances in all directions. Consider a two-dimensional aerofoil advancing into still air. Its net effect is to create downwards momentum by giving vanishingly small ultimate velocities (typically v, say) to a virtually unlimited airstream (mass flow rate $\dot{m}$), because its effects, however weak, spread indefinitely through the stream if it is unbounded. This is why the perturbation velocity tends to zero at great distances. At the same time, if there were a finite amount of new kinetic energy created by the passage of the aerofoil, this would imply a drag on the aerofoil, to balance the energetics. However, if we keep the momentum creation rate, of order $\dot{m}v$, finite while letting $\dot{m} \to \infty$, $v \to 0$, the kinetic energy creation rate, of order $\frac{1}{2}\dot{m}v^2$, tends to zero. This explains why there is no drag, once viscous effects are ignored; the force on the aerofoil must be in such a direction that an advancing aerofoil does no work on the fluid. The situation is different in a blade-row in a turbomachine because each blade has access to only a finite stream of fluid. Example 12.6 concerned this case and showed how $\mathbf{U}$ had to be replaced by $\mathbf{v}_m$, the mean velocity.

12.13 Flow past wings of finite span

We close with a qualitative discussion of the more difficult problem of the three-dimensional flow past aeroplane wings, which are inevitably of finite span. The flow becomes three-dimensional simply because the pressures below the wing are mainly higher than the pressures above the wing, while the pressure beyond the wing tip takes intermediate values. The flow past the wings is therefore modified from its plane, two-dimensional form by gaining a velocity component outwards towards the tips under the wing and inwards away from the tips above the wing. Beyond

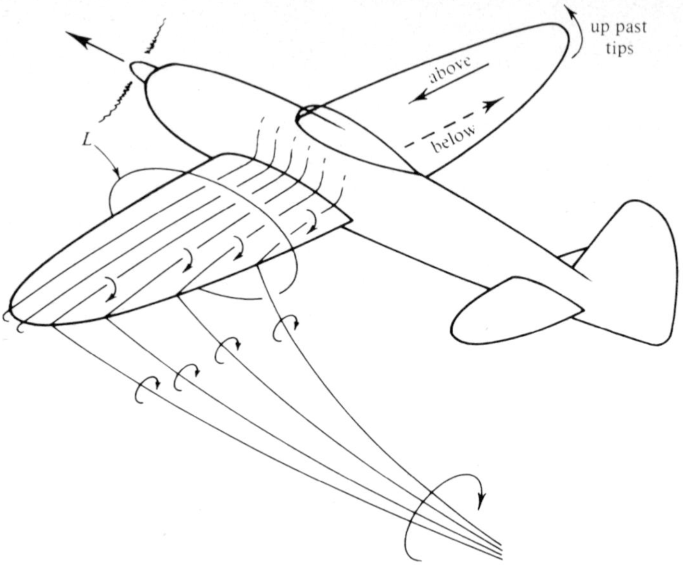

Fig. 12.16. Flow past a lifting wing. Spanwise velocities are shown at the starboard wing, vorticity lines are shown above and behind the port wing.

the tips an upwards flow develops. These transverse velocities are indicated on the right-hand (starboard) wing of the aeroplane shown in figure 12.16.

When the resulting oblique streams above and below the wing rejoin behind the trailing edge of the wing they obviously have incompatible spanwise or transverse velocities. Therefore there is a vortex sheet or shear layer created in the wake of the wing (superposed upon any wake of viscous origin emanating from the boundary layer). In this shear layer the vorticity is oriented mainly in the flight direction, to be consistent with the discontinuity in spanwise velocity. This is described as vorticity *shed* by the wing. Because of the continuity of vorticity lines, with div curl $\mathbf{v} \equiv 0$, these vorticity lines can be traced back into the boundary layers on the wing where the vorticity is mainly spanwise. Figure 12.16 shows schematically the vorticity lines above the left-hand (port) wing. The pattern is different under the wing because the spanwise vorticity there is mainly in the opposite direction. The vorticity shed is most intense near the tips, for there the discontinuity in spanwise velocity behind the wing is obviously greatest. The vortex sheet trails behind because, as was already remarked, there cannot be flow *through* a vortex sheet.

The vortex sheet rapidly wraps itself up as suggested in the figure into a more concentrated line vortex which persists for a long distance behind each wing tip. With very large modern passenger aircraft these vortex wakes present a grave hazard to other aviation until they die out.

Consider the circulation Γ round a loop such as L lying in a plane perpendicular to the spanwise axis of the aeroplane. Because of the shedding of vorticity this circulation Γ falls off as the wing-tip is approached, and the lift per unit span correspondingly falls below the value it would have had under two-dimensional conditions. Essentially some of the 'lifting pressure' is escaping round the wing tips. The higher the span/chord ratio, or *aspect ratio*, of the wing, the less deleterious are these three-dimensional effects.

The vortex motions behind the wing contain a great deal of kinetic energy. When they are created by an aeroplane moving forward through initially still air the necessary energy can only come from work done by a component of force on the aeroplane in its direction of motion, i.e. there must be a drag on the aeroplane (and propulsive thrust to overcome it). This so-called *induced drag* is quite additional to any drag associated with viscous effects. How does this extra drag come about?

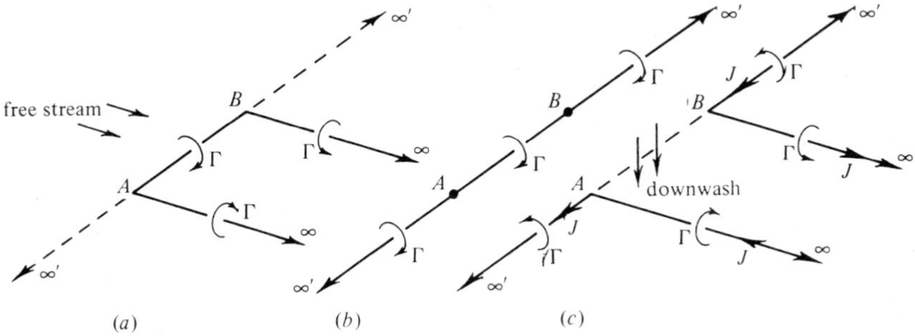

Fig. 12.17. (*a*) Horseshoe vortex. (*b*) Infinite-span vortex. (*c*) Vortices needed to change infinite-span vortex into horseshoe vortex, i.e. (*b*) + (*c*) = (*a*).

To explain induced drag simply, it is best to use a crude but not misleading model of the vorticity pattern, the *lifting line approximation*. This is illustrated in figure 12.17(*a*). All the vorticity is assumed to be shed at the wing tips A and B and the wing chord is assumed so small compared with the span that all the net spanwise vorticity in the boundary layers can be treated as bunched together into a straight line vortex AB of circulation Γ. This is what is called a *bound vortex*, for it still contains a solid body capable of carrying a net pressure reaction. (A free vortex, unable to sustain a reaction, would just blow away.) Continuity of vorticity flux demands that the tip vortices $A\,\infty$, $B\,\infty$ also have circulation Γ. It is usual to treat them also as being straight lines, oriented parallel to the flight direction. The complete vortex is called the *horseshoe vortex*.

The flow past the wing is essentially a solution of Laplace's equation which represents the superposition of a uniform stream upon the laplacian flow field associated with the bent vortex $\infty\, AB\, \infty$, of strength Γ. Superposition is legitimate because Laplace's equation is linear. There is a complete mathematical analogy between the irrotational flow associated with a vortex-line configuration and the vacuum magnetic field of line electric currents, flowing in the same pattern as the vortex lines. The Biot–Savart law is one way of predicting the action-at-a-distance of these line currents, i.e. of calculating the magnetic field 'induced' by the currents. By the same token, the flow associated with the line vortices can be calculated by a law of Biot–Savart form, and is usually described in terms of an 'induced' velocity field, even though it is not appropriate to speak in simple cause-and-effect terms, as if a vortex *caused* a certain velocity elsewhere. The line-vortex (a singularity, if modelled as having vanishing diameter) and the other velocities are merely co-existing aspects of the whole phenomenon.

The difference between the two-dimensional flow over a wing of virtually infinite span and the three-dimensional flow is describable in terms of replacing the parts $\infty' A$ and $B\,\infty'$ of an infinite-span vortex (see figure 12.17(*b*)) by the trailing vortices $A\,\infty$, $B\,\infty$. We can regard the three-dimensional flow as the result of superposing on to the two-dimensional flow of a uniform stream past the infinite span vortex $\infty' AB\,\infty'$ a correction flow field associated with the extra two line vortices shown in figure 12.17(*c*). These have the effect of altering the vortices beyond the wing tips to be trailing instead of spanwise. It is easy, from the analogy with magnetism, to form a qualitative notion of the extra velocity patterns attributable to the finiteness of the span, i.e. to the correction vortices shown in figure 12.17(*c*). If these lines carried currents J as shown, a magnetic field that was directed vertically downwards would appear along the line AB. The directions for J are chosen to give a circulation of magnetic field consistent with the velocity circulation round the vortices. By analogy, the induced, extra velocities are vertically downwards in the vicinity of AB. This effect is called *downwash*. It is calculated in problem 12.20.

The result is that wing AB behaves as if flying in an airstream that is inclined downwards. As a consequence the orientation of the lift force is altered, because at each point it is perpendicular to the effective airstream direction. It is therefore canted backwards, and now includes a component opposed to the aeroplane's motion, i.e. *a drag*. This is the origin of induced drag. In a situation where the circulation is determined by the Zhukovskii condition at each section along the span (an effect which is suppressed once we make Γ constant along AB) the downwash reduces the effective angle of incidence at each section and therefore also reduces the lift and circulation. The loss of lift has already been remarked upon.

In books on aerofoil theory it is shown how the magnitude of the change in lift and the drag can be calculated according to the crude model given here, or by better ones which allow for vorticity-shedding all along the trailing edge. Low aspect ratio and delta wings call for even more sophisticated treatments. Generalisations of these ideas permit the calculation of the performance of fans, propellers and windmills.

12.14 A final valediction

The exploitation of the analogy between electromagnetism and fluid mechanics provides a very appropriate note on which to end this book, especially since the ideas of line, surface and volume integrals, div, grad and curl, have all come together in this final synthesis in the context of three-dimensional fluid dynamics. The author hopes that the reader who has persevered to the end with the intermingled treatment of electrical and mechanical topics that this book has adopted now shares his view that these two aspects of applied vector field theory illuminate each other so instructively that the combined study of both together, along with the relevant new mathematical concepts, leads to a much deeper and more imaginative understanding of both than would the study of either alone.

Problems 12

12.1 A bathplug vortex is almost the same as a free vortex, but has a small radial flow, which turns each horizontal plane of flow virtually into one long spiral streamline, along which $p + \frac{1}{2}\rho v^2 = \text{const}$. Treating the streamlines as circles verify that this is consistent with the balance between radial presence gradient $\partial p/\partial r$ and 'centrifugal force' $\rho v^2/r$ if the velocity v is appropriately related to the radius r. Then check that this relation is consistent with Kelvin's theorem applied to a circular, concentric loop drawn in the fluid, shrinking slowly inwards.

*12.2 This problem concerns unsteady rotational flow:
A large shallow circular open tank of radius R contains water of uniform depth h_0 rotating with a uniform angular velocity Ω_0 that is small enough for the free surface to be treated as plane. At $t = 0$ a central drain is opened and water is withdrawn at a steady volumetric rate $\pi R^2 k$. Except in a small region near the centre, the water surface continues to remain virtually horizontal while v_r (the radial) and v_θ (the swirling) components of velocity may be treated as independent of vertical position. The flow is symmetric about the vertical central axis.

Show that the radial position r of a fluid particle initially at radius r_0 is given subsequently by the relation $r^2(h_0 - kt) = h_0 r_0^2 - ktR^2$. From Kelvin's theorem find an expression for v_θ for this fluid particle. Hence show that the velocity components expressed as a function of r and t are:

$$v_r = -\frac{k(R^2 - r^2)}{2r(h_0 - kt)} \quad \text{and} \quad v_\theta = \frac{\Omega_0}{rh_0}\{ktR^2 + r^2(h_0 - kt)\}.$$

At a given instant, at what radius do maximum or minimum values of circulation $2\pi r v_\theta$, swirl velocity v_θ or angular velocity v_θ/r occur? Show that the minimum swirl velocity reaches a maximum value when the tank is half empty. Discuss your results.

12.3 (*a*) If the density of a fluid depends purely on pressure p, and Φ is $\int dp/\rho$ integrated from some datum pressure p_0, show that ρ^{-1} grad p = grad Φ. Infer that Kelvin's theorem holds, if no other force acts on the fluid. Identify Φ in the case where the entropy is constant. (Do not assume perfect gas properties.)
(*b*) Show that the condition $\rho = f(p)$ is also necessary for hydrostatic equilibrium of a fluid of non-uniform density at rest under any conservative force field (per unit *mass*) such as gravity. (When the condition is not met, natural convection ensues. The question whether the equilibrium is stable depends on further considerations.)
(*c*) Heavy compressible fluid is set into motion round the closed circuit

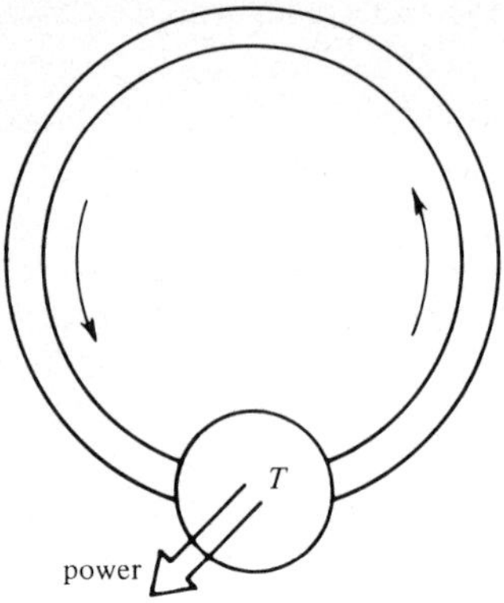

Fig. 12.18

shown in figure 12.18. The pressure drop across the turbine T causes the density on the left to be greater than on the right. The resulting unbalanced gravity force causes the fluid to continue to circulate, producing power from the turbine.

Investigate the feasibility of this alleged power system considering both isothermal and adiabatic conditions (and perhaps others!).

*12.4 (*a*) The *scalar* convection equation is of the form

$$D\theta/Dt = \lambda\nabla^2\theta$$

where θ is a property (e.g. temperature) of a travelling fluid particle whose conservation is modified by the presence of the relevant diffusivity λ (e.g. K/C in the case of temperature – see problem 9.12(*e*)). The corresponding *vector* convection equation for a vector property $\mathbf{u}$ is

$$\partial\mathbf{u}/\partial t = \text{curl}\,(\mathbf{v}\times\mathbf{u}) + \lambda\nabla^2\mathbf{u}$$

where $\mathbf{v}$ is fluid velocity and λ a relevant diffusivity. Show that if div $\mathbf{v}$ and div $\mathbf{u}$ are zero

$$D\mathbf{u}/Dt = (\mathbf{u}\cdot\text{grad})\,\mathbf{v} + \lambda\nabla^2\mathbf{u},$$

and note the contrast with the scalar equation. Show that vorticity $\boldsymbol{\omega}$ (in a viscous fluid) and magnetic field $\mathbf{B}$ (in a conducting fluid) obey this equation for $\mathbf{u}$ and identify λ in each case. (For the viscous case you will need to know the Navier–Stokes equation.) Study the way in which the contrast with scalar diffusion disappears in two-dimensional plane flow for problems with $\mathbf{u}$ in the direction of no variation.
(*b*) What is λ in (i) an inviscid fluid, (ii) a perfectly conducting fluid? In the inviscid fluid, $\iint \boldsymbol{\omega}\cdot d\mathbf{a}$ for a travelling loop is constant (Kelvin's theorem). What is the analogous result for a perfectly conducting fluid? (Section 4.8 is relevant.)
(*c*) A perfectly conducting fluid of variable density ρ undergoes a motion

which is symmetrical about an axis. Swirl, radial and axial velocities may all be present. The fluid contains a magnetic field B whose field lines are all circles in planes perpendicular to the axis, concentric with the axis. Show that for any travelling fluid element $B/\rho r = \text{const.}$, where r denotes distance from the axis.

12.5 Reconsider problem 10.7 and interpret the term $(\boldsymbol{\omega}\cdot\text{grad})\mathbf{v}$ (which equals $\omega\partial\mathbf{v}/\partial s$, where s is measured along vorticity lines) in relation to the vorticity stretching and tilting mechanisms.

12.6 Consider a steady, rotational, inviscid, compressible flow in which $\boldsymbol{\omega}$ and $\mathbf{v}$ are parallel. Is this consistent with Euler's equation and Helmholtz's theorem?

By setting $\boldsymbol{\omega} = \alpha(\rho\mathbf{v})$, where α is some scalar, and exploiting the identity div curl $\equiv 0$, investigate how α varies along streamlines. Is your result consistent with Kelvin's theorem applied to a travelling cross-section of a streamtube?

12.7 (*a*) In example 12.2 use Bernoulli's equation to check that the pressure drop along the streamline which starts at level y_1 and finishes at level y_2 is independent of y_1. Why must this be the case?
(*b*) Repeat example 12.2 for the case where the contraction of the y-dimension is still $1:\frac{1}{2}$ but there is no contraction in the other direction, i.e. the flow is two-dimensional. Show that the reduction in velocity variation is less than in example 12.2.
(*c*) Repeat (*b*) using the stream function ψ and the relation $\nabla^2\psi = f(\psi)$ (where $\nabla^2 = \partial^2/\partial x^2 + \partial^2/\partial y^2$, x being measured along the duct), showing that

$$f(\psi) = \psi/a^2 + A/a, \text{ here,}$$

if $\psi = 0$ is the streamline along the side from which y_1 and y_2 are measured.
*(*d*) Verify the vorticity-stretching effect in problem 12.2. (Vertical vorticity $= (1/r)(\mathrm{d}/\mathrm{d}r)(rv_\theta)$.)

*12.8 By taking the divergence of the identity $\mathbf{v}\times(\boldsymbol{\omega}\times\mathbf{v}) \equiv \mathbf{v}^2\boldsymbol{\omega} - (\boldsymbol{\omega}\cdot\mathbf{v})\mathbf{v}$, show that in steady, inviscid, constant-density flow $\mathbf{v}\cdot\text{grad}\,(\boldsymbol{\omega}\cdot\mathbf{v}) = 2\,\boldsymbol{\omega}\cdot\mathbf{a}$, in which $\mathbf{v}$ = velocity, $\boldsymbol{\omega}$ = vorticity, $\mathbf{a}$ = acceleration. Is this result consistent with the growth of streamwise vorticity as the fluid enters the bend shown in figure 12.8?

12.9 A shallow circular closed tank is full of liquid. Tank and liquid are spinning at uniform angular velocity about the vertical axis of the tank, when the tank is suddenly held stationary. Discuss the influence of secondary flows associated with the viscous boundary layers on the top and bottom of the tank in bringing the fluid to rest more quickly than would be the case in a tank so deep that the top and bottom had little influence.

12.10 A certain fluid motion has $v_x = 4y - 2x$, $v_y = 2y - 2x$, $v_z = 0$. Is it (i) irrotational, (ii) incompressible? Find the form of the velocity potential ϕ

and/or the stream function ψ if they exist, and the equations of the streamlines. What kind of curves are they?

A loop in the x, y plane moves with this motion. Does its area A vary? Use Stokes's theorem to express Γ, the circulation round it. Does the behaviour of Γ satisfy Kelvin's theorem?

12.11 (*a*) In steady, two-dimensional, constant-density, rotational flows show that $\boldsymbol{\omega} \times \mathbf{v} = \omega \operatorname{grad} \psi$, in which $\boldsymbol{\omega}$ = vorticity (of magnitude ω). The stream function ψ is independent of z. Show that if ω = const.

$$\omega\psi + p/\rho + \tfrac{1}{2}\mathbf{v}^2 = \text{uniform}.$$

Can you relate this to Coriolis effects?

*(*b*) For the flow $\psi = A(x^2 + y^2) + B \tan^{-1} y/x$, show that ω = const. if A, B = const. What do the streamlines look like and what kind of symmetry does the flow possess? (Polar coordinates may be helpful.)

Because ψ is multi-valued (from the $\tan^{-1}$ term) it follows from (*a*) that p would be multi-valued (which is absurd) unless the flow field were cut by at least one barrier across which there is a jump in p. This is essentially what is done in a radial-flow turbomachine rotor. The moving blades support these pressure differences and thereby allow power exchange with the fluid stream. In this case the flow is steady (if the blade thickness is neglected) but the pressure is unsteady in time at each point. How would it vary?

Consider the simple case of straight radial blades. How fast must they rotate? (The velocity component of a point on the blade normal to the blade must be equal to the component of fluid velocity normal to the blade.) What parameter in the expression for ψ measures the flow rate through the machine per unit axial length? If there are N blades show that the pressure difference across each is $8\pi\rho AB/N$. Relate this result to the orthodox energy/momentum theory for turbomachine rotors. (More general theory allows for variation in the axial dimension of the rotor.)

12.12 A steady two-dimensional, constant-density flow has the stream function $\psi = \mathrm{e}^x + \mathrm{e}^y$. Find the vorticity and show that it is consistent with Kelvin's theorem. If the point P_1, (x_1, y_1), is on the streamline $\psi = \psi_1$, what streamline is the point P_2, $(x_1 + k, y_1 + k)$, on? Deduce that all streamlines take the same shape and sketch the flow pattern.

Consider also the flow given by $\psi = \cos x \cos y$. Does this satisfy Kelvin's theorem? Sketch the flow pattern for $|x|$ and $|y| < \pi/2$.

*12.13 (*a*) Show that $\mathbf{j} \times \mathbf{B} = (\mathbf{B} \cdot \operatorname{grad}) \mathbf{B}/\mu_0 - \operatorname{grad} (\mathbf{B}^2/2\mu_0)$ in the absence of $\partial \mathbf{D}/\partial t$ and magnetisation. Noting the analogy with the momentum integral (see section 12.10) prove that the total magnetic force $\iiint \mathbf{j} \times \mathbf{B}\, d\tau$ on the contents of a closed surface can be expressed as the surface integral

$$\oiint \{(\mathbf{B} \cdot \mathbf{da})\, \mathbf{B}/\mu_0 - (\mathbf{B}^2/2\mu_0)\, \mathbf{da}\}.$$

Verify that this is the force that would be produced by a fictitious stress system applied to the surface consisting of a *normal* (tensile) stress $(B_n^2/\mu_0 - B^2/2\mu_0)$, and a shear stress $B_n B_t/\mu_0$ in the direction of B_t, where B_n and B_t are the normal and tangential components of **B** at each point. (These stresses are known as *Maxwell stresses* and enable the total magnetic force on a volume to be calculated from a knowledge merely of the field at its surface. The result is still correct even when the volume contains magnetic material and virtual current curl **M** replaces actual current **j**. But note that Maxwell stresses only indicate the total force on a volume, *not* the state of the real stresses at a particular point. For example, a body devoid of current and magnetisation, exposed to a magnetic field is not thereby put into a state of stress, despite the Maxwell stresses at its surface. Maxwell stresses are somewhat like the Poynting vector; only their non-uniformity is significant; any arbitrary balanced stress system can be added to them with impunity.)
(*b*) Which Maxwell stress is responsible for propelling the induction motor referred to in section 11.9?
(*c*) Show that Maxwell stress also accounts for the magnetic pressure due to skin-effect (see section 11.4) by considering a surface layer which includes all of the skin-effect layer.
(*d*) Show that the normal and shear Maxwell stresses are $(B^2/2\mu_0)\cos 2\theta$ and $(B^2/2\mu_0)\sin 2\theta$, where θ is the angle between **B** and the normal to the surface. What value of θ gives a pure shearing stress?
(*e*) Reconsider problem 11.20(*b*) in the light of Maxwell stresses.

12.14 The moving fluid sample which is enclosed by a control surface S at some particular instant is, at a time dt later, enclosed by a slightly different control surface S'. If the motion is a steady flow, express the difference in momentum contents of S and S' in the form of a surface integral $\oiint \rho(\mathbf{v}\,dt\cdot d\mathbf{a})\mathbf{v}$, describing the characteristics of the fluid lying between S and S'. By equating the total forces acting on the fluid sample (including pressure) to the rate of change of total momentum of the sample, deduce the momentum integral theorem of section 12.10.

*12.15 For an irrotational flow with constant density ρ and velocity potential ϕ, show that in relation to the contents of a control surface, the following surface integrals apply:

total momentum $= \rho \oiint \phi\, d\mathbf{a}$,
total moment of momentum about the origin $= \rho \oiint \phi\mathbf{r} \times d\mathbf{a}$, and
total kinetic energy $= \frac{1}{4}\rho \oiint \operatorname{grad}(\phi^2)\cdot d\mathbf{a}$.

12.16 (*a*) Extend the result of section 12.10 to include the effect of gravity.
(*b*) Liquid of constant density ρ flows steadily at volumetric flow rate Q down an outlet pipe of cross-sectional area A as shown in figure 12.19. The outlet terminates with a diverging section made of thin metal, which doubles the cross-sectional area of the pipe. Only the divergent part of the

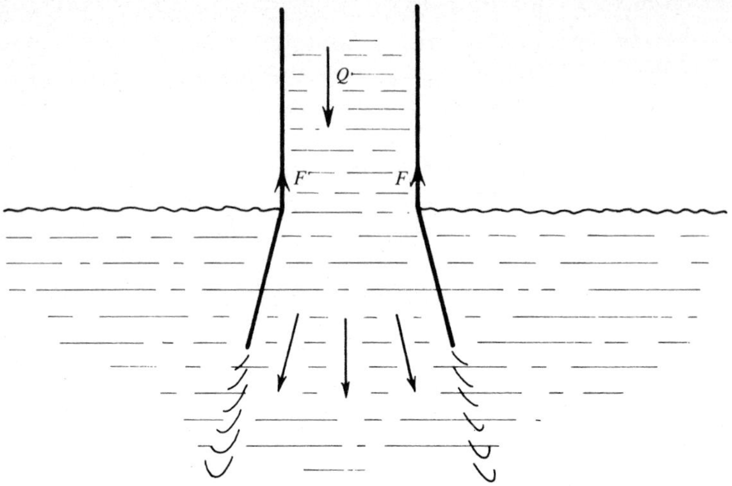

Fig. 12.19

pipe is submerged below the free surface of the liquid in the large pool that receives the discharge. It may be assumed that Bernoulli's equation (including gravity but ignoring transverse velocity components) applies in the diverging section, that variations of velocity or pressure over horizontal cross-sections of the pipe are negligible, and that the pressure at outlet is the same as that prevailing in the virtually static fluid alongside the pipe.

Calculate the total vertical force F necessary to support the divergent section of pipe. (Try various choices of control surface.)

12.17 A two-dimensional free jet of liquid of velocity v and thickness d impinges on a rigid plate inclined at an angle θ to the direction of motion and splits into two unequal jets travelling in opposite directions along the plate. If the liquid's density ρ is constant, its free surface is exposed to uniform ambient pressure everywhere and the reaction on the plate is normal to it (viscosity being negligible) calculate:

(*a*) the velocities in the jets on the plate,
(*b*) their thicknesses and
(*c*) the reaction on the plate per unit width.

Show that the component of the reaction normal to the original jet is greatest when $\theta = \pi/4$. (Neglect gravity.)

12.18 The rate of flow of water from an experiment is measured by allowing it to fall as a free jet into an open weigh tank and finding the increase in apparent weight of the tank as the water level rises from one height to another in a measured time. There are two kinds of errors in this method:

(*a*) Errors due to the changing momentum of the water in the jet entering the water surface at various heights.

(*b*) Errors due to the changing amount of water contained in the jet between the experiment and the water surface in the tank.

Making suitable, clearly stated assumptions, investigate the degree to which these errors are mutually compensatory.

*12.19 At large distances the potential flow past a two-dimensional aerofoil becomes indistinguishable from potential flow with circulation past a circular cylinder. (See problem 10.12(*d*).) Verify that the terms neglected in section 12.12, when integrated round a loop L which is a large circle of radius r, centred on the cylinder or aerofoil, give zero contribution as $r \to \infty$.

Show also that the two contributions associated respectively with pressure variation and momentum flux in the integral for $\mathbf{F}$ in terms of $\mathbf{U}$ and $\mathbf{v}'$ are equal as $r \to \infty$.

12.20 By analogy with the Biot–Savart law, infer that the induced velocity at a point P 'due to' a line vortex of circulation Γ is given by

$$\frac{\Gamma}{4\pi}\int \frac{\mathbf{r} \times \mathrm{d}\mathbf{r}}{r^3}$$

integrated along the line in the direction consistent with the right-hand screw rule in relation to Γ, where $\mathbf{r}$ is the vector $\overrightarrow{PQ}$, Q being at the line element $\mathrm{d}\mathbf{r}$. What simple statement is the analogue of Ampère's law?

Calculate the downwash velocity at the centre-span point in figure 12.17 if the span is $2a$.

Appendix I: A note on differentials, Taylor's theorem and mathematical modelling

The essence of continuum field theory, whether in fluid or solid mechanics, thermodynamics, electromagnetism, or diffusion theory, is mathematical modelling based on the central idea of the calculus, namely, that mathematical statements about the changes of quantities in a given vicinity in some space (normally three dimensions and time) become greatly simplified in the limit as the sizes of the vicinity and of the changes are all allowed to decrease to zero. (But we stop before molecular lumpiness manifests itself.) This happens because various ratios of small changes settle down to identifiable values (differential coefficients) and other quantities, of a higher order of smallness, finally make a vanishing contribution. The art of doing this mathematical modelling process depends on being able to know which terms must be retained because they will contribute to the differential coefficients and which should be abandoned, preferably at the earliest possible stage, as they will not contribute in the limit.

Typically one is expressing conservation, or some other physical constraint, applied to variables P, etc., as they vary in space and time, and one needs an expression for δP, the change in P consequent upon a change of the independent variables, x and y say (to take a simple case) to $(x + \delta x)$ and $(y + \delta y)$. The δ-notation indicates a small change. Granted that P is smoothly varying (differentiable) in the vicinity, and expandable as a series, Taylor's theorem gives

$$P(x + \delta x, y + \delta y) - P(x, y) =$$

$$\delta P = \frac{\partial P}{\partial x}\delta x + \frac{\partial P}{\partial y}\delta y + \frac{1}{2!}\left\{\frac{\partial^2 P}{\partial x^2}(\delta x)^2 + 2\frac{\partial^2 P}{\partial x \partial y}(\delta x \delta y) + \frac{\partial^2 P}{\partial y^2}(\delta y)^2\right\} \ldots$$

After the limiting process has been applied to the physical statement, involving δP and other relevant quantities, one usually finds that the terms such as $(\delta x)^2$ or higher powers of small quantities finally make no contribution, whereas the first-order terms in δP, δx, etc. contain the information which survives in the limit. A simple example is where x and y are constrained by some relation so that $(\delta y/\delta x)$ tends to some identifiable value $(\mathrm{d}y/\mathrm{d}x)$ in the limit as δx, δy, etc. tend to zero. Then

$$\frac{\delta P}{\delta x} = \frac{\partial P}{\partial x} + \frac{\partial P}{\partial y}\frac{\delta y}{\delta x} + \frac{1}{2}\left(\frac{\partial^2 P}{\partial x^2}(\delta x) + \text{etc.}\right) + \text{higher order terms,}$$

and when δx, etc. tend to zero we have

$$\frac{\mathrm{d}P}{\mathrm{d}x} = \frac{\partial P}{\partial x} + \frac{\partial P}{\partial y}\frac{\mathrm{d}y}{\mathrm{d}x} + 0.$$

In the face of this kind of behaviour, most books on mathematics applied to physical science anticipate the limiting process and omit the higher order terms right from the start, e.g. they write

$$\delta P \doteqdot \frac{\partial P}{\partial x}\delta x + \frac{\partial P}{\partial y}\delta y,$$

with the approximation sign $\doteqdot$. But the = sign is often used instead, even though it is not strictly true until the equation has been divided by some increment δx, say, and the limit has been taken. (Failure to divide by δx before taking the limit merely gives $0 = 0$, of course!) Another common notation (the one used in this book) is to write

$$\mathrm{d}P = \frac{\partial P}{\partial x}\mathrm{d}x + \frac{\partial P}{\partial y}\mathrm{d}y,$$

using the = sign, where $\mathrm{d}P$, etc. are called *differentials*. They have the convenient property that $\mathrm{d}P \div \mathrm{d}x$ is in fact equal to $(\mathrm{d}P/\mathrm{d}x)$, the limiting value of the ratio $(\delta P/\delta x)$. But what exactly *are* these differentials?

The strict view is to regard the relation $P(x, y)$ as defining a curved surface in a (P, x, y) space. At a point Q, (P_0, x_0, y_0), on the surface, the equation of the tangent plane is

$$(P - P_0) = (\partial P/\partial x)(x - x_0) + (\partial P/\partial y)(y - y_0),$$

in which $\partial P/\partial x$ and $\partial P/\partial y$ are evaluated at Q. From the equation for $\mathrm{d}P$, it is apparent that $\mathrm{d}P$, etc. can be interpreted as finite excursions of P etc. *on the tangent plane*. This idea can be generalised to more dimensions if there are more independent variables.

Comparing the equation for $\mathrm{d}P$ with Taylor's series shows that if $\delta x = \mathrm{d}x$ and $\delta y = \mathrm{d}y$ the difference between δP and $\mathrm{d}P$ is a small quantity of order $(\delta x)^2$. Thus the difference between the tangent plane and the actual surface becomes less and less, the smaller the excursions, and any terms associated with this difference make no contribution as δx, etc. tend to zero, as already remarked. So the scientist engaged on mathematical modelling soon operates the calculus by treating small excursions on the surface as if they were on the tangent plane (i.e. linearly related) and then the linear relations between differentials became effectively statements about the actual variations of the quantities, although the implications of the differential relations interpreted this way are then strictly true only in the limit. The process is most simply thought of as 'anticipating the limit'. This book adopts such an approach, i.e. it treats differentials as small increments.

Another very familiar manifestation of the same ideas occurs in the integration of a function P of one variable x. The area under the curve of P against x is the value of the sum $\Sigma\, \bar{P}\delta x$ where δx is one of a series of successive increments of x as it rises monotonically from x_1 to x_2, and $\bar{P}$ is an appropriate value near $P(x)$ and $P(x + \delta x)$ such that $\bar{P}\delta x$ is the area under the curve of P against x in this range, i.e. $\bar{P}$ is the value at $(x + \theta\delta x)$ (where θ is between 0 and 1). Taylor's theorem gives

$$\bar{P} = P(x) + \theta\delta x(\mathrm{d}P/\mathrm{d}x) + \text{higher order terms (h.o.t.)}$$

and

$$\text{area under curve} = \sum \bar{P}\delta x = \sum P\delta x + (\delta x)^2 \text{ terms} + \text{h.o.t.}$$

In the limit all δx tend to zero, the number of contributions increases indefinitely and the limit of $\Sigma P\delta x$ (if it exists) is written $\int P\,\mathrm{d}x$, the curly 'ess' sign replacing the sigma, and $\mathrm{d}x$ replacing δx, by convention. While $\Sigma P\delta x$ usually tends to a finite limit, the $(\delta x)^2$ and higher terms become smaller and finally vanish, with the result that

$$\text{area under curve} = \text{the '}\textit{integral}\text{'} \int P\,\mathrm{d}x.$$

Knowing that this is the behaviour, the scientist normally models by anticipating the limit. He works with increments $\mathrm{d}x$ (rather than δx) and sums contributions as integrals ($\int$) rather than finite sums (Σ). This book adopts this posture in relation to line, surface and volume integrals. It is a crucial part of many arguments that the distinction between P and $\bar{P}$ can be ignored in anticipation of the limit. (See for instance the argument in section 11.1 as to why the result curl $\mathbf{E} = -\partial\mathbf{B}/\partial t$ follows from Faraday's law and Stokes's theorem.) In effect one is saying that $\mathrm{d}x$, $\mathrm{d}\mathbf{a}$, etc. are so small that variation of the integrand over them can be ignored, for the variation contributes terms which vanish in the limit. It is just this fact which allows the modelling process to be fruitful.

Another related ingredient in the modelling process is the geometrical approximation that only becomes exact in the limit. Treating the area element in figure 3.9 as a rectangle $R^2\,\mathrm{d}\lambda\,\mathrm{d}\phi\cos\lambda$ is an example. Another familiar simple example occurs with plane polar coordinates. Consider two adjacent points P, (r, θ), and Q, $(r + \delta r, \theta + \delta\theta)$, on some curve C, as shown in figure A.1(a). PN is perpendicular to

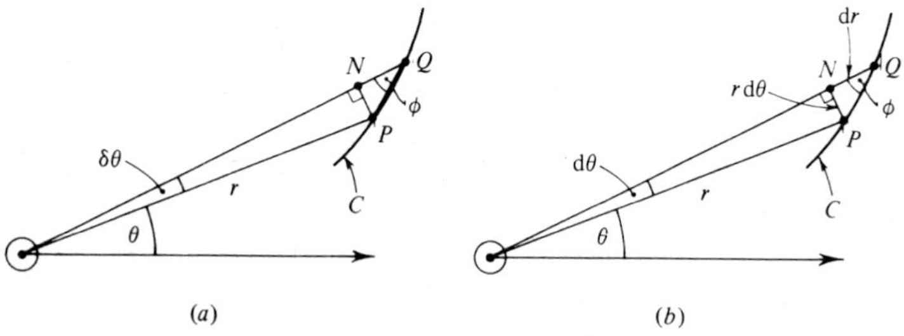

Fig. A.1

OQ and equals $r\sin\delta\theta$, while $QN = r + \delta r - r\cos\delta\theta$. From Taylor's theorem we have

$$PN = r\delta\theta - \tfrac{1}{6}r(\delta\theta)^3 + \text{h.o.t.}$$

$$QN = \delta r + \tfrac{1}{2}r(\delta\theta)^2 + \text{h.o.t.}$$

If there was interest in the angle ϕ, we would write

$$\tan\phi = \frac{PN}{QN} = \frac{r\delta\theta - \frac{1}{6}r(\delta\theta)^3 \ldots}{\delta r + \frac{1}{2}r(\delta\theta)^2 \ldots} \to r\frac{\mathrm{d}\theta}{\mathrm{d}r}$$

in the limit as P and Q tend to coincidence and ϕ becomes the angle between the radius vector and the tangent at P.

What the practised scientist or applied mathematician does is to anticipate the limit and label the diagram as in figure A.1(*b*) with differentials $\mathrm{d}r$, $\mathrm{d}\theta$, etc., omitting the higher-order terms which he knows will not finally contribute. In effect, PN is treated as though it were an arc of a circle of radius r, subtending $\mathrm{d}\theta$ at the centre O, and OPN is treated as an isosceles triangle with $OP = ON$ so that QN is the increase in r.

Another way in which geometrical approximation enters is through the ignoring of the small changes of direction of a vector over a line, surface or volume element, because any correction terms would not contribute in the limit. The point was referred to in section 1.8. In effect this simply means that the integrand includes some variable resolution factor $\cos\theta$ whose variation over each element may be ignored in anticipation of the limit.

Obviously some skill and experience is essential, especially in more complicated problems, and the beginner should always include higher-order terms if in any doubt, particularly if the equation is one where the surviving lowest-order terms on which the limiting process is to be based are higher than first order. An example appears in section 12.3.

We conclude with a particular modelling example which brings out some of these points and reveals why certain higher-order terms fail to contribute in the limit in more complicated situations.

Consider an unsteady, two-dimensional heat-conduction problem, characterised by the scalar variable T (temperature) and the vector variable $\mathbf{Q}$ (heat flux intensity), which depend on x, y and t. The task is to express conservation of energy, given that the non-uniform material has a heat capacity C per unit volume which depends on position (x, y) and temperature T, but not directly on time.

We shall first discuss the problem cautiously at some length, recording contributions higher than the lowest essential order, and then confirm the correctness of the direct method which discards the non-essential terms at an earlier stage.

Consider the rectangular space element shown in figure A.2 in which x ranges from x_0 to $(x_0 + \Delta x)$ and y ranges from y_0 to $(y_0 + \Delta y)$. The thickness in the z-direction is unity. Consider a time interval from t_0 to $(t_0 + \Delta t)$. Let suffix 0 denote values at (x_0, y_0, t_0). For values intermediate between x_0 and $(x_0 + \Delta x)$, etc. we shall use $(x_0 + \delta x)$, etc. and treat δx, etc. as variables. The quantities of interest comprise Q_x and Q_y (components of $\mathbf{Q}$), T and C. Taylor's theorem gives:

$$Q_x = Q_{x0} + \left(\frac{\partial Q_x}{\partial x}\right)_0 \delta x + \left(\frac{\partial Q_x}{\partial y}\right)_0 \delta y + \left(\frac{\partial Q_x}{\partial t}\right)_0 \delta t + \text{h.o.t.},$$

and a similar expression for Q_y,

$$\delta T = T - T_0 = \left(\frac{\partial T}{\partial x}\right)_0 \delta x + \left(\frac{\partial T}{\partial y}\right)_0 \delta y + \left(\frac{\partial T}{\partial t}\right)_0 \delta t + \text{h.o.t.},$$

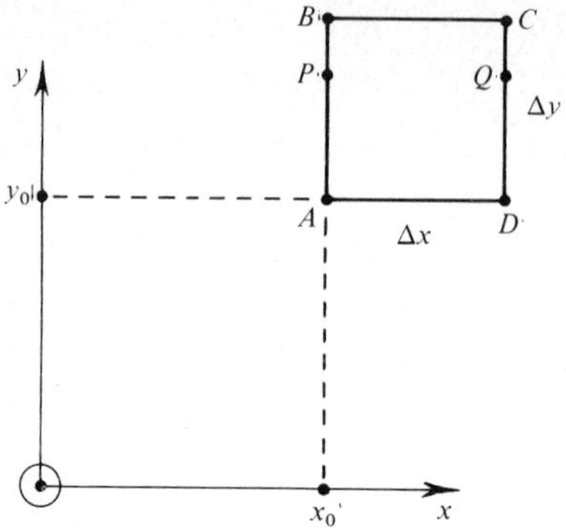

Fig. A.2

$$\frac{\partial T}{\partial t} = \left(\frac{\partial T}{\partial t}\right)_0 + \left(\frac{\partial^2 T}{\partial t \partial x}\right)_0 \delta x + \left(\frac{\partial^2 T}{\partial t \partial y}\right)_0 \delta y + \left(\frac{\partial^2 T}{\partial t^2}\right)_0 \delta t + \text{h.o.t.},$$

and $$C = C_0 + \left(\frac{\partial C}{\partial x}\right)_0 \delta x + \left(\frac{\partial C}{\partial y}\right)_0 \delta y + \left(\frac{\partial C}{\partial T}\right)_0 \delta T + \text{h.o.t.},$$

where δT is given above. First consider the heat entering face AB (where $\delta x = 0$) during Δt:

$$\text{heat in through } AB = \int_0^{\Delta y} \int_0^{\Delta t} Q_x \, \mathrm{d}(\delta t)\mathrm{d}(\delta y) (\text{with } \delta x = 0)$$

$$= Q_{x0} \Delta t \Delta y + \left(\frac{\partial Q_x}{\partial y}\right)_0 \Delta t \frac{(\Delta y)^2}{2} + \left(\frac{\partial Q_x}{\partial t}\right)_0 \frac{(\Delta t)^2}{2} \Delta y + \text{h.o.t.}$$

if we use the Taylor series. The heat escaping through CD during Δt has extra terms because $\delta x = \Delta x$ there:

$$\text{heat out through } CD = \int_0^{\Delta y} \int_0^{\Delta t} Q_x \, \mathrm{d}(\delta t)\mathrm{d}(\delta y) (\text{with } \delta x = \Delta x)$$

$$= \text{heat through } AB + \left(\frac{\partial Q_x}{\partial x}\right)_0 \Delta x \Delta t \Delta y + \text{further h.o.t.}$$

Similarly,

$$\text{heat out through } BC = \text{heat in through } AD + \left(\frac{\partial Q_y}{\partial y}\right)_0 \Delta x \Delta t \Delta y + \text{h.o.t.},$$

and we see that the *net* heat leaving the volume element during Δt equals

$$\left\{\left(\frac{\partial Q_x}{\partial x}\right)_0 + \left(\frac{\partial Q_y}{\partial y}\right)_0\right\} \Delta t \Delta x \Delta y + \text{h.o.t.}$$

Note that the lowest order of essential quantities is *third* in this problem, even though lower-order quantities appeared earlier. Next we turn to the heat deposited within the element during Δt. This is the integral

$$I = \int_0^{\Delta t}\int_0^{\Delta x}\int_0^{\Delta y} C\frac{\partial T}{\partial t}\,\mathrm{d}(\delta y)\,\mathrm{d}(\delta x)\,\mathrm{d}(\delta t),$$

in which C is expressed as a function of x, y and t by eliminating δT. The first terms of this integral may be evaluated, after some manipulation, to yield

$$I = C_0\left(\frac{\partial T}{\partial t}\right)_0 \Delta t\Delta x\Delta y$$

$$+\frac{(\Delta x)^2}{2}\Delta t\Delta y\left\{\left(\frac{\partial C}{\partial x}\right)_0+\left(\frac{\partial C}{\partial T}\right)_0\left(\frac{\partial T}{\partial x}\right)_0+C_0\left(\frac{\partial^2 T}{\partial t\,\partial x}\right)_0\right\}+\frac{(\Delta y)^2}{2}\Delta t\Delta x \text{ term}$$

$$+\frac{(\Delta t)^2}{2}\Delta x\Delta y\left\{\left(\frac{\partial C}{\partial T}\right)_0\left(\frac{\partial T}{\partial t}\right)_0^2+C_0\left(\frac{\partial^2 T}{\partial t^2}\right)_0\right\}+\text{h.o.t.}$$

Energy conservation requires

$$\left\{\left(\frac{\partial Q_x}{\partial x}\right)_0+\left(\frac{\partial Q_y}{\partial y}\right)_0\right\}\Delta t\Delta x\Delta y + C_0\left(\frac{\partial T}{\partial t}\right)_0\Delta t\Delta x\Delta y+\text{h.o.t.} = 0.$$

Dividing by Δt, Δx and Δy and letting them tend to zero gives

$$\frac{\partial Q_x}{\partial x}+\frac{\partial Q_y}{\partial y}+C\frac{\partial T}{\partial t} = 0,$$

evaluated at (x_0, y_0, t_0). This is recognisable as div $\mathbf{Q} = -C\partial T/\partial t$, which is the sort of statement which is frequently written down without much ado throughout this book, essentially on the basis of a minimal argument such as the following, expressed in relation to an element $\mathrm{d}x$, $\mathrm{d}y$ and time interval $\mathrm{d}t$ (where we now express increments as differentials):

The quantity (heat out of *CD* – heat into *AB*) depends on the differences between corresponding points such as *P* and *Q* on *AB* and *CD* associated with the different value of x. (*N.B.* y and t for *P* and *Q* are the same at each instant.) The only essential contribution therefore comes from x-variation and so we write

$$\text{heat out of } CD - \text{heat into } AB = \{(\partial Q_x/\partial x)\mathrm{d}x\}\,\mathrm{d}y\,\mathrm{d}t$$

in which $\{(\partial Q_x/\partial x)\mathrm{d}x\}$ is the essential change in x-wise heat flux in going from *AB* to *CD*, and $\mathrm{d}y$ is the area of these faces. Thus the net heat flow from the element is $(\partial Q_x/\partial x + \partial Q_y/\partial y)\mathrm{d}x\,\mathrm{d}y\,\mathrm{d}t$.

As regards the heat deposition term, it should be obvious that allowing for variations in the local, instantaneous heat deposition density $C\partial T/\partial t$ will generate higher-order terms that will not contribute after the final limit operation. We therefore ignore these terms and write directly

$$\text{heat deposition rate} = (\mathrm{d}x\,\mathrm{d}y)(C\partial T/\partial t)\mathrm{d}t$$

in which $\mathrm{d}x\,\mathrm{d}y$ is the volume. The conservation equation follows immediately.

It should be evident to the reader how much economy of effort is achieved by the direct method, once it has been mastered.

As a final point, the reader should note that for brevity this book adopts the use of the d-prefix also for quantities such as $\mathrm{d}a$, $\mathrm{d}\tau$, $\mathrm{d}\Gamma$, $\mathrm{d}\Phi$ which are not increments of variables and which are not first-order small quantities in general; $\mathrm{d}a$ and $\mathrm{d}\Gamma$ are second order and $\mathrm{d}\tau$ and $\mathrm{d}\Phi$ are third order.

Appendix II: Suggested lecture demonstrations and coursework tasks

The demonstrations which the author has found useful in connection with the material covered in this book vary greatly in their level of complication; some involve only simple resources, others require access to well-equipped laboratories and workshops. They are listed in the same order as the topics in the book, and the relevant section number is recorded in brackets.

Many of the more obvious models that can be used are not listed. For a three-dimensional subject like vector analysis, plane drawings on a blackboard or projection screen are a poor representation of such essentially three-dimensional concepts as right-handed axes, alternative surfaces spanning the same loop, the relation between grad ϕ and constant-ϕ surfaces, etc. Models made from wire, string, paper, cardboard, perspex, balloons, etc. are much better and are easily constructed.

1. *Not all physical entities that have magnitude and direction are vectors* (1.3)

A suitably labelled rectangular box or block can easily be used to demonstrate that successive finite rotations through 90° in different orientations do not produce the same result when applied in different orders, and therefore cannot be vectors.

2. *Vector moment* (1.11)

A three-dimensional model is essential if students are to grasp the significance of the components of the vector moment $\mathbf{r} \times \mathbf{F}$. An enlarged version of figure 1.8 is satisfactory.

3. *Laplacian field plotting* (8.4)

(*a*) A commercially available field-plotter, used with teledeltos paper to provide two-dimensional solutions of Laplace's equation makes a very satisfactory lecture demonstration or coursework task, particularly if it is used to generate equipotentials for two dual problems which can then generate a complete curvilinear square mesh (see section 8.8). The failure of the 'squares' to be square at boundary singularities is easily demonstrated.

(*b*) Given the resources, it is well worth showing two-dimensional laplacian solutions with the aid of a Hele–Shaw apparatus, which is commercially available as well as being fairly easy to make. It is even possible to project an enlarged image on a screen. The author employs several models which demonstrate seepage flows through and under dams, using water with dye to reveal streamlines and a gap of 0.5 mm. The most simple and satisfactory is that appearing in figure 8.8.

(*c*) It is also worth demonstrating solutions of laplacian problems by using the membrane analogy applied to soap films. This works best for problems where the dependent variable is specified all round the boundary, rather than its normal derivative. (By applying a pressure difference across the film, solutions to Poisson's equation may be exhibited, also. See section 9.2.)

4. *Unsteady diffusion equation analogue* (9.4 and 11.2)

The nature of the dynamic thermal or electromagnetic skin effect phenomenon cannot be captured by static diagrams, and a working model is very illuminating. It clearly reveals the slowing down of penetration according to the $\sqrt{t}$ law and, in oscillatory cases, the reduction in penetration associated with higher frequencies and the decay within one wavelength.

The analogue is a lumped, hydraulic model. A bank of, say, 25 vertical glass tubes of 2.5 mm bore are connected at their lower ends by a short length of capillary tube (10 mm long, 1 mm bore is satisfactory) between each pair of adjacent tubes. The end capillary leads to a 'boundary condition' tank in which the level of water can be varied at will by a plunger. Methylated spirits is a better fluid to use than water. In constructing this apparatus it is important not to make the response too slow, especially if it is to be used in lectures.

5. *Demonstrations of the persistence of fluid rotation and non-rotation* (10.2 and 12.2)

(*a*) Blowing out a candle (and failing to suck it out) makes a good lecture demonstration, both entertaining and instructive. A vacuum cleaner or equivalent device capable of sucking or blowing, with a cigarette to provide smoke for flow-visualisation, can demonstrate the natural asymmetry between sucked and blown flow.

(*b*) A cylindrical glass jar with a small, sealable neck permits simple, instructive demonstrations. The author has used a bottle about 200 mm high and 200 mm in diameter. Some strands of wool may be used to reveal the motion of the water in the bottle.

(i) First one leaves a small amount of air in the bottle and shows how easy it is to spin the fluid about a vertical axis, in the manner of one washing milk bottles. (This is a very complicated phenomenon.)

(ii) Then one fills the bottle completely and meets total failure at getting the water to spin by the same technique – *conservation of non-rotation*! Most audiences find this a revelation.

(iii) Finally one spins the fluid as in (i) before turning it over until the axis is vertical again. Though the bottle is inverted, the water is still spinning the same way as it was previously – *conservation of rotation*! (As a contrast it is worth first showing a wheel spinning about a vertical axle and then inverting the axle.)

6. *Demonstration of curl as a property of fields* (10.9 and 11.1)

Essentially this demonstration aims to convey the geometrical quality of the relation $\oint \mathbf{v} \cdot d\mathbf{r} = \text{curl}\, \mathbf{v} \cdot d\mathbf{a}$ for a small loop $d\mathbf{a}$ i.e. the idea that at each point there is an orientation for $d\mathbf{a}$ which gives maximum circulation while all orientations normal to that give no circulation and intermediate ones demonstrate the familiar 'cosine-quality' of vectors. The loop is a search coil 80 mm in diameter with, say, 300 turns of 28 s.w.g. copper wire connected to a suitable large sensitive ammeter, through a small rectifier if necessary. The source of the field is an a.c. magnet, run off the mains, producing an oscillating and highly rotational electric field in the vicinity. To be effective, strong e.m.f.s must be detectable when the search coil is held 200–300 mm from the magnet. Then the relationship between circulation of **E** and orientation is clearly demonstrable.

7. *Electromagnetic skin-effect* (11.3)

The most easily demonstrated effect is the $\pi/4$ change in the phase between the alternating current and voltage along a wire as the frequency is raised from low to high values, by means of a suitable variable-frequency oscillator. The voltage and current wave forms can be displayed on a two-beam oscilloscope. Using a steel or iron wire enables lower frequencies to be used, as the skin-effect is then more marked. A steel wire of 1.0 mm diameter shows the effect well as the frequency rises from around 50 Hz up to 50 kHz.

8. *Electromagnetic waves* (11.5 *et seq.*)

Demonstrations with two microwave frequencies are convenient:

(*a*) *X-band (c. 10 GHz)* The basic equipment consists of a horn transmitter producing a narrow, plane-polarised beam and a horn receiver, which is arranged to produce an audible signal proportional to microwave power received. These then permit the following demonstrations:

(i) Narrowness and straightness of beam.
(ii) 'Opaqueness' of metal sheet reflector, 'transparency' of dielectric sheet.
(iii) Angle of reflection off sheet obeys usual law.
(iv) Opaqueness of metal grating (when aligned so as to short-circuit **E**-field of wave) demonstrates polarisation. (Grating consists of bars and spaces of approximately 2 mm width.)
(v) 45° prism full of liquid paraffin shows a total internal reflection.
(vi) Wavelength can be demonstrated by interferometry (see figure A.3). *AB* is a fixed reflecting plate, *CD* a reflecting plate which can be moved parallel to itself and *EF* is a partial reflector (the metal grating in (iv) may be used).

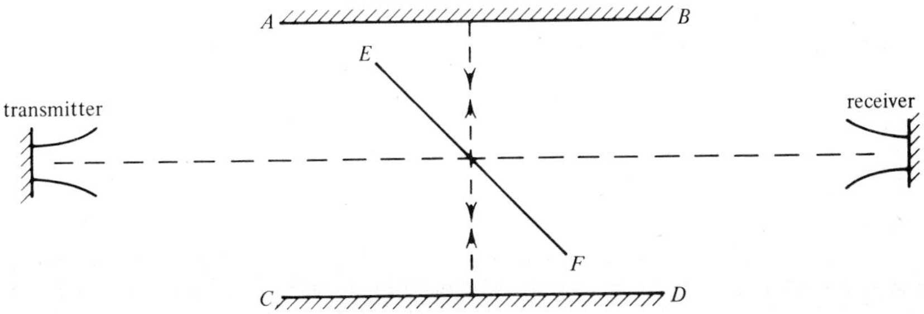

Fig. A.3. Microwave interferometer.

Moving *CD* between positions that give a minimum received signal readily reveals the semiwavelength.

(*b*) *800 MHz* This frequency is convenient for showing behaviour in wave-guides. The presence of electric fields and magnetic fields is revealed by movable dipole and loop probes. These can be connected to a detector which activates a low voltage d.c. circuit for illuminating a small bulb on the probes to aid location of maximum fields.

Typical experiments can include a rectangular wave-guide 150 mm × 300 mm about 1.5 m long, excited by an aerial near its closed end, so as to adopt the TE_{01} mode. The fact that the tangential **E** is zero at the walls and that the magnetic field is greater at the side walls and is zero in the middle is easily revealed. (See problem 11.11.)

A circular guide 350 mm in diameter can be used to show the axisymmetric E_{01} mode with **E** radial and **H** tangential at the walls.

By closing most of the open end with a reflecting plate, standing wave nodes and antinodes inside the guide may be shown. This may also be done with the plate held at some distance from the open end of the wave-guide.

9. *Vorticity, Kelvin and Helmholtz theorems* (12.2 *et seq*.)

Of the many films available on fluid mechanics, one of the most educative is the 40-minute film 'Vorticity' produced by Educational Services Inc. for the U.S. National Committee on Fluid Mechanics Films. It is strongly recommended that this film be used to supplement any lecture course on the topics in this book. Other films in the same series are also relevant. For instance, the film 'Flow-visualisation' has a sequence which makes very clear the difficult distinction between streamlines, particle paths and dye-lines in unsteady flow. The author's own film in the series ('Magnetohydrodynamics') has sequences which show how vorticity is the key to understanding the influence of force fields acting on a fluid.

10. *Coursework tasks*

As well as being required to do suitably selected problems, students should be given some extended tasks of a more empirical nature. There are many possibilities, which fall into three categories:

(*a*) *Experimental tasks* Obvious examples include the plotting of two-dimensional laplacian fields by the use of the conducting or Hele–Shaw analogies (see 3(*a*) and 3(*b*) above). In the fluid-mechanical field, comparing the drag on, say, a cylinder transverse to a stream with the integrated effect of pressures measured round the cylinder is an instructive exercise, as is the integration of axial velocity in a pipe (measured by a pitot traverse) to yield flow rate, for comparison with a direct measurement. If possible students should do their own microwave experiments (see 8 above).

(*b*) *Drawing-office tasks* These should certainly include an exercise on solving a two-dimensional laplacian problem by sketching curvilinear squares. Another good exercise is to construct two-dimensional field lines by (i) the use of isoclines (compare problem 3.24) and (ii) the use of superposed stream functions, exploiting the intersections of the known field lines of simpler fields which can be superposed to yield the field in question if the field lines are drawn for equal increments of the stream functions. For a simple example, consider the flow between a line source and a line sink. The field lines (for equal increments of the stream functions) for the source or sink alone consist of equally-spaced radial lines. The intersections of those lines from the source and the sink which yield a constant value for the total stream function (when the two fields are superposed) can be joined together to generate the field lines, as in figure A.4. (In this case they are coaxal circles.) A more interesting example is the combination of a line source with a uniform stream.

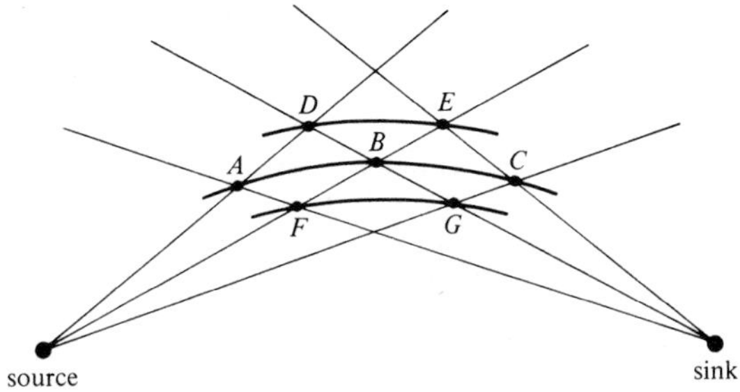

Fig. A.4. The straight lines are the field lines for the source or the sink alone, drawn for equal increments of the stream function. *ABC* is part of a field line for the field resulting from superposing the source and sink fields. *DE*, *FG* are parts of other field lines, similarly.

Equipotentials may be generated by a similar technique.

A quite different kind of task is to solve the one-dimensional unsteady diffusion (or heat conduction) equation by Schmidt's method (see figure A.5):

The differential equation

$$\frac{\partial T}{\partial t} = \alpha \frac{\partial^2 T}{\partial x^2}$$

is replaced by its finite-difference equivalent

$$\frac{T' - T}{\Delta t} = \frac{\alpha(T_+ + T_- - 2T)}{(\Delta x)^2},$$

in which T at a point changes to T' after time Δt, and T_+ and T_- are the values of T at the two positions Δx away from the point. If we choose $2\alpha\Delta t = (\Delta x)^2$ then

$$T' = \tfrac{1}{2}(T_+ + T_-),$$

and the evolution of a temperature profile may easily be plotted graphically as in figure A.5.

(*c*) *Computing tasks* It is desirable that students should also get experience of solving field problems on a computer, either by numerical integrations (e.g. of magnetic fields, found from the Biot–Savart law) or by finite-difference solutions of the field equations such as Laplace's equation. It should be well within the capability of a second-year student to program and run the solution of Laplace's equation in two-dimensions in a domain with straight boundaries (parallel to or at 45° to the

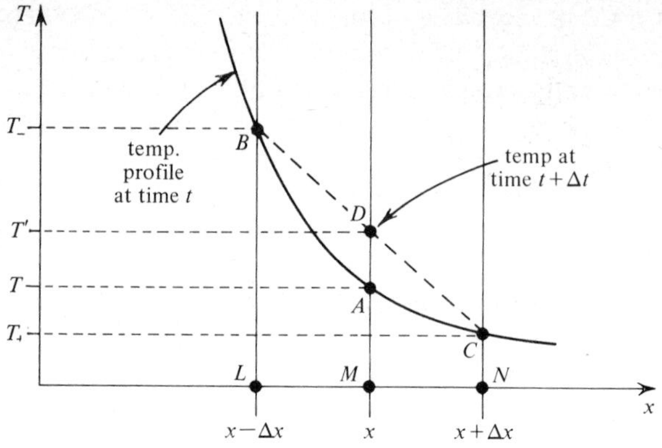

Fig. A.5. Schmidt's method: A, B, C, record the temperature levels at three adjacent points, L, M, N, at time t. The temperature at M for time $t + \Delta t$ is given by D, the intersection of BC with the vertical through M.

axes) upon which the dependent variable is specified. (This choice of boundaries avoids the 'unequal star' problem.) The finite-difference equation to be solved is

$$\sum \phi_i = 4\phi_0 \quad \text{(if } \phi \text{ is the unknown)}$$

where the left-hand side is the sum of the values of ϕ at the four points in the square mesh nearest to the point at which the value ϕ_0 applies. The calculation proceeds by iteration for the values of ϕ at all the interior points in turn, using the formula in the form:

$$\text{adjustment} = (\phi_0)_{\text{new}} - (\phi_0)_{\text{old}} = \frac{1}{4}\left(\sum \phi_i\right)_{\text{old}} - (\phi_0)_{\text{old}}.$$

The iteration should cease when the changes of ϕ fall within a specified tolerance. To accelerate the process, use the 'over-relaxation' formula:

$$\text{adjustment} = (\phi_0)_{\text{new}} - (\phi_0)_{\text{old}} = w\left\{\frac{1}{4}\left(\sum \phi_i\right)_{\text{old}} - (\phi_0)_{\text{old}}\right\},$$

in which w-values around 1.5 are effective for problems involving up to 100 mesh points.

As a variation, Poisson's equation, $\nabla^2\phi = -1$, say, may be solved, with $\phi = 0$ all round the boundary, or students can be asked to computerise Schmidt's method (see (*b*) above). Other interesting, simple tasks include solving the (linear) one-dimensional wave equation by the method of characteristics, suitably explained. Computing provides a very good way of getting students to understand the fundamental differences in behaviour of elliptic, parabolic and hyperbolic equations.

Answers to problems

1.1 scalars: kinetic energy, pressure, mass density, electric charge, frequency, voltage, equilibrium black body radiation intensity.
vectors: momentum, angular velocity, acceleration, weight, light-beam intensity.

1.2 Route is sum of five displacement vectors, two of which (across roads) are known. Take them first. Remaining three must be collinear for shortest route.

1.3 $\cos\theta = \Sigma \cos\alpha_1 \cos\alpha_2$.

1.4 $ak + bl + mc = 0$.

1.5 $\pm(\mathbf{i} - 2\mathbf{j} + \mathbf{k})/\sqrt{6}$.

1.6 $8/\sqrt{101}$.

1.7 vector product $= \pm 2\mathbf{a} \times \mathbf{b}$.

1.8 (i) P fixed (bisects O_2A where $\overrightarrow{O_1A} = \mathbf{a}$). (ii) Locus is line parallel to O_1O_2 at a distance $|\mathbf{a}|/O_1O_2$ from it, in plane through O_1O_2 perpendicular to $\mathbf{a}$. (*N.B.* Here $\mathbf{a}$ and O_1O_2 must be perpendicular.) (iii) Locus is sphere with centre = midpoint of O_1O_2, radius $= (k + \frac{1}{4}O_1O_2{}^2)^{1/2}$.

1.9 (*c*)

$$\begin{vmatrix} x - x_1 & y - y_1 & z - z_1 \\ x_1 - x_2 & y_1 - y_2 & z_1 - z_2 \\ x_2 - x_3 & y_2 - y_3 & z_2 - z_3 \end{vmatrix} = 0, \quad \text{a linear equation in } x, y, z.$$

If $P_1P_2P_3$ collinear, identically zero because third row is then a multiple of second row. Plane is then undefined.

1.10 $\sin^2\theta = 1 - \cos^2\theta$.

1.11 If inflexion, speed is proportional to $e^{-\mu\theta}$ where θ = sum of *positive* values of angles turned through between start, finish and points of inflexion.

1.12 (*b*) $(-11\mathbf{i} - 8\mathbf{j} + 9\mathbf{k})/\sqrt{266}$.

1.13 Radius of curvature $= (1 + \sin^2 t)^{-1/2}$.

1.15 Only if the body is rigid.

1.16 (*a*) $V = \mathbf{a}\cdot\mathbf{r}$.

1.18 Direction cosines are $(2/\sqrt{17}, 3/\sqrt{17}, -2/\sqrt{17})$, max. moment $= \sqrt{17}$.

2.1 (II) (ii) is consistent. A may be taken positive *or* negative. The choice is arbitrary. To know which is the *standard* choice, a comparison with some standard, identified charge such as the electron (arbitrarily chosen as negative) would be necessary.

2.2 (*a*) $E = 0$ as seen by moving observer.
(*b*) Motion is cycloidal, i.e. like that of a point on the rim of a rolling wheel.

3.2 $\frac{1}{2}(1 - 1/\sqrt{2})$.

3.9 (*b*) $\phi = \frac{1}{2}x^2y^2 + \frac{1}{2}z^2$, $\int_0^P F\cdot d\mathbf{r} = 1$.

3.10 (*c*) $V = -\dfrac{q}{4\pi\epsilon_0}\log\left[\dfrac{(x-a)^2+y^2}{(x+a)^2+y^2}\right]$, or $\dfrac{px}{2\pi\epsilon_0(x^2+y^2)}$ for doublet.

Doublet equipotentials are cylinders of radius $p/(4\pi\epsilon_0 V_0)$, axis $x = p/(4\pi\epsilon_0 V_0)$, $y = 0$.

3.14 Both $\frac{1}{6}\sqrt{3}$. (*N.B.* $da = (\sqrt{3})\, dx\, dy$).

3.15 Orthogonal if $(\partial\mathbf{r}/\partial u)\cdot(\partial\mathbf{r}/\partial v) = 0$. True for latitude and longitude.

3.16 $\pi R^2(A - 3B/4)$ downwards.

3.18 $\pi r^2 p/\sqrt{2}$.

3.20 (*a*) 2π (e.g. by exploiting problem 3.19(*a*)).
(*b*) 4π, (*c*) $2\pi/3$ (i.e. one-sixth of complete solid angle).

3.23 (*b*) $\cos\theta = \frac{2}{3}$.

3.24 (i) $x^2 + y^2 = \text{const.}$, (ii) $xy = \text{const.}$,
(iii) and (iv) $y = \text{const.} \times e^{x^2/2}$.

4.1 No magnetic potential. (Can only exist in regions devoid of current.)

4.3 $\mathbf{r}\cdot d\mathbf{a}/r^3$.

4.5 B is $[M/TQ]$; J is $[Q/T]$; μ_0 is $[ML/Q^2]$; ϵ_0 is $[T^2Q^2/ML^3]$.

4.6 $E = -d\alpha/dt$, $v = -Q\alpha/M$.

4.7 L/R.

4.9 Net electrical energy *from* system $= L_0 J_0{}^2 - \frac{1}{2}L_0 J_0{}^2 =$ mechanical work *into* system.

4.10 (*b*) Half each from decrease in stored magnetic energy and mechanical work against $\mathbf{j} \times \mathbf{B}$ forces.

4.11 (*b*) again.

5.1 (*a*) $1/\sqrt{3}, 1/\sqrt{2}, 1/\sqrt{2}, 1/\sqrt{2}$. (*b*) $1+\sqrt{2}-3\pi/4$.

5.3 $\frac{3}{4}, \frac{1}{4}$ of K.E. is lost, $\frac{3}{32}$ goes into P.E.

5.4 (*b*) $(637/768)\pi h^3 k \tan\alpha$; $(127+2048/\sqrt{3})h/1274$.

5.7 $\oiint \rho\mathbf{v}\cdot d\mathbf{a} = 0$, $\oiint (\rho h_0\mathbf{v}+\mathbf{Q})\cdot d\mathbf{a} = 0$ (or P).

5.9 Approximately 10^{20} volts. Its enormous magnitude confirms the impossibility of the operation.

5.11. $\frac{1}{2}\epsilon_0\mathbf{E}^2$, $\frac{1}{2}k\epsilon_0\mathbf{E}^2$.

5.15 (i) Excess temp $= R(a^2-r^2)/4K$, (ii) $\pi a^4 CR/8K$.

6.1 $\pi\epsilon_0 dV_0/\log_e 2$, increased by factor k if dielectric constant k.

6.2 $c = (\mu_0\epsilon_0)^{-1/2}$, $C/D = -c$.

6.3 (*a*) $B = \frac{1}{2}\mu_0 r\, dq/dt$, $\partial\mathbf{B}/\partial t = 0$.

6.4 $B = \mu_0 x\dot{q}/2\pi r$, $\oint \mathbf{B}\cdot d\mathbf{r}/\mu_0 = \int_0^x \dot{q}\cdot dx =$ (iii) or (iv).

6.5 $r>R: H=\frac{1}{2}jR^2/r$, $B = \frac{1}{2}\mu_0 jR^2/r$; $r<R: H=\frac{1}{2}jr$, $B = \frac{1}{2}\mu jr$. **B** and **M** discontinuous.

6.6 (*a*) $(k\mu_0\epsilon_0)^{-1/2}$, (*b*) $(\mu\epsilon_0)^{-1/2}$.

6.8 $\mu = (3/\pi)10^{-3}$; $\mu/\mu_0 = 760$; $L = 1/\pi$ henry.

6.10 (*b*) $\tan\theta_i = k\tan\theta_o$, (*c*) $4Q/5$; $11/15$.

6.11 (*b*) $\tan\theta_i = (\mu/\mu_0)\tan\theta_o$. As $\mu\to\infty$, $\theta_o\to 0$ unless $B_n = 0$.

(*c*) (i) 45° to negative y-direction, (ii) as for (i),

(iii) 30° to negative y-direction.

7.2 (*a*) $al+bm+cn = 0$. Divide by $\sqrt{(a^2+b^2+c^2)}$. $|\text{grad}\, C| = 1$. perp. dist. $= C = A/\sqrt{(a^2+b^2+c^2)}$.

(*c*) (i) $-1/r$ (ii) $\log_e r$.

7.3 (*b*) $q = \tan p$, i.e. $p = f(q)$.

7.5 $dp/dr = \rho_m a$, force $= (\rho_m-\rho_c)a\epsilon$ outwards.

7.14 (*a*) Yes, for div $\mathbf{v} = 0$. (*b*) Yes, for div $\mathbf{v} = 0$.

7.16 (*b*) $Q_x = -K(y-2xy+y^2)$, $Q_y = -K(x-2y-x^2+2xy)$,

$R = 2K(y-x+1)$, total release rate $= 2K/3$, heats from short faces $= K/6$ each, from long face $= K/3$.

E is uniform (to be conservative).

$\sigma \propto (y-x+1)$ (ohmic heating $= \sigma\mathbf{E}^2$ per unit vol.).

7.17 (*a*) $n = -3$, i.e. inverse-square law.

(*b*) $\epsilon_0 j \cdot \text{grad}\, \tau$. (*c*) $\iint n\mathbf{v} \cdot \mathbf{da}$.

7.18 (*a*) 4π, (*b*) 4.

8.1 (*a*) $2(y^2z^2 + z^2x^2 + x^2y^2)$. (*b*) $-2 \sin(x+y)$. (*c*) 0.

8.2 $\phi = 2k\alpha\beta\gamma/\sqrt{(3)}a$.

8.3 (*a*) (i) $\phi = xyz$; (ii) $\psi = \sin x \sin y$;

(iii) $\phi = e^x \sin y$, $\psi = -e^x \cos y$;

(iv) $\phi = \frac{1}{2}(x^2 + y^2)$, flux $= 1$ in (ii) or $e^{\pi/2}$ in (iii).

8.4 (*b*) $A = -E_0$, $B = a^2E_0(k-1)/(k+1)$, $C = -2E_0/(k+1)$.

(*c*) $A = \frac{1}{2}\mu_0 J$, $B = \frac{1}{2}\mu_0 JR^2$.

8.5 $aq/\pi\sqrt{\{(x^2 - y^2 - a^2)^2 + 4x^2y^2\}}$, max. at closest point of approach.

8.6 (*a*) $m = -1$. (*b*) See section 8.10.

8.8 $\psi = \rho gkxy/2H + \text{const.}$

8.9 (*a*) $\psi = KEe^{-\pi x/a} \sin(\pi y/a)$.

8.10 (*a*) Infinite (because of the contributions at the corners). The Fourier series for ψ diverges at $x = 0$, $y = \pm a/2$.

(*b*) $T_{max} = (4Q_0a/K\pi^2)(1 - \frac{1}{9} + \frac{1}{25} \ldots)$.

8.11 Resistance $= 2.5\tau/t$. See figure A.6.

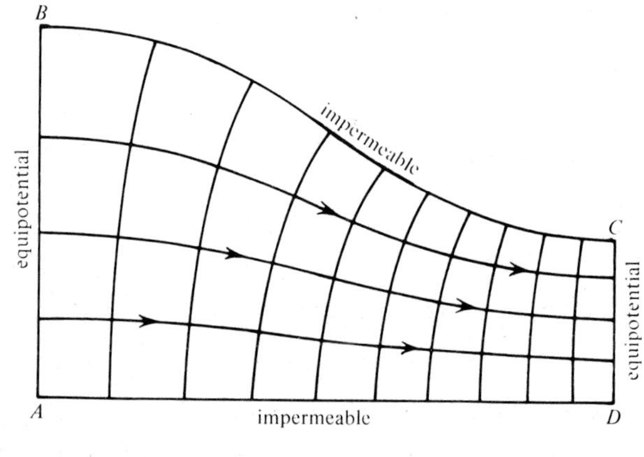

Fig. A.6. The solution to problem 8.11.

8.14 150%.

8.15 (*c*) Coaxal circles. (*d*) Half.

9.1 $V = (Ad^2/\epsilon_0\pi^2)\sin(\pi x/d)$.

9.4 (*a*) $\pi\sqrt{(3D/C)}$. (*b*) $\pi\sqrt{(D/C)}$.

9.5 $\frac{1}{2}\pi\sqrt{(K/C)}$.

9.7 (*a*) C/Km^2. (*b*) $\dfrac{4T_0}{\pi}\left\{e^{-(\pi^2K/a^2C)t}\cos\dfrac{\pi x}{a} - \frac{1}{3}e^{-(9\pi^2K/a^2C)}\cos\dfrac{3\pi x}{a} + \text{etc.}\right\}$.

9.8 (*a*) $\dfrac{d^2\theta}{dz^2} = -\dfrac{z}{2}\dfrac{d\theta}{dz}, \quad \theta = \text{erf}_c(\tfrac{1}{2}z)$.

(*b*) $\dfrac{d^2\theta}{dz^2} = -\dfrac{1}{2}\dfrac{d}{dz}(\theta z), \quad \theta = \dfrac{1}{2\sqrt{\pi}}e^{-z^2/4}$.

9.9 (i) $Ax + kA^2t/m$. (ii) $A/t^{1/3} - mx^2/6kt$.

9.10 $V_0 = \left(\dfrac{81j^2md^4}{32\epsilon_0{}^2e}\right)^{1/3}$.

9.11 (*a*) $DT/Dt = 0$. (*b*) $D\rho/Dt = acfe^{(g-b)t}$,

$\rho_P = acf(e^{(g-b)t} - 1)/(g-b) + a$.

9.12 (*a*) $\oiint p\mathbf{v}\cdot d\mathbf{a} = \iiint \text{div}\, p\mathbf{v}\, d\tau$.

(*c*) Expansion work and fall in kinetic energy.

9.13 $(\mathbf{u}\cdot\text{grad})\mathbf{v}$ is perp. to $\mathbf{v}$, $(\mathbf{u}\cdot\text{grad})\mathbf{u}$ is perp. to $\mathbf{u}$, $(\mathbf{v}\cdot\text{grad})\mathbf{v}$ is parallel to $\mathbf{v}$.

9.15 $x = e^{t_1-t_0}, \quad y = 1 + t_1 - t_0e^{t_1-t_0}$.

9.17 (*a*) $p = \frac{1}{2}\rho\Omega^2r^2 - \rho gz + \text{const.}$ Yes.

(*b*) $p = -\frac{1}{2}\rho A^2/r^2 - \rho gz + \text{const.}$

9.21 Waves travel backwards relative to traffic for dv/dn is negative.

9.22 $a_0\delta\rho/\rho_0$.

9.23 $a_0\rho_0u, \quad ul/a_0$.

9.24 (*a*) $a/2L, \quad 2a/2L, \quad 3a/2L$, etc.

(*b*) $y = \dfrac{8\epsilon}{\pi^2}\left(\sin\dfrac{\pi x}{L}\cos\dfrac{\pi at}{L} - \frac{1}{9}\sin\dfrac{3\pi x}{L}\cos\dfrac{3\pi at}{L} + \frac{1}{25}\sin\dfrac{5\pi x}{L}\cos\dfrac{5\pi at}{L}\ldots\right)$.

9.25 $\{A/\sin(\omega L/a_0)\}\sin(\omega x/a_0)\sin\omega t$.

9.26 $a_0/2d, \ (\sqrt{2})a_0/2d, \ (\sqrt{3})a_0/2d, \ a_0/d$. (*N.B.* 1 or 2 of X, Y and Z may be constant.)

9.27 (*b*) 2.405, 5.52, 8.05 times $(T/\rho)^{1/2}/2\pi R$.

10.1 (*a*) (i) 0; (ii) $\mathbf{k}(\partial g/\partial x - \partial f/\partial y)$. (*b*) $\mathbf{v} = \mathbf{\Omega}\times\mathbf{r}, \quad \text{curl}\,\mathbf{v} = 2\mathbf{\Omega}$.

10.2 Streamlines are rectangular hyperbolas. $\boldsymbol{\omega} = \mathbf{0}$.

10.4 (*b*) $\text{curl}\,\mathbf{j} \times \mathbf{B} = \mathbf{0}$.

10.9 $\mathbf{u}$ is conservative with potential $\frac{1}{2}(x^2 + y^2) - 3xy$; $\mathbf{v}$ is solenoidal.

10.17 (*b*) $4(e^2 - 1)/(e^2 + 1)$.

10.19 (*d*) $q = 0$ if $\theta = \pi/2$, $\mathbf{j} = \mathbf{0}$ if $\theta = 0$.

11.2 (*a*) $B_x = -\frac{1}{2}\mu_0 jy, \quad B_y = \frac{1}{2}\mu_0 jx \quad (r < a);$

$B_x = \frac{1}{2}\mu_0 jya^2/(x^2 + y^2), \quad B_y = \frac{1}{2}\mu_0 jxa^2/(x^2 + y^2) \quad (r > a).$

(*b*) $E_x = \frac{1}{2}\mu_0 Ny(\mathrm{d}J/\mathrm{d}t), \quad E_y = -\frac{1}{2}\mu_0 Nx(\mathrm{d}J/\mathrm{d}t).$

(*c*) $E_z = \Omega B(x \cos \Omega t + y \sin \Omega t).$

11.3 (*a*) p, B and H are all constant along current flow lines.

(*b*) $(\mathbf{B} \cdot \text{grad})\mathbf{B}$ is non-zero because $\mathbf{B}$ changes direction along $\mathbf{B}$-lines.

11.4 (*a*) At 10^4 Hz: in ratio $0.022 \times 10^{-5} : 1.75 \times 10^{-5} : 1$; $\delta = 3.18$ m. At 10^9 Hz: in ratio $0.022 : 1.75 : 1$ (i.e. $\mathbf{j}$ and $\partial\mathbf{P}/\partial t$ comparable, and theory inadequate).

11.5 (*b*) $E = e^{-x\sqrt{(m\mu_0\sigma)}}$.

11.6 71.

11.9 (*b*) Pressure $= 2B_0{}^2/\mu_0$; duration T. (*c*) momentum $= B^2 T/\mu_0$,

$B^2/\mu_0 c$ per unit vol.

11.10 $2/(1 + \sqrt{k})$.

11.13 Voltage $= bc\mu_0 J$., no current after time a/c.

11.16 (*b*) $\mathbf{S}$ is $\frac{1}{2}r\epsilon_0 E\, \mathrm{d}E/\mathrm{d}t$ radially inwards.

(*c*) Twice the wave frequency. Energy uniform when $2\pi ct/\lambda = n\pi/4$ (n odd).

11.18 Breakdown limits $\mathbf{E}$; 2.65×10^9; $0.38\,\mathrm{m}^2$.

11.19 (*a*) $\mathbf{H}$.

11.20 (*a*) curl $\mathbf{M}$ and div $\mathbf{M}$ will not in general vanish inside a magnet.

12.3 (*a*) $\Phi = h$ (enthalpy).

12.4 (*b*) (i) and (ii) $\lambda = 0$; magnetic flux linked constant.

12.6 $\alpha =$ const. along streamlines.

12.7 (*b*) Velocity variation falls from $1 : 2.718$ to $1 : 1.30$.

12.10 Incompressible, $\psi = 2y^2 - 2xy + x^2 +$ const., ellipses, $A =$ const., $\Gamma = -6A$.

12.11 (*b*) Spiral streamlines, p rises linearly in time with regular discontinuous drops, blade angular velocity $= -2A$, flow rate per unit length $= 2\pi B$.

12.12 $\psi = \psi_1 e^k$, irrespective of x_1, y_1, given ψ.

12.13 (*b*) Shear stress (i.e. field is not purely radial).

(*d*) $\theta = \pi/4$.

12.16 (*b*) $\rho Q^2/8A$.

12.17 (*a*) Velocity unchanged. (*b*) $d \cos^2 \dfrac{\theta}{2}$, $d \sin^2 \dfrac{\theta}{2}$. (*c*) $\rho v^2 d \sin\theta$.

12.18 They do compensate exactly, within reasonable assumptions.

12.20 Circulation = strength of vortex linked; $\Gamma/2\pi a$.

Index